Circular Economy and Sustainability

This book series aims at exploring the rising field of Circular Economy (CE) which is rapidly gaining interest and merit from scholars, decision-makers and practitioners as the global economic model to decouple economic growth and development from the consumption of finite natural resources. This field suggests that global sustainability can be achieved by adopting a set of CE principles and strategies such as design out waste, systems thinking, adoption of nature-based approaches, shift to renewable energy and materials, reclaim, retain, and restore the health of ecosystems, return recovered biological resources to the biosphere, remanufacture products or components, among others.

However, the increasing complexity of sustainability challenges has made traditional engineering, business models, economics and existing social approaches unable to successfully adopt such principles and strategies. In fact, the CE field is often viewed as a simple evolution of the concept of sustainability or as a revisiting of an old discussion on recycling and reuse of waste materials. However, a modern perception of CE at different levels (micro, meso, and macro) indicates that CE is rather a systemic tool to achieve sustainability and a new eco-effective approach of returning and maintaining waste in the production processes by closing the loop of materials. In this frame, CE and sustainability can be seen as a multidimensional concept based on a variety of scientific disciplines (e.g., engineering, economics, environmental sciences, social sciences). Nevertheless, the interconnections and synergies among the scientific disciplines have been rarely investigated in depth.

One significant goal of the book series is to study and highlight the growing theoretical links of CE and sustainability at different scales and levels, to investigate the synergies between the two concepts and to analyze and present its realisation through strategies, policies, business models, entrepreneurship, financial instruments and technologies. Thus, the book series provides a new platform for CE and sustainability research and case studies and relevant scientific discussion towards new system-wide solutions.

Specific topics that fall within the scope of the series include, but are not limited to, studies that investigate the systemic, integrated approach of CE and sustainability across different levels and its expression and realisation in different disciplines and fields such as business models, economics, consumer services and behaviour, the Internet of Things, product design, sustainable consumption & production, bio-economy, environmental accounting, industrial ecology, industrial symbiosis, resource recovery, ecosystem services, circular water economy, circular cities, nature-based solutions, waste management, renewable energy, circular materials, life cycle assessment, strong sustainability, and environmental education, among others.

Catherine De Wolf • Sultan Çetin
Nancy Bocken
Editors

A Circular Built Environment in the Digital Age

Springer

Editors
Catherine De Wolf
Circular Engineering for Architecture,
Institute of Construction & Infrastructure
Management, Department of Civil,
Environmental and Geomatic
Engineering
ETH Zurich
Zurich, Switzerland

Sultan Çetin
Management in the Built Environment,
Architecture and the Built Environment
Delft University of Technology
Delft, The Netherlands

Nancy Bocken
Maastricht Sustainability Institute, School
of Business & Economics
Maastricht University
Maastricht, The Netherlands

This work has been supported by Maastricht University, ETH Zurich, and Delft University of Technology.

ISSN 2731-5509 ISSN 2731-5517 (electronic)
Circular Economy and Sustainability
ISBN 978-3-031-39674-8 ISBN 978-3-031-39675-5 (eBook)
https://doi.org/10.1007/978-3-031-39675-5

This Springer imprint is published by the registered company Springer Nature Switzerland AG
The registered company address is: Gewerbestrasse 11, 6330 Cham, Switzerland

Paper in this product is recyclable.

Foreword

We need an economic system based on the insight that the Earth is a closed system. In a closed system with defined boundaries, materials, just like space, must be considered as limited and treated as such. They should be used accordingly in such a manner that their availability is not limited to one single, temporary application but allows them to circulate in our systems indefinitely.

This insight is of particular relevance for the architecture, engineering, and construction sectors, which, according to the UNEP Global Status Report for Building and Construction, account for 40–50% of material consumption of our global economy and 37% of greenhouse gas emissions. Without decisive action, the severe pressure on resources caused by the construction industry will only aggravate. According to the International Institute for Sustainable Development, roughly half of the building stock needed by 2060 has yet to be built while several critical boundaries of our planetary system have already been surpassed.

For all these reasons, we need a new circular building culture in which we design, develop, build, and rebuild in such a way that "limited-edition" materials are available indefinitely and resources are used responsibly. Buildings need to be conceived as documented and deconstructible material depots, from which, after their useful life, all materials can be recovered for future (building) projects. In addition, we need to reduce our consumption by applying strategies to use fewer materials and resources, and use materials longer. In short, we need to narrow, slow, close, and regenerate our material loops, as the authors of this book articulate.

Yet the complexity of the building process, the fragmentation of the construction value chain, the variety of actors involved, and the long-time horizon of building projects are factors that make the realisation of circularity in the built environment particularly challenging. We therefore urgently need new knowledge and tools that allow us to navigate and, more importantly, shape this highly complex environment in new ways. Data and digitalisation can provide us with exactly these new tools and bring transparency into the building process by bridging the information and knowledge gaps that exist between disciplines, processes and time.

This publication, for the first time, connects and describes the many digital innovations which can drive circularity in the built environment. Its authors come from a global community of scientists and entrepreneurs who are at the forefront of shaping a new building paradigm.

In this book, a groundbreaking vision for the future of the built environment emerges, fueled by cutting-edge technologies that offer new possibilities for circular and regenerative design and construction. This book evokes a vision for a new era of the built environment in which geographical data will enable us to place our building structures and activities more responsibly in the wider context of a place or city; a vision where AI-informed computational design tools will support architects in creating fully circular structures that require less material for construction, can be maintained and adapted during their useful life, and can be easily deconstructed once they reached their expiration date. Additive manufacturing and robotics will bring new possibilities for realising and decomposing these circular structures, while maintenance and material management will be supported by blockchain-enhanced digital twins and material passports. All these new tools will need to converge into one integrated digital ecosystem whereby, as Alexander Koutamanis formulates in this book, "information should be treated not as a product of integration but as the integrator of all activities."

New tools alone will not be enough to bring the urgently needed transformation of our system. We also need a new consciousness for the consequences of our actions. We seem to be far from Jean-Jacques Rousseau's remark (from his 1754 essay "On the Origin of the Inequality of Mankind") that "the fruits of the earth belong to all, but the earth to none." We need to develop a new building culture that, among other things, will reconcile the health of our planetary system and the interests of future generations with the temporary needs of people, society, and businesses. Digitalisation in the many areas of the built environment as described in this book has the potential not only to see new buildings via augmented reality but to shape a circular built environment through an augmented consciousness.

Co-founder of Turntoo
The Netherlands

Sabine Oberhuber

Architect, Founder of RAU, Turntoo, Madaster
The Netherlands

Thomas Rau

Sabine Oberhuber is one of the first pioneers of the circular economy and co-founder of Turntoo, the first company in the Netherlands focusing on the transition to a circular economy. She has developed some of the first business models and strategies for circularity and has helped shape the thinking about the transition to a circular economy. Turntoo works with leading companies and public bodies to develop circular business models and processes to reduce or eliminate material waste, realising breakthrough concepts such as Light as a Service, in cooperation with Philips. Turntoo also assists municipalities with circular city strategies and area development.

Thomas Rau is an architect, entrepreneur, innovator and founder of RAU, Turntoo and Madaster. The architectural firm RAU has developed innovative concepts and set the tone in the field of environmentally conscious, climate-neutral and energy-efficient building at an early stage. RAU is now the undisputed authority in the Netherlands on plus-energy building and circular value creation in architecture. Thomas Rau was nominated for the Circular Economy leadership Award of the World Economic Forum and received the Circular Hero Award by the Dutch Ministry of Infrastructure.

In 2016, Thomas Rau and Sabine Oberhuber published the best-selling book *Material Matters*, in which they describe the critical building blocks for achieving a circular economy. In 2016, they initiated Madaster, the cadastre for materials. Madaster is active in eight countries and considered as the leading solution for creating and registering material passports for the built environment. For its potential for systemic change, Madaster won the Digital Top 50 Award for Social Impact which is awarded by Google, McKinsey and Rocket Internet.

Acknowledgements

We would like to extend our heartfelt gratitude to all those who have immensely contributed to the quality of this book. Their valuable insights and expertise have enriched our understanding and presentation of the subject matter. We are grateful to the authors for delivering excellent and thought-provoking chapters and perspectives. We are tremendously grateful to Jennifer Bartmess for her invaluable expertise and dedication while reviewing and copyediting the chapters. We would also like to thank the anonymous reviewers who participated in the blind peer-review process for their careful evaluation and thoughtful comments. Our sincerest appreciation goes to our team members who provided feedback on the technologies discussed in this book. Additionally, we would like to thank our families and friends for their unwavering support. Finally, we express our gratitude to the readers of this book. It is our sincere hope that this work helps them navigate through the emerging field of a circular built environment in the digital age and that this knowledge will contribute to a regenerative future.

Open access of this book was made possible with the financial support from the TU Delft Open Access Fund, the Maastricht University Library's Open Access Book Fund, the Swiss National Science Foundation (SNSF) Project Fund under grant agreement No 200021E_215311, and the European Union's Horizon 2020's European research Council (ERC) funding scheme under grant agreement No 850159 (project Circular X).

Introduction

> "The ability of technological advancement to do more and more with less and less until eventually you can do everything with nothing."
> R. Buckminster Fuller in *Nine Chains to the Moon*

By 2050, two-thirds of the world's population will be living in cities, and by 2030, three billion people will need new housing (UN Habitat 2023). As a growing sector, especially due to increasing urbanisation and the need for new housing, the architecture, engineering, and construction (AEC) industry is responsible for depleting resources, generating waste, and emitting greenhouse gases at a tremendous scale and speed. Construction and demolition processes are highly resource-intensive, accounting for more than 40% of the total raw materials extracted worldwide and generating over 35% of total waste – additionally, the building sector is responsible for approximately 39% of global energy-related greenhouse gas emissions (Schrör 2011; Allwood et al. 2011; Yuan et al. 2012; Di Maria et al. 2018; Abergel et al. 2019; Eurostat 2020; and summarised by Çetin et al. 2021). The industry's linear model of production is at the core of the problem: resources are extracted, buildings are used, and then materials are disposed of when a building is no longer needed. If we continue this linear model to meet the unprecedented and growing demand for constructing new buildings, we will deplete the Earth's resources and pave the way for an even greater climate catastrophe.

Industry 5.0 technologies and circular economy principles hold untapped potential for achieving a sustainable built environment. This book aims to address the urgent need for a sustainable built environment by leveraging this potential. Industry 5.0 is focused on creating a sustainable, human-centric, and resilient future using advanced technologies. While many industries have already begun implementing circular economy principles through digitalisation, the AEC industry is lagging behind.

To address the challenges the construction industry faces, we can apply circular economy strategies, such as service life enhancement of materials, rehabilitation, dis- and reassembly, design for reuse, and implementation of regenerative design. The goal is to make the built environment part of the solution rather than part of the problem. We must urgently shift from a linear take-make-waste model to a circular one in which we use resources wisely and prevent them from becoming waste. Adopting digital innovation is crucial to achieving this paradigm shift, but currently there is a lack of understanding of the potential synergies between the circular economy and digital transitions. This "twin transition" could be leveraged to tackle the unprecedented challenges facing the industry. This book highlights the importance of these synergies and explores how digital technology can help accelerate the circular transition of the built environment.

But first, what is a circular model? The circular economy has become a popular concept among scholars, NGOs, business professionals, and policymakers to address issues of sustainable development (Geissdoerfer et al. 2017; Kirchherr et al. 2017). The concept has evolved through the work of designers and architects such as McDonough and Braungart (2002), who coined the cradle-to-cradle concept, Benyus (1997), who led the work on Biomimicry and designing according to nature's principles, and Stahel (2010), who advocated new business models focused on performance and services rather than mere product sales. Policies focused on waste reduction, pollution prevention, and resource efficiency have existed in many parts of the world for decades but have only recently culminated in circular economy policies with the opportunity to create a broader more comprehensive framework to address resource issues (Bocken et al. 2017). Yet as several organisations, regions, and countries across the globe are increasingly bringing circular economy explicitly into their visions, goals, and policies, it has become an important concept to help transfer thinking about the future of the society, economy, and sustainable innovation.

In this book, we regard the circular economy as an important lever to support sustainable development and secure the resources to sustain our current and future generations by minimising the resource inputs and waste, emissions, and energy leakage of products over time (Bocken et al. 2021). This may be achieved through four distinct resource strategies (see Çetin et al. 2021; Konietzko et al. 2020):

- Narrowing the loop: using fewer resources through increased efficiency in the production and design process
- Slowing the loop: using and consuming less by extending product lifespans, and avoiding unnecessary consumption
- Closing the loop: reusing materials or resource-efficient post-consumer recycling
- Regenerating the loop: focusing on leaving the environment (and society) in a better state than before

Clarifying the relevance of a circular model brings us to a second question: how is digital technology relevant? There are many opportunities to combine digital with circular principles in existing buildings, new buildings, and even demolition projects (Çetin et al. 2021). Digitalisation can solve practical challenges related to material scarcity and carbon emissions reductions. Digital technologies could also help with de- and re-constructing buildings more quickly, economically, and intelligently. For example, we can use the advances we have seen in recent years in digital fabrication to start a digital *de-fabrication* design approach, augmented by other digital innovations (such as matchmaking algorithms, extended reality, blockchain, etc.). This would disrupt the current construction sector's value chain and reverse the current architectural design approach towards de- and re-fabrication for effectively reusing building materials. Digital transformation holds great potential for helping the transition to a circular economy.

Emergent digital innovations make this a timely topic. Digital technologies – from building information modelling (BIM) software to Internet of Things (IoT) sensors, artificial intelligence (AI) algorithms, blockchain technology, or digital fabrication technologies – can help architects, engineers, and construction professionals optimise design, construction, monitoring, decision-making, maintenance, and performance of building systems to improve efficiency and reuse, reduce waste, and even provide safer working conditions.

This book addresses the pressing need for a comprehensive overview of the technologies that are most relevant to the circular building industry and how they can be used to achieve circularity goals. Each chapter focuses on a particular digital technology and its application for the circular economy, providing at least one practical example and discussing potential future developments. Leading experts in the field of digitalisation and circular economy offer their insights into how emerging digital technologies can be used to address circular economy strategies in the built environment.

The book is divided into three distinct parts, each with a unique focus, for a more in-depth exploration of each technology and its application, as well as a clear framework for understanding how these technologies can work together to drive the circular economy transition.

Part I explores the role of data, with topics including digital representation (BIM and digital twins), geographic information systems (GIS), scanning technologies and reality capture, AI, machine learning, and material passports in enabling digital innovation for circularity. Part II delves into design and fabrication, covering technologies such as computational tools, additive manufacturing and robotic fabrication, and extended reality. Part III examines business, management, and governance, exploring topics such as reverse logistics, blockchain technology, digital logbooks, circular business models, and regeneration.

Finally, the book concludes with a discussion of how these digital tools can be combined to create practical, economic, and policy-driven solutions to drive the twin transition of shifting to a circular economy and digital transformation in the built environment. By presenting a comprehensive exploration of these technologies and their potential applications, this book aims to inspire and accelerate the knowledge needed to drive the circular economy transition.

Catherine De Wolf
Sultan Çetin
Nancy Bocken

References

Abergel T, Dulac J, Hamilton I, Jordan M, Pradeep A (2019) Global status report for buildings and construction—towards a zero-emissions, efficient and resilient buildings and construction sector. United Nations Environment Programme & International Energy Agency. https://wedocs.unep.org/20.500.11822/30950

Allwood JM, Cullen JM, Carruth MA, Cooper DR, McBrien M, Milford RL, Moynihan MC, Patel AC (2011) Sustainable materials: with both eyes open. UIT Cambridge, Cambridge

Benyus JM (1997) Biomimicry: Innovation inspired by nature. HarperCollins, Australia

Bocken N, Stahel W, Dobrauz G, Koumbarakis A, Obst M, Matzdorf P (2021) Circularity as the new normal: Future fitting Swiss business strategies. PwC Switzerland. https://www.wwf.ch/sites/default/files/doc-2021-01/Circularity-as-the-new-normal_whitepaper-EN.pdf

Bocken NM, Olivetti EA, Cullen JM, Potting J, Lifset R (2017) Taking the circularity to the next level: a special issue on the circular economy. J Industr Ecol 21(3):476–482

Çetin S, De Wolf C, Bocken N (2021) Circular digital built environment: an emerging framework. Sustainability 13, 6348, https://doi.org/10.3390/su13116348

Di Maria A, Eyckmans J, Van Acker K (2018) Downcycling versus recycling of construction and demolition waste: Combining LCA and LCC to support sustainable policy making. Waste Manage 75:3–21. https://doi.org/10.1016/j.wasman.2018.01.028

Eurostat (2020) Waste statistics. https://ec.europa.eu/eurostat/statistics-explained/index.php/Waste_statistics#Total_waste_generation. Accessed 21 Apr 2023

Geissdoerfer M, Savaget P, Bocken NM, Hultink EJ (2017) The circular economy: a new sustainability paradigm? J Clean Prod 143:757–768. https://doi.org/10.1016/j.jclepro.2016.12.048

Kirchherr J, Reike D, Hekkert M (2017) Conceptualizing the circular economy: An analysis of 114 definitions. Resour Conserv Recycl 127:221–232. https://doi.org/10.1016/j.resconrec.2017.09.005

Konietzko J, Bocken N, Hultink EJ (2020) A tool to analyze, ideate and develop circular innovation ecosystems. Sustainability 12(1):417. https://doi.org/10.3390/su12010417

McDonough W, Braungart M (2002) Cradle to cradle: remaking the way we make things. North Point, New York

Schrör H (2011) Generation and treatment of waste in Europe 2008. Environment and Energy Eurostat Statistics in Focus 44:2011

Stahel W (2010) The performance economy, 2nd ed. Palgrave Macmillan.

UN Habitat (2023) Housing. https://unhabitat.org/topic/housing. Accessed 21 Apr 2023

Yuan H, Chini AR, Lu Y, Shen L (2012) A dynamic model for assessing the effects of management strategies on the reduction of construction and demolition waste. Waste Manage 32(3):521–531

Introductory Perspective

For a long time, the built environment was mostly based on natural 'zero carbon' materials, such as dimension stone (Roman aqueducts), clay (houses in Saana), backed clay (bricks), plants (trees and bamboo), materials which decay without harm to Nature, or become food for other organisms – bacteria, insects, worms – at the end of their use. The industrial revolution in the UK mass-produced iron, steel, and cement, extending society's limits beyond natural materials into a high-carbon material domain.

A notable symbol of the beginnings of the industrial revolution is the Iron Bridge, built in 1779. This remarkable monument stands as a testament to where the Industrial Revolution originated and dominates the small town that shares its name. The first known instance of mass steel production is credited to China in the third century AD. They employed techniques similar to what is now known as the Bessemer Process, which later enabled bulk steel production. In 1855, Henry Bessemer obtained British patents for a pneumatic steelmaking process, using blasts of air to remove impurities from molten steel. In 1824, Joseph Aspdin of Leeds, Yorkshire, England, secured a patent for a material produced from a synthetic mixture of limestone and clay, which came to be known as cement.

Today, the construction industry is the biggest consumer of material resources as well as a major polluter. Cement and steel are the backbones of infrastructure and buildings. Cement production creates 2.3 billion tonnes of carbon dioxide per year, while iron and steel production releases some 2.6 billion tonnes – 6.5% and 7.0% of global CO_2 equivalent emissions, respectively (Fennell et al. 2022). Their future will depend on public policies of environmental protection, market supplies of natural resources, and the availability of landfills.

In the twentieth century, steel-reinforced concrete and steel structures were increasingly used, together with technical progress in industrialised building methods and new materials: plastic for pipes and cables, chemicals for joints and insulation.

In the late 1960s, when I studied architecture and urban planning at ETH Zurich, our focus was on design and engineering issues, building regulations and zoning laws. The World population was 2.5 billion people, partly living in big cities of the Northern hemisphere, New York, London, Tokyo, Moscow, and Paris. The Digital Age consisted of computation centres where big machines were fed by punch cards. Neil Alden Armstrong had just set foot on the Moon, using computers with 128-bit chip technology. In 1950, Eduard Stiefel at ETH rented the Zuse Z4, a relay computer developed in Germany by Konrad Zuse, making it the world's first commercial digital computer. The Z4 was the first computer at a continental European university, it remained at ETH Zurich for 4 years.

In the twenty-first century, the Digital Age introduced CAD, digital twins, Internet of Things, and 'speaking' elevators mainly in industrialised regions, which enabled more efficient construction methods but also created digital outcasts in the population.

Today, eight billion people live on our planet. China, India, and Africa are with over one billion people, with Africa having the fastest population growth. Major cities are on both sides of the equator: Tokyo, Jakarta, Chongqing, Delhi, Shanghai, Seoul, Mumbai, Manila, New York City, Sao Paolo, Beijing, and Mexico City. Concrete is the second-most-consumed product on the planet, after clean water: world production of cement is 530 kg per person per year, and of steel, 240 kg (but only part of which goes into the built environment). Changes to building codes and in the education of architects, engineers, and contractors could reduce demand for cement and steel by only 26%, according to the International Energy Agency (IEA 2019).

The objective of the Circular Economy is "doing more with less over longer periods" by maintaining the value, purity, and utility of stocks (of natural, human, cultural, financial, and manufactured objects and materials), with a focus on the sustainable use of these stocks. In other words, the Circular Economy is about the design and construction as well as the use phase, the smart operation, and maintenance of the built environment. A circular economy in construction is also a solution towards the environmental impacts of buildings.

The Digital Age can contribute to increasing the efficiency of building and construction activity and support the standardisation of materials and dimensions to facilitate the efficient reuse of components and material resources (urban mining). But remember that in industrialised regions, the annual volume of new construction is only about 2% of the stock volume!

Therefore, the biggest contribution of the Digital Age will be in improving the sustainable use of the stocks of infrastructure and buildings by extending their service lives and improving their operation and maintenance phase – the heart of the Circular Economy.

Walter R. Stahel

References

Fennell P, Driver J, Bataille C, Davis SJ (2022) Going net zero for cement and steel. Nature 603:577
IEA (2019). Material Efficiency in Clean Energy Transitions (International Energy Agency)

Walter R. Stahel, a Swiss architect and sustainability advocate, is renowned for co-founding the Product Life Institute in Geneva and pioneering the concept of the circular economy. His book, *The Performance Economy*, introduced selling goods as services to encourage resource efficiency. Stahel collaborates with the Ellen MacArthur Foundation and the European Commission. He received accolades such as a doctorate honoris causa from the Université de Montréal and a senior research fellowship at the Circular Economy Research Centre in Paris. In 1989, Stahel and Orio Giarini published *The Limits to Certainty*, emphasising that finite resources were a limit to the unsustainable linear industrial economy and proposed a circular economy that utilises resources effectively and considers ecological factors.

Contents

About the Editors

Catherine De Wolf is an Assistant Professor of Circular Engineering for Architecture at the Swiss Federal Institute of Technology of Zurich (ETH Zurich), where she conducts research on digital innovation towards a circular built environment. She has a dual background in civil engineering and architecture and completed her PhD at the Massachusetts Institute of Technology (MIT). She is a faculty member in the ETH Centre for Augmented Computational Design in Architecture, Engineering and Construction (Design++), the ETH AI Center, the Swiss Federal Laboratories for Materials Science and Technology (Empa), the Circular Future Cities group of the Future Cities Lab, and the National Centre of Competence in Research on Digital Fabrication (DFAB). She has worked in academia (e.g., University of Cambridge, University of Technology Delft, Nanjing University), in industry (e.g., Arup, Elioth, Ney & Partners), and in governmental institutions (e.g., Joint Research Centre of the European Commission). She also co-founded the circular construction company Anku, the Structural Engineers 2050 Commitment, and the Digital Circular Economy (DiCE) Lab.

Sultan Çetin is a PhD candidate at the Delft University of Technology. She received bachelor's degrees in Architecture and Civil Engineering in Turkey and a master's degree from Eindhoven Technology of University. Sultan worked as an Architect on international projects in Russia, Azerbaijan, and the Netherlands for several years. Since 2019, she has been conducting doctoral research on digitalisation for a circular building industry. She founded the Digital Circular Economy (DiCE) Lab with Catherine De Wolf in 2021. After the disastrous earthquakes in Turkey in 2023, she founded a network platform called RIAR (Research Industry Alliance for Recovery) to connect experts to find recovery solutions.

Nancy Bocken is a Professor in Sustainable Business and Circular Economy at Maastricht University, Maastricht Sustainability Institute (MSI) and is a leading researcher on topics such as sustainable business models, business experiments for sustainability, circular economy, and sufficiency. She is also a Fellow at Cambridge

Institute for Sustainability Leadership, advisor to TNO (Dutch association for applied scientific research) and a Board member of the Philips Foundation. Before going into academia, she held positions in the logistics, banking, and consulting sectors. She holds a PhD from the Department of Engineering, University of Cambridge and is a co-founder of her own circular and sustainable business, HOMIE.

Part I
Data

Chapter 1
From Building Information Modelling to Digital Twins: Digital Representation for a Circular Economy

Alexander Koutamanis

Abstract Building information modelling (BIM) has ushered in the era of symbolic building representation: building elements and spaces are described not by graphical elements but by discrete symbols, each with properties and relations that explicitly integrate all information. Digital twinning promises even more: a digital replica in complete sync with the building and its behaviour. Such technologies have obvious appeal for circularity because they accommodate the rich information it requires and link circularity goals to other activities in AECO (architecture, engineering, construction and operation of buildings).

Present implementations of BIM may fall short of the promise, and digital twinning may be hard to achieve, but they remain crucial not only for circularity but for all AECO disciplines. To realise the potential of such representations, information should be treated not as a product of integration but as the integrator of all activities. Similarly, digitalisation should be at the core of business models and deployment plans, not an additional or even optional layer at a high cost. This calls for a coherent approach that includes the full capture of building information, supports the detailed exploration of circular operations, uses the results to constrain decisions and actions and does so throughout the life cycle.

Keywords Information · Digitalisation · Representation · Building information modelling (BIM) · Digital twinning

1.1 Building Information Modelling and Digital Twinning

Rhetoric has three modes of persuasion: pathos, ethos and logos. Circularity is derived from pathos: appeals to emotions and ideals, expressing beliefs about the environment and materiality. It is reinforced by ethos: arguments from authorities and other credible sources, such as scientists and industry leaders. When it comes to

A. Koutamanis (✉)
Delft University of Technology, Delft, the Netherlands
e-mail: a.koutamanis@tudelft.nl

C. De Wolf et al. (eds.), *A Circular Built Environment in the Digital Age*, Circular Economy and Sustainability, https://doi.org/10.1007/978-3-031-39675-5_1

implementing circularity, however, it is the logos that matters most: the reasoning that underlies business models, material flow calculations, feasibility assessments, implementation requirements, deployment plans, etc. Information is the basic resource for making such analyses and projections reliable and transparent: valid, meaningful data that describe past and future states of the world, providing input to and accommodating output from decision processes.

This chapter focuses on the critical, fundamental role of information in the context of circularity. It explains the two most relevant general-purpose technologies, building information modelling (BIM) and digital twinning, and links them to passports and logbooks proposed specifically for circularity. It then moves on to current and proposed uses of the technologies in AECO (architecture, engineering, construction and operation of buildings), including with respect to circularity, and concludes with guidelines for developing circularity business models and practical applications.

1.1.1 BIM

BIM is a frequently misrepresented and therefore misunderstood technology. Many poor definitions describe not the phenomenon itself but its applications and effects (Sacks et al. 2018), often from the perspective of existing analogue practices. The production of drawings and other conventional documents to incrementally improve efficiency or reduce errors takes up a disproportionate amount of the BIM literature but does not explain how BIM is structured and how its structure helps to achieve certain objectives. Instead, it makes BIM appear as a mere step in AECO computerisation. The truth is more revolutionary: BIM marks the transition to *symbolic representation* (Koutamanis 2022). While earlier technologies like computer-aided design (CAD) focused on the graphic implementation mechanisms of building representations, BIM makes explicit the symbols described by these mechanisms.

Symbolic representation is already the norm in many computer applications. In a digital text, the capital 'A' is not a group of three strokes, as in handwriting, but the Unicode symbol U+0041, explicitly entered through a keyboard and stored as such, regardless of how it appears on the screen. Any change to the symbol does not come from changing the three strokes but from changing the properties of the symbol (e.g. a different font or size) or switching to a different symbol (e.g. U+1D434 for the mathematical capital 'A'). Symbolic representation underlies a lot of machine intelligence. In digital texts, knowing each letter allows computers to recognise words and sentences and subsequently understand grammar and syntax.

Similarly, in BIM, a window is not the group of line segments one sees in a graphic view like a floor plan but a symbol explicitly entered in a specific location of a wall. One can reposition the window in the wall, but changing its type or even its size may require switching to a different symbol. The interfaces of BIM software tend to depart from facsimiles of analogue drawing, which confuse users into

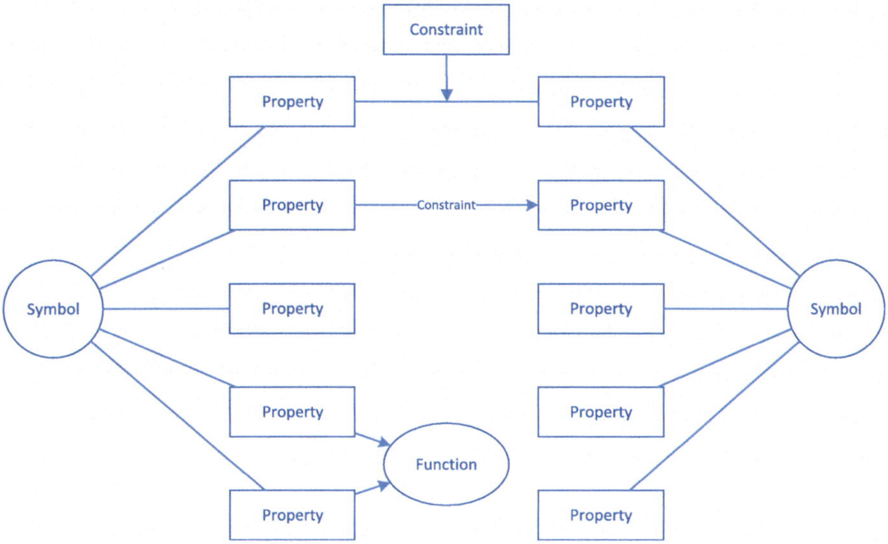

Fig. 1.1 Symbols, properties and connections

thinking that they are drawing and obscure the symbolic structure of the model. We should think of BIM models not as 2D or 3D drawings with additional data but as *graphs* of interconnected symbols. In fact, connections are between specific symbol properties (Fig. 1.1): the co-termination of two walls links the endpoints of their axes, while the orientation of a wall is inherited by the windows it hosts.

External constraints, such as the maximum height of a roof in planning regulations, are also linked to relevant symbol properties, while other constraints affect relations between two symbols, such as when windows are not allowed in certain wall parts. As a result, all primary information resides in the properties and relations of the symbols in a model. This allows for the derivation of further information through functions, e.g. calculations of fire resistance on the basis of the material composition of a building component. It also supports the production of various views of the model, including conventional drawings. As for machine intelligence, the potential is already evident in the *behaviours* of symbols: a window sticks to the hosting wall, and the shape of a room follows the bounding building elements.

Integration, a key selling point of BIM, comes from this symbolic structure. With all information residing in symbols, there are no multiple representations from different disciplines that must be combined to obtain a full description. Instead, all actors have access to different symbols, properties and relations in a model, in adjustable *worksets* that give them specific rights and responsibilities. This integration of information and its dynamic relation to authorship and custodianship also mean that information processing and AECO activities can be accommodated in BIM. The same holds for continuity through phases and stages: a symbolic representation can contain the entire history of a building.

BIM is often called 'object-oriented'. This is misleading because the term has a different meaning in computer science but also because we should not equate symbols with real things. In English, the letter 'a' corresponds to five different sounds (phonemes). Knowing how to pronounce the letter depends on the context (the word). When considering representations in building, the correspondence between symbols and things can be even fuzzier. A window may be considered a discrete component, but a wall is an assemblage with variable composition and indeterminate form. Its material layers often continue into other walls, forming construction networks that are not captured by wall symbols in BIM. A main reason for this is geometric bias: continuous walls are segmented into separate symbols by the geometry of their axes.

Despite such fuzziness and resulting ambiguities, the symbolic representation underlying BIM remains the obvious choice for AECO computerisation, with a potential similar to that of the Latin alphabet or the Hindu-Arabic numerals. The graph of symbols and their relations is a transparent, consistent and efficient foundation for any application. The capacity for integration and continuity means that information efforts can be consolidated into a single representation that caters for all aspects, goals and disciplines.

1.1.2 Digital Twinning

While the use of BIM has yet to reach a satisfactory level or achieve significant efficiencies, AECO has already adopted a new buzzword: digital twinning. In contrast to BIM, digital twinning has yet to consolidate into a recognisable technology. Quite frequently, any virtual model seems to qualify as a digital twin, purely on the basis of intent. However, a digital twin is more than a model: it is a digital replica of something physical. It describes the form, behaviour and performance of the thing, including uses, users and direct context – all that is required for precise and accurate analyses and forecasts of future states of the physical twin.

Information in a digital twin is dynamic and reciprocal: sensors in the physical twin that monitor temperature, light, sound, occupancy, vibration, etc., send their data to the digital twin, where they become attached to relevant properties of the appropriate symbols. The products of the digital twin travel in the reverse direction, guiding actuators in operational adaptations, e.g. the functioning of heating systems, and informing users through displays (Fig. 1.2). In other words, the twins are connected in both directions in near real time and are capable of communication and synchronisation (Chen 2017; Liu et al. 2018). Consequently, we can distinguish between representations (static models, as in BIM), shadows (representations which are updated by data from the physical things) and twins (full two-way synchronisation) (Fuller et al. 2020; Sepasgozar 2021).

Digital twins of buildings are invariably based on BIM (Boje et al. 2020; Sacks et al. 2020; Begić and Galić 2021; Mêda et al. 2021; Shahat et al. 2021; Tagliabue et al. 2021; Alibrandi 2022; Shaharuddin et al. 2022). At the same time, it is stressed

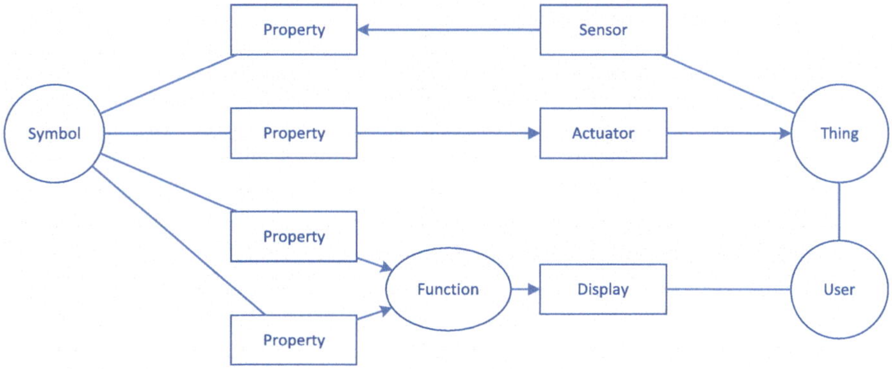

Fig. 1.2 Connections between symbols in a digital twin and things in a physical twin

that digital twinning is more than BIM, as it includes sociotechnical and process aspects, especially in use (Boje et al. 2020; Davila Delgado and Oyedele 2021; Sepasgozar 2021). This makes it significantly more demanding than as-built BIM in terms of reliability, precision and completeness. Furthermore, it is questionable whether BIM can accommodate and process the big data produced by sensors in the built environment. Rather than a foundation, BIM is a predecessor to digital twinning, based on the same symbolic approach to representation (Boje et al. 2020; Koutamanis et al. 2021).

More than on BIM, digital twinning relies on the Internet of Things (IoT): the networks that connect sensors, actuators and displays in a building, making it 'smart', i.e. automating certain operations, such as opening doors and regulating ventilation systems. In addition to such local automation, the IoT also collects data from all sources to capture the history and the overall conditions in a building. This improves local operations by connecting them to global goals and constraints. The IoT is not just an enabler but a necessity because digital twinning presupposes a building heavily populated by IoT for bidirectional communication and synchronisation, including feedback to users and operators (Farsi et al. 2020; Fuller et al. 2020; Lu et al. 2020; Sepasgozar 2021). The collection of data for digital twinning could be much more extensive than in most smart buildings, resulting in a lack of suitable physical twins and possibly rendering digital twinning a pipe dream. Alternatively, one could tolerate low-fidelity solutions as early deployment stages and encourage incremental development (Mêda et al. 2021). However, experience with BIM maturity levels suggests that such tolerance is self-defeating because it provides alibis for not taking the trouble to use the technology properly while continuing processes that actually undermine it. The degree of validation and verification required in digital twinning makes any attempt to pass off static models as twins as misguided as calling 2D drawings BIM.

1.1.3 Passports and Logbooks

BIM and digital twinning are general-purpose technologies. There are also stand-alone information technologies specifically developed for circularity in AECO. These are referred to by terms such as *building* or *material passport* or *logbook*. Chapter 5 by Honic et al. in this book describes the potential of such technologies and relevant life cycle and standardisation challenges in detail. Therefore, from the perspective of this chapter, it suffices to emphasise that BIM, as an integrated information environment, is more than a useful source of data (Durmišević 2018; Bertin et al. 2020). There is a significant overlap between BIM and material or building passports (Charef and Emmitt 2021), even when the latter are based on other sources for product composition breakdown.

The advantage of BIM is that it makes materials situated and connected to life cycle processes (Honic et al. 2019). This supports design for deconstruction and disassembly (Minunno et al. 2018; Xing et al. 2020; Marzouk and Elmaraghy 2021; O'Grady et al. 2021) and other circularity goals. Translating manufacturers' disassembly instructions into simulations in BIM improves legibility and completeness, especially concerning resources that may be available or required. It also verifies the disassembly procedures and validates designs with respect to them. Including the location of a component among its metaproperties in a passport does not offer the same advantages.

In conclusion, passports and logbooks are amenable to the integrating power of BIM and digital twins, which can accommodate product information (Kebede et al. 2022), life cycle energy data (Shah et al. 2023) and other key information in their properties and relations. In BIM, information collections such as material passports can become views of the model, similarly to bills of quantities. Linking their goals and constraints to all activities in design, construction and operation through BIM returns connections to information sources that help make material flow registration and analysis realistic and reliable (Miatto et al. 2022).

1.2 BIM in the Built Environment

There is general agreement that digital uptake in AECO is slow and limited, even though investment in digitisation may not be that low (Turk 2021; Koutamanis 2022). Nevertheless, BIM was received with unprecedented willingness and optimism as a solution to major inefficiencies and malperformances (Sacks et al. 2018; Ernstsen et al. 2021), but rapid adoption was not accompanied by a scope wide and coherent enough to effect fundamental changes. There are persistent complaints about BIM costs, complexity and social and organisational aspects that contrast with its arguably unrealistic promotion (Miettinen and Paavola 2014; Oesterreich and Teuteberg 2019) and put smaller enterprises at a disadvantage (Dainty et al. 2017; Murguia et al. 2023). BIM is commonly deployed in hybrid situations, where it overlaps with other technologies (Davies 2017). This conflicts with the holistic character of BIM and reduces its potential. As AECO remains attached to existing, document-based practices, BIM is generally restricted to office use and the production of such documents. Out of the office, the reliance of AECO on low-cost human labour does little to promote digitalisation.

Even in office use, BIM has not always facilitated innovation. Its emphasis on integration and interoperability is not linked to models of labour division and specialisation (Turk 2020). It is also questionable that complex assemblages such as buildings can be broken down into hierarchical ontologies by merely observing real-world buildings and following pre-existing, paper-based standards (Koutamanis et al. 2021). Unfortunately, such limitations are seldom experienced, as most applications and models tend to remain selective, partial and restricted to specific tasks, such as clash detection between load-bearing structures and building services.

BIM has yet to make its presence felt beyond design and construction, in the costly and resource-intensive use stage (Gao and Pishdad-Bozorgi 2019; Abideen et al. 2022; Benn and Stoy 2022; Durdyev et al. 2022; Matos et al. 2022; Pinti et al. 2022; Tsay et al. 2022). Making and especially maintaining as-is models appears to be beyond the scope or capacities of most organisations, which are already overwhelmed by the amount of existing information and the multiplicity of channels through which they exchange information.

1.3 BIM and Digital Twinning for a Circular Economy

BIM, while not perfect, remains preferable to its predecessors and indicative of the symbolic direction building representations are taking. Implemented properly, it offers information integration and continuity, unambiguous interpretation by both humans and machines and full and reliable support of complex analyses. This supports goals such as circularity and the information-intensive processes they require.

At the same time, present limitations in BIM create interest in technological advances. Digital twinning promises the additional capacity to accommodate and process all states of the physical twin, past and present (Rafael Sacks et al. 2020). This helps transform static evaluations into dynamic life cycle processes, combining, e.g. end-of-life assessment with adaptable planning (Chen et al. 2021). This transition from static to dynamic is demanding but seems justified by feasibility evaluations, which confirm a significant potential for improved life cycle assessment and control (Tagliabue et al. 2021).

Neither BIM nor digital twinning are goals for AECO; they are means towards domain-specific performances. Moreover, circularity may be viewed as an imposed, external societal constraint. As with any such constraint, it may conflict with established practices and be poorly served by existing tools, which are attuned to other priorities. To remove such obstacles, the general capacities of digital twinning, BIM and digitalisation should be taken for granted, and attention should be on specific, critical issues (Çetin et al. 2021). General intentions, such as reducing inefficiencies, improving communication, optimising design performance or just providing visualisations (Wong and Fan 2013; Akinade et al. 2017; Minunno et al. 2018; Charef and Emmitt 2021), can be relevant but do not amount to a specific, coherent approach.

1.3.1 Registration of Relevant Information

The first step in a coherent approach to circularity with BIM or digital twins is to learn to rely on symbolic representation. Any full model or twin can easily cover circularity information needs without additional investment, but in practice representations can be selective or opportunistic and hence incomplete or inconsistent. Deferring the information burden to any particular goal and its stakeholders (as with passports) is not a viable option. Instead, all AECO stakeholders should insist on joint, permanent working environments, not disconnected repositories or documentation for different phases. There can be no half-hearted BIM or digital twin deployment: economising on investment means severely limited potential and low returns.

The first reason why a digital solution cannot be made for circularity solely is *cost:* the value of what it supports can hardly be justified by the returns, certainly in the perception of most AECO stakeholders with different priorities. General-purpose solutions such as BIM are clearly preferable because they support most such priorities. If circular goals can be added to them, then circularity stakeholders can reap the benefits, while others are stimulated to include circularity in their considerations.

The perennial question in AECO is not so much who makes a BIM model but who maintains it, especially in the life cycle of a building. If this does not happen collaboratively by conjoining the core processes of all actors, and preferably automatically, there is little hope for success. Collaborative solutions also lower the participation threshold for smaller enterprises and offer enticing benefits in terms of digital support and room for fruitful specialisation. In return, the enterprises contribute to the completeness and up-to-dateness of information simply by using it.

The second reason for a lack of digital solutions for circularity is *selectivity:* any information solution motivated primarily or exclusively by circularity inevitably remains restricted to circularity factors and aspects. It may even suffer from inattentional blindness, which causes omissions of important data simply because we concentrate on other matters (Chabris and Simons 2010). One can naturally work with conscious concentration towards a full, inclusive solution, but then the results would amount to something akin to BIM or digital twins, i.e. a comprehensive solution that could only justify costs and improve returns by being open to other goals and priorities, too.

1.3.2 Exploration of Circular Operations

The second step towards circularity with BIM or digital twins is to utilise their capacities for exploring deconstruction and disassembly (Akanbi et al. 2019; van den Berg et al. 2021). In the same way that we simulate construction processes, we can also simulate the expected maintenance, refurbishment, renovation and deconstruction processes with the accuracy and precision required for feasibility, effectiveness and efficiency. This provides direct support for construction-related circularity goals (narrow and regenerate through efficiency improvement) and a useful background for others (slow and close through reliable life cycle projections). It also stresses the necessity of detail and realism. For deconstruction in particular, we should acknowledge that it is not a mere reversal of construction. As Van den Berg explains in Chap. 11 in the relevant chapter in this volume, information is a key issue in organising reverse logistics. As-is representations are essential for the identification and harvesting of reusable resources from existing buildings because as-built models (i.e. construction documentation) are neither sufficient nor reliable enough. Closing loops requires certainty about the state of components and materials, as well as about their physical context, which has changed from an accommodating construction site to a finished, functioning building. This calls for solutions that are full and realistic, including all details of deconstruction in space and time, e.g. how cranes and scaffolds would function in the existing building. Van den Berg (Chap. 11) describes a number of focused explorations and demonstrations that must graduate from opportunistic demonstrations of potential to standard facilities in BIM and digital twinning.

1.3.3 Constraining Design, Construction and Operation

Based on the second step, we should explicitly describe circularity dependencies and constraints in properties and relations of symbols (e.g. constraints on interfacing between components for effective deconstruction). Relations are of particular importance in this respect because they link interfacing between components to symbol behaviours. They can ensure that the building design and construction allow for deconstruction (Sanchez et al. 2021), e.g. avoid additions that spoil interfaces designed for disassembly, such as equalising layers of in situ concrete over demountable floor slabs. If symbols refuse to accept such additions to their properties or relations, similarly to a door not accepting positioning outside a wall, the scope for human error becomes much smaller. This is particularly important in the use phase, where changes are only too frequently improvised, in both refurbishment and maintenance. The representation can also anticipate circularity operations, such as the replacement of some components when they fall below a certain performance level, by including among the symbol triggers that adjust the timing of loops.

1.3.4 Life Cycle Registration and Guidance

The final step is an extension of the previous three: use 4D symbolic representations to monitor the detailed history of a building, preferably in near-real time. As symbol properties and relations can register the activities and effects of maintenance, refurbishment, etc., material flows are measured and managed not by questionable proxies but with primary, precise and accurate data (Minunno et al. 2018; Chen and Huang 2020; Marzouk and Elmaraghy 2021). Up-to-date information is essential for the planning of circularity operations: narrowing, slowing, closing and regenerating can be based not just on initial assumptions and projections but on constantly refinable and dynamic decision frameworks that include permanent validation and verification facilities for making sense of the existing building conditions for deconstruction (Van den Berg, Chap. 11). The bidirectional relation between digital twins and buildings is clearly advantageous in this respect, as it covers not only monitoring but also adaptations in the behaviour of the physical twins, e.g. adjusting the heating and ventilation of a building in order to reduce the extent of material ageing in specific components.

1.4 Current Applications of BIM and Digital Twinning to Circularity

Judging the efficacy of a technology or approach requires realistic applications that can be analysed with respect to both means and ends. However, most publications on circularity and digital twinning, as well as many on circularity and BIM, are programmatic or aspirational. They focus on aspects such as technology and platform development, enablers and challenges (Copeland and Bilec 2020; Fuller et al. 2020; Ganiyu et al. 2020; Rafael Sacks et al. 2020; Davila Delgado and Oyedele 2021; Sepasgozar 2021; Shahat et al. 2021; Ammar et al. 2022; Charef 2022). Actual case studies are thin on the ground and mostly presented as plans or untested prototypes. The best examples illustrate that highly specific subjects and goals are beneficial for both the setup of a digital twin and analyses in it (Funari et al. 2021). Laboratory case studies, however limited, represent useful steps forward, especially for learning and testing (Rocca et al. 2020; Marzouk and Elmaraghy 2021).

The narrow scope of digital twinning case studies is inherent to any early stage. With the sensitising of architects, engineers, authorities and clients to environmental issues and the life cycles of materials, ambition and attention inevitably become dispersed over a wide range of subjects and possibilities, from key applications in AECO to promising digital technologies (Hillebrandt et al. 2019; Çetin et al. 2021), arguably at the cost of coherence, consistency and effectivity. There is no uniform solution that applies to all aspects and goals. Each component, material or building has different potential, not just generically but in every instance and situation.

However, even advanced and convincing cases with a narrow and well-defined scope, such as bridge maintenance, still fall short of a full digital twin (Mahmoodian et al. 2022). Other studies are hampered by the small samples available, as longitudinal or long-term data are required for consistent and reliable results (Rita et al. 2022). This is particularly true of attempts to go beyond the microscale of materials and elements and extend to the macroscales of neighbourhoods and cities, so as to identify and promote synergies (Bejtullahu and Morishita-Steffen 2021). Such extensions inevitably shift attention from new designs to the existing stock. Existing buildings, especially historical ones, involve knowledge not easy to codify in systems developed for today. So, it is not only information we are lacking, it is also decision-making and design tools (Durmišević 2018; Bianchini et al. 2021).

One of the key problems with case analysis is that evaluation tends to be weak, based on opinion rather than objective criteria. Information collected through questionnaires, interviews and similar means (Charef and Emmitt 2021; Çetin et al. 2022) should not be taken at face value. It contains opinions, subjective estimates and uncorroborated reported results that indicate belief or strategic support for potential rather than tangible, verifiable results. As time-use studies demonstrate, personal estimates can be heavily biased by goals and emotions: stressed people overestimate how they spend their time and produce sums of more than 24 hours per day (Robinson and Godbey 1999). This calls for yet another use of BIM and digital twins: the collection of reliable, comprehensive and consistent data, which can be processed through generally accepted methods towards case analyses and benchmarks. Without such objective information processing, it is impossible to arrive at clear evidence that not only convinces but also shows what can be improved and how.

1.5 Business Models for BIM and Digital Twinning in a Circular Built Environment

Business models address organisational aspects, such as who, what, when and how in key tasks that contribute towards delivering desired results and outcomes. Information is of critical significance here, especially in product-as-service, bundling, dematerialisation, life extension and similar models that depend on fine-tuning or combinations (Charter and McLanaghan 2018; McCausland 2022). These require transitions from production-driven to customer-centred approaches and changes in collaboration patterns and supply chain structures (Qi et al. 2022; Wang et al. 2022; Xiang et al. 2022). Whether the business model follows an innovation or a resource strategy (Bocken and Ritala 2022), rich information is a prerequisite for reliability and feasibility (Shah et al. 2023). Projected states and indicators must be substantiated and monitored, so that lessons learned are fed back to related decisions.

The same organisational aspects and their goals are critical for the utilisation of information technologies. Despite the key role of information, the digitalisation of products and processes is not always included in digital twinning business models, which often retain legacy conditions and practices (Deckert et al. 2022). Digitalisation is still treated as external to core processes: a layer to be superimposed when needed. Consequently, the business case for digitalisation and information is hampered by investment and operation costs that are deemed too high, despite the promise of substantial efficiency improvement.

In AECO, digitalisation has yet to develop into a connecting tissue between all stakeholders and actors, as in other economic areas (Floridi 2014). Attachment to analogue practices and their information carriers remains too strong, regardless of changes in the objectives of projects, enterprises or society. This contrasts sharply not only with other industries but even more with daily life. The same AECO practitioners who are reluctant to fully embrace integrated digital information solutions in their professional activities make extensive use of social media, e-commerce, e-banking, etc., in their private lives. The result is that AECO computerisation is characterised by isolated islands, not the networks necessary for business value. BIM, digital twinning and all other forms of digital information are treated as the product of integration rather than the *integrator* that enables better collaboration and performance (Davila Delgado and Oyedele 2021).

This does not imply lack of attempts at new business models that build on digitalisation. On the contrary, there are many proposals from which we can learn. Looking at business models related to digital twinning (as the most demanding case) across application areas, industries and countries (Kumar et al. 2022), certain characteristics emerge:

- The emphasis is on potential (rather than effectiveness), particularly for competitiveness, which requires venturing beyond legacy solutions and comfort zones.
- Control applications appear to offer easier deployment than production applications, but in both cases the main promise is value co-creation through support for decision-making and management of operations and services (West et al. 2021).
- Differences between industries are largely due to legacy practices and industry structures (Morelli et al. 2022). There appears to be no uniform solution for universal transformation.
- Importance is attached to platforms, autonomous stakeholders operating on them and networks emerging from the interaction between stakeholders and platforms (Rocca et al. 2020).

In summary, digital twinning seems not easily attainable in practice, especially for subjects like buildings, which undergo many, often invisible changes in their protracted lifespans and require a high level of detail to capture both contexts and user experiences.

Some therefore argue that the business case should be motivated by a clear goal such as the reduction of energy consumption. This guides the development of business value towards measurable results while serving wider societal goals like sustainability and improving the lives of users and consumers. They also stress that data strategies should be imposed top-down, as part of business value, rather than left to the willingness or ability of stakeholders and actors (Apte and Spanos 2021).

Such arguments sound autocratic but nevertheless produce clear solutions in a notoriously fragmented and backward-looking industry like AECO. Judging from the half-hearted commitment and relatively low investment in computerisation, business models involving BIM or digital twinning need to include the technologies in their core and give them the primary role of integrator. Developing add-on business models for digitalisation on top of circularity models is self-defeating because it makes information technologies an option, moreover an expensive one, with tenuous connections to goals and values. So long as stakeholders are under the impression that circularity in the built environment is feasible without a radical digital reform of practically all processes, there is little hope for wide and effective deployment.

Digitalisation should be specified according to general principles, rather than specific objectives such as circularity, so as to ensure inclusiveness and completeness. This provides the necessary context for explaining how different aspects can support each other in the business model, e.g. how maintenance activities contribute to the fine-tuning of timely deconstruction, thereby alleviating the burden of fact-finding in circularity monitoring and assessment. Conversely, circularity constraints guide maintenance towards not only timely replacement but also higher performance in the building.

1.6 Discussion

One thing we no longer need to justify or defend is digitalisation. Everyone is aware of its importance and pervasiveness. The fact that information is key to digitalisation is sometimes less obvious, let alone that information is the integrator of human interactions. Goals like circularity are not only highly demanding in information, they also require radical changes in all related industries. These characteristics make circularity clearly dependent on the digital transformation of the whole of AECO, in the same way that digitalisation has transformed communications, entertainment, social contacts, etc. While such transformation is feasible, the problem with digitalisation in AECO is not lack of potential but low priority. So long as it is seen as a mere means to basic tasks, it cannot deliver its full promise. In turn, this reduces willingness to invest in digitalisation and hence the performance of digital solutions.

To break this vicious circle, brave plans are necessary. Circularity has to assume fully integrated digital information for the built environment and include it in the core of its processes as the connecting tissue between aspects, stakeholders and actors. In other words, the first, critical step is that AECO commits to BIM and applies it to all aspects and tasks. This ensures reliable and effective support for circularity, as well as a wide scope for it, for two key reasons. Firstly, being successful with just a few components or materials does not justify the circularity claims and investments – for circularity to be truly effective, it must apply widely to the built environment. Secondly, to achieve that, circularity must be present in all aspects, become embraced by the corresponding disciplines and made part of their goals and methods. Keeping it separate, as an additional layer, turns it into an afterthought and an option.

This information environment cannot be initiated by any single aspect or goal. Circularity may endorse it, but it is the whole of AECO that must sustain it throughout the life cycle. This sounds like a tall order, but thankfully BIM, properly and consistently applied, is a good starting point. Its limitations are not trivial but not such that they preclude effectiveness and efficiency in any discipline or the collaboration between disciplines. What AECO needs is more experience with working in such an environment – experience that can be invaluable in further transitions, e.g. to the enticing prospect of digital twinning.

1.7 Key Takeaways

- BIM has considerable potential to integrate information processing, thus providing comprehensive and situated information that covers most circularity needs.
- BIM seamlessly links circularity to other activities in design, construction and operation.
- Digital twinning promises even more: digital replicas in full synchronisation with the physical twin and its past, present and future states.
- The successful deployment of powerful technologies such as BIM and digital twinning requires significant investment, commitment and consistency.

References

Abideen DK, Yunusa-Kaltungo A, Manu P, Cheung C (2022) A systematic review of the extent to which BIM is integrated into operation and maintenance. Sustainability 14:8692. https://doi.org/10.3390/su14148692

Akanbi LA, Oyedele LO, Omoteso K, Bilal M, Akinade OO, Ajayi AO, Davila Delgado JM, Owolabi HA (2019) Disassembly and deconstruction analytics system (d-das) for construction in a circular economy. J Clean Prod 223:386–396. https://doi.org/10.1016/j.jclepro.2019.03.172

Akinade OO, Oyedele LO, Omoteso K, Ajayi SO, Bilal M, Owolabi HA, Alaka HA, Ayris L, Henry Looney J (2017) BIM-based deconstruction tool: towards essential functionalities. Int J Sustain Built Environ 6:260–271. https://doi.org/10.1016/j.ijsbe.2017.01.002

Alibrandi U (2022) Risk-informed digital twin of buildings and infrastructures for sustainable and resilient urban communities. ASCE-ASME J Risk U A 8:04022032. https://doi.org/10.1061/AJRUA6.0001238

Ammar A, Nassereddine H, AbdulBaky N, AbouKansour A, Tannoury J, Urban H, Schranz C (2022) Digital twins in the construction industry: a perspective of practitioners and building authority. Front Built Environ 8. https://doi.org/10.3389/fbuil.2022.834671

Apte PP, Spanos CJ (2021) The digital twin opportunity. MIT Sloan Manag Rev 63:15–17

Begić H, Galić M (2021) A systematic review of construction 4.0 in the context of the BIM 4.0 premise. Buildings 11:337. https://doi.org/10.3390/buildings11080337

Bejtullahu F, Morishita-Steffen N (2021) From resilient and regenerative materials to a resilient and regenerative built environment. In: Andreucci MB, Marvuglia A, Baltov M, Hansen P (eds) Rethinking sustainability towards a regenerative economy. Springer International Publishing, Cham, pp 263–278

Benn M, Stoy C (2022) BIM for CREM: exploring the benefit of building information modelling for facility management in corporate real estate management. Buildings 12:400. https://doi.org/10.3390/buildings12040400

Bertin I, Mesnil R, Jaeger J-M, Feraille A, Le Roy R (2020) A BIM-based framework and databank for reusing load-bearing structural elements. Sustainability 12:3147. https://doi.org/10.3390/su12083147

Bianchini C, Attenni M, Potestà G (2021) Regenerative design tools for the existing city: Hbim potentials. In: Andreucci MB, Marvuglia A, Baltov M, Hansen P (eds) Rethinking sustainability towards a regenerative economy. Springer International Publishing, Cham, pp 23–43

Bocken N, Ritala P (2022) Six ways to build circular business models. J Bus Strategy 43:184–192. https://doi.org/10.1108/JBS-11-2020-0258

Boje C, Guerriero A, Kubicki S, Rezgui Y (2020) Towards a semantic construction digital twin: directions for future research. Autom Constr 114:103179. https://doi.org/10.1016/j.autcon.2020.103179

Çetin S, De Wolf C, Bocken N (2021) Circular digital built environment: an emerging framework. Sustainability 13:6348

Çetin S, Gruis V, Straub A (2022) Digitalization for a circular economy in the building industry: multiple-case study of Dutch social housing organizations. RCR Adv 15:200110. https://doi.org/10.1016/j.rcradv.2022.200110

Chabris CF, Simons DJ (2010) The invisible gorilla : and other ways our intuitions deceive us. Crown, New York

Charef R (2022) The use of building information modelling in the circular economy context: several models and a new dimension of BIM (8d). Clean Eng Technol 7:100414. https://doi.org/10.1016/j.clet.2022.100414

Charef R, Emmitt S (2021) Uses of building information modelling for overcoming barriers to a circular economy. J Clean Prod 285:124854. https://doi.org/10.1016/j.jclepro.2020.124854

Charter M, McLanaghan S (2018) Business models for a circular economy. In: Charter M (ed) Designing for the circular economy. Routledge, London, pp 89–101

Chen Y (2017) Integrated and intelligent manufacturing: perspectives and enablers. Engineering 3:588–595. https://doi.org/10.1016/J.ENG.2017.04.009

Chen Z, Huang L (2020) Digital twin in circular economy: remanufacturing in construction. IOP Conf Ser Earth Environ Sci 588:032014. https://doi.org/10.1088/1755-1315/588/3/032014

Chen C, Zhao Z, Xiao J, Tiong R (2021) A conceptual framework for estimating building embodied carbon based on digital twin technology and life cycle assessment. Sustainability 13:13875. https://doi.org/10.3390/su132413875

Copeland S, Bilec M (2020) Buildings as material banks using RFID and building information modeling in a circular economy. Procedia CIRP 90:143–147. https://doi.org/10.1016/j.procir.2020.02.122

Dainty A, Leiringer R, Fernie S, Harty C (2017) BIM and the small construction firm: a critical perspective. Build Res Inf 45:696–709. https://doi.org/10.1080/09613218.2017.1293940

Davies K (2017) Making friends with frankenstein: hybrid practice in BIM. Eng Constr Archit Manag 24:78–93. https://doi.org/10.1108/ECAM-04-2015-0061

Davila Delgado JM, Oyedele L (2021) Digital twins for the built environment: learning from conceptual and process models in manufacturing. Adv Eng Inform 49:101332. https://doi.org/10.1016/j.aei.2021.101332

Deckert C, Kalefeld J, Kutz M (2022) Business model innovation for the internet of things. Springer International Publishing, Cham, pp 321–333

Durdyev S, Ashour M, Connelly S, Mahdiyar A (2022) Barriers to the implementation of building information modelling (BIM) for facility management. J Build Eng 46:103736. https://doi.org/10.1016/j.jobe.2021.103736
Durmišević E (2018) Reversible building design. In: Charter M (ed) Designing for the circular economy. Routledge, London, pp 344–359
Ernstsen SN, Whyte J, Thuesen C, Maier A (2021) How innovation champions frame the future: three visions for digital transformation of construction. J Constr Eng Manag 147:05020022. https://doi.org/10.1061/(ASCE)CO.1943-7862.0001928
Farsi M, Daneshkhah A, Hosseinian-Far A, Jahankhani H (eds) (2020) Digital twin technologies and smart cities. Springer International Publishing, Cham
Floridi L (2014) The 4th revolution: how the infosphere is reshaping human reality. Oxford University Press, New York/Oxford
Fuller A, Fan Z, Day C, Barlow C (2020) Digital twin: enabling technologies, challenges and open research. IEEE Access 8:108952–108971. https://doi.org/10.1109/ACCESS.2020.2998358
Funari MF, Hajjat AE, Masciotta MG, Oliveira DV, Lourenço PB (2021) A parametric scan-to-fem framework for the digital twin generation of historic masonry structures. Sustainability 13: 11088. https://doi.org/10.3390/su131911088
Ganiyu SA, Oyedele LO, Akinade O, Owolabi H, Akanbi L, Gbadamosi A (2020) BIM competencies for delivering waste-efficient building projects in a circular economy. DIBE 4:100036. https://doi.org/10.1016/j.dibe.2020.100036
Gao X, Pishdad-Bozorgi P (2019) BIM-enabled facilities operation and maintenance: a review. Adv Eng Inform 39:227–247. https://doi.org/10.1016/j.aei.2019.01.005
Hillebrandt A, Riegler-Floors P, Rosen A, Seggewies J-K (2019) DETAIL. München
Honic M, Kovacic I, Rechberger H (2019) Improving the recycling potential of buildings through material passports (MP): an Austrian case study. J Clean Prod 217:787–797. https://doi.org/10.1016/j.jclepro.2019.01.212
Kebede R, Moscati A, Tan H, Johansson P (2022) Integration of manufacturers' product data in BIM platforms using semantic web technologies. Autom Constr 144:104630. https://doi.org/10.1016/j.autcon.2022.104630
Koutamanis A (2022) Building information – representation and management: principles and foundations for the digital era. TU Delft OPEN, Delft
Koutamanis A, Dainty A, Kvan T, Turk Ž (2021) The enigma of BIM. Archit Struct Construct 3:1. https://doi.org/10.1007/s44150-021-00017-6
Kumar V, Leng J, Akberdina V, Kuzmin E (eds) (2022) Digital transformation in industry: digital twins and new business models. Springer International Publishing, Cham
Liu Z, Meyendorf N, Mrad N (2018) The role of data fusion in predictive maintenance using digital twin. AIP Conf Proc 1949:020023. https://doi.org/10.1063/1.5031520
Lu Q, Xie X, Heaton J, Parlikad AK, Schooling J (2020) From BIM towards digital twin: strategy and future development for smart asset management. Springer International Publishing, Cham, pp 392–404
Mahmoodian M, Shahrivar F, Setunge S, Mazaheri S (2022) Development of digital twin for intelligent maintenance of civil infrastructure. Sustainability 14:8664. https://doi.org/10.3390/su14148664
Marzouk M, Elmaraghy A (2021) Design for deconstruction using integrated lean principles and BIM approach. Sustainability 13:7856. https://doi.org/10.3390/su13147856
Matos R, Rodrigues H, Costa A, Rodrigues F (2022) Building condition indicators analysis for BIM-FM integration. Arch Comput Methods Eng 29:3919–3942. https://doi.org/10.1007/s11831-022-09719-6
McCausland T (2022) Digital twins. Res Technol Manag 65:69–71. https://doi.org/10.1080/08956308.2022.1999637
Mêda P, Calvetti D, Hjelseth E, Sousa H (2021) Incremental digital twin conceptualisations targeting data-driven circular construction. Buildings 11:554. https://doi.org/10.3390/buildings11110554

Miatto A, Sartori C, Bianchi M, Borin P, Giordano A, Saxe S, Graedel TE (2022) Tracking the material cycle of italian bricks with the aid of building information modeling. J Ind Ecol 26: 609–626. https://doi.org/10.1111/jiec.13208
Miettinen R, Paavola S (2014) Beyond the BIM utopia: approaches to the development and implementation of building information modeling. Autom Constr 43:84–91. https://doi.org/10.1016/j.autcon.2014.03.009
Minunno R, O'Grady T, Morrison GM, Gruner RL, Colling M (2018) Strategies for applying the circular economy to prefabricated buildings. Buildings 8:125. https://doi.org/10.3390/buildings8090125
Morelli G, Musso F, Murmura F, Bravi L (2022) Knowledge analysis on the industry 4.0 diffusion in italian manufacturing: opportunities and threats. Springer International Publishing, Cham, pp 195–214
Murguia D, Vasquez C, Demian P, Soetanto R (2023) BIM adoption among contractors: a longitudinal study in Peru. J Constr Eng Manag 149:04022140. https://doi.org/10.1061/(ASCE)CO.1943-7862.0002424
O'Grady TM, Brajkovich N, Minunno R, Chong H-Y, Morrison GM (2021) Circular economy and virtual reality in advanced BIM-based prefabricated construction. Energies 14:4065. https://doi.org/10.3390/en14134065
Oesterreich TD, Teuteberg F (2019) Behind the scenes: understanding the socio-technical barriers to BIM adoption through the theoretical lens of information systems research. Technol Forecast Soc Change 146:413–431. https://doi.org/10.1016/j.techfore.2019.01.003
Pinti L, Codinhoto R, Bonelli S (2022) A review of building information modelling (BIM) for facility management (FM): implementation in public organisations. Appl Sci 12:1540. https://doi.org/10.3390/app12031540
Qi Q, Tao F, Nee AYC (2022) Chapter 1 – From service to digital twin service. In: Tao F, Qi Q, Nee AYC (eds) Digital twin driven service. Academic Press, pp 1–31
Rita LPS, Silva JR, Junior RRF (2022) Best regional practices for digital transformation in industry: the case of the industry 4.0 program in Portugal. Springer International Publishing, Cham, pp 163–181
Robinson JP, Godbey G (1999) Time for life : the surprising ways americans use their time. Pennsylvania State University Press, University Park
Rocca R, Rosa P, Sassanelli C, Fumagalli L, Terzi S (2020) Integrating virtual reality and digital twin in circular economy practices: a laboratory application case. Sustainability 12:2286. https://doi.org/10.3390/su12062286
Sacks R, Eastman C, Lee G, Teicholz P (2018) BIM handbook. Wiley, Hoboken
Sacks R, Brilakis I, Pikas E, Xie HS, Girolami M (2020) Construction with digital twin information systems. Data-centric Eng 1:e14. https://doi.org/10.1017/dce.2020.16
Sanchez B, Rausch C, Haas C, Hartmann T (2021) A framework for BIM-based disassembly models to support reuse of building components. Resour Conserv Recycl 175:105825. https://doi.org/10.1016/j.resconrec.2021.105825
Sepasgozar SME (2021) Differentiating digital twin from digital shadow: elucidating a paradigm shift to expedite a smart, sustainable built environment. Buildings 11:151. https://doi.org/10.3390/buildings11040151
Shah D, Kathiriya H, Suthar H, Pandya P, Soni J (2023) Enhancing the building's energy performance through building information modelling—a review. Springer Nature Singapore, Singapore, pp 247–253
Shaharuddin S, Abdul Maulud KN, Syed Abdul Rahman SAF, Che Ani AI (2022) Digital twin for indoor disaster in smart city: a systematic review. Int Arch Photogramm Remote Sens Spat Inf Sci XLVI-4:315–322. https://doi.org/10.5194/isprs-archives-XLVI-4-W3-2021-315-2022
Shahat E, Hyun CT, Yeom C (2021) City digital twin potentials: a review and research agenda. Sustainability 13:3386. https://doi.org/10.3390/su13063386

Tagliabue LC, Cecconi FR, Maltese S, Rinaldi S, Ciribini ALC, Flammini A (2021) Leveraging digital twin for sustainability assessment of an educational building. Sustainability 13:480. https://doi.org/10.3390/su13020480
Tsay GS, Staub-French S, Poirier É (2022) BIM for facilities management: an investigation into the asset information delivery process and the associated challenges. Appl Sci 12:9542. https://doi.org/10.3390/app12199542
Turk Ž (2020) Interoperability in construction – mission impossible? DIBE 4:100018. https://doi.org/10.1016/j.dibe.2020.100018
Turk Ž (2021) Structured analysis of ICT adoption in the european construction industry. Int J Constr Manag 1–7. https://doi.org/10.1080/15623599.2021.1925396
van den Berg M, Voordijk H, Adriaanse A (2021) BIM uses for deconstruction: an activity-theoretical perspective on reorganising end-of-life practices. Constr Manag Econ 39:323–339. https://doi.org/10.1080/01446193.2021.1876894
Wang X, Kumar V, Kumari A, Kuzmin E (2022) Impact of digital technology on supply chain efficiency in manufacturing industry. Springer International Publishing, Cham, pp 347–371
West S, Stoll O, Meierhofer J, Züst S (2021) Digital twin providing new opportunities for value co-creation through supporting decision-making. Appl Sci 11:3750. https://doi.org/10.3390/app11093750
Wong K, Fan Q (2013) Building information modelling (BIM) for sustainable building design. Facilities 31:138–157. https://doi.org/10.1108/02632771311299412
Xiang F, Fan J, Ke S, Zuo Y (2022) Chapter 2 – Digital twin-driven service collaboration. In: Tao F, Qi Q, Nee AYC (eds) Digital twin driven service. Academic Press, pp 33–58
Xing K, Kim KP, Ness D (2020) Cloud-BIM enabled cyber-physical data and service platforms for building component reuse. Sustainability 12:10329. https://doi.org/10.3390/su122410329

Alexander Koutamanis studied architecture in Greece and received his PhD from Delft University of Technology, where he is associate professor of computational design. He has also worked as a designer and consultant in practice. His research interests include representation, recognition, information, interaction, affordances, briefing and decision support. Recent publications include an open textbook on building information management, an edited volume on construction and demolition waste recycling (including a chapter on BIM) and journal papers on dimensions and potential in BIM.

Chapter 2
Geographic Information Systems for Circular Cities and Regions

Tanya Tsui, Wendy Wuyts, and Karel Van den Berghe

Abstract A geographic information system (GIS) stores, manipulates, analyses, and visualises spatial data. GIS enables the mapping of building elements and components and can optimise the location of facilities for circular activities, thus contributing to the closing of material loops and the spatial development of circular cities and regions. This chapter presents use cases of GIS in the circular built environment, with examples from academia, industry, and government. Academics use GIS data for urban mining studies to estimate the location and availability of secondary construction materials. Businesses in industry use GIS analysis to inform the facility location of circular construction hubs and (reverse) logistics. Governments use GIS to monitor and assess the circular spatial development potential of their (industrial) territories. In order to integrate GIS into circular economy solutions, improvements need to be made in making spatial data available and in presenting findings that emerge from it. Finally, present enthusiasm for GIS tools should be balanced by a deeper understanding of the connection between digital tools and governance decisions.

Keywords Geographic information systems · GIS governance tools · Spatial data · Spatial analysis · Circular cities

T. Tsui (✉) · K. Van den Berghe
Faculty of Architecture and the Built Environment, Delft University of Technology, Delft, The Netherlands
e-mail: t.p.y.tsui@tudelft.nl

W. Wuyts
Department of Civil Engineering and Manufacturing, Norwegian University of Science and Technology, Gjøvik, Norway

Omtre AS, Hønefoss, Norway

C. De Wolf et al. (eds.), *A Circular Built Environment in the Digital Age*, Circular Economy and Sustainability, https://doi.org/10.1007/978-3-031-39675-5_2

2.1 What Is GIS?

A geographic information system (GIS) is a system for managing, analysing, and visualising geographic data. Geographic data integrates location data (where things are) with all types of descriptive information (what things are like there). GIS is utilised in multiple technologies, processes, techniques, and methods, and it is associated with various disciplines, including engineering, planning, management, logistics, telecommunications, and business. The ubiquity of GIS can be attributed to the fact that a large variety of problems are affected by their location and thus can incorporate the use of location data (Goodchild 2010; Chang 2018).

2.2 GIS in the Built Environment

Within the built environment, GIS is used as a tool to create, share, and analyse spatial data. Spatial data related to the built environment can be created from processing data sources such as satellite images. This can be seen in the creation of high-resolution 3D models of cities using photogrammetry (ArcGIS 2023a), LiDAR (laser imaging, detection, and ranging) (TU Delft 2023), and Google Street View data (Spotr 2023; Chap. 3 by Gordon et al. on scanning technologies). These models are especially relevant for the built environment: 3D data is essential for various analyses, including estimations of wind load, solar exposure, and temperature changes in city blocks or neighbourhoods.

Spatial data can also be visualised and shared, allowing stakeholders to track and maintain elements in buildings, infrastructure, and even cities. Tracking is often achieved in combination with other digital technologies, such as building information modelling (BIM). This can be seen in the tracking and tracing of urban infrastructure (ESRI 2020), as well as city management systems that crowdsource citizens' maintenance requests for their municipality on a map (Liu 2021). The tracking of elements in the built environment can be seen in the development of digital twins – virtual models of the built environment that store detailed information on urban elements, such as buildings, greenery, and infrastructure.

Finally, spatial data can be analysed to create insights and aid decision-making in the built environment. This can be seen in transportation optimisation (Santi et al. 2014), site selection of services (Kontos et al. 2005), analysing energy potential (van den Dobbelsteen et al. 2011), and urban morphology (Spatial Morphology Group 2023).

2.3 GIS for a Circular Built Environment

In recent years, researchers from industrial ecology, economic geography, and urban planning have highlighted the importance of space as a major factor in the study of circular economy (Wuyts et al. 2022; Bahers et al. 2022; Bucci et al. 2022). Creating, sharing, and analysing spatial data in GIS can therefore bring further insights and aid decision-making in a circular built environment. This section provides a brief overview of GIS for a circular built environment through creating, mapping, sharing, and analysing spatial data. Section 2.4 provides detailed examples of these methods, introducing cases from academia, industry, and government.

GIS's capability in creating and mapping spatial data can be used to visualise material flows and the availability of secondary material and land. Existing spatial data can be used to map out the availability of secondary materials embedded in buildings today in a process known as 'urban mining' (Van den Berghe and Verhagen 2021). An example can be seen in Fig. 2.1. Flows of secondary materials can be mapped using transportation data or waste management data. The mapping of stocks and flows can highlight hotspots with a high concentration of available secondary resources or the presence of material reuse. This information can then be used to assist the planning of the location of material reuse actors – facilities that collect, store, and redistribute construction waste to be reused in new construction sites, thus narrowing and closing material loops (see Sect. 2.4.1).

Sharing spatial data in GIS platforms allows for tracking buildings, infrastructure, and waste flows. Tracking the conditions of buildings and infrastructure allows

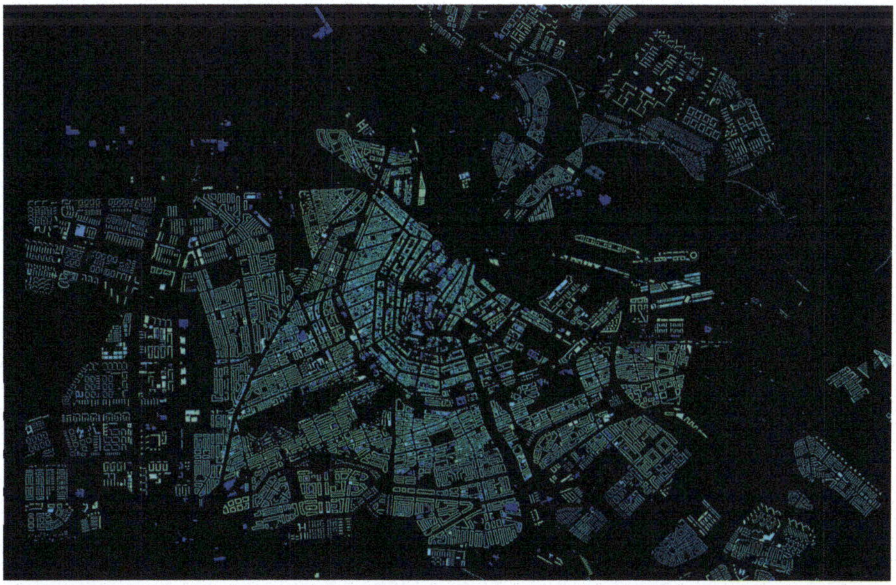

Fig. 2.1 Mapping of copper availability in Amsterdam, using open data on residential buildings (Waag 2016)

maintenance and repair works to be conducted in a coordinated and timely manner, which ensures a longer life cycle, resulting in the slowing of material loops. Digital platforms, which allow for the exchange of secondary materials, often have a mapping element that allows users to know the locations of stakeholders (Rotor 2023; Superuse 2023) (see also Sects. 2.4.2.2 and 2.4.2.3). By tracking waste material flows, governments can also monitor their level of circularity and transportation emissions, giving policymakers a better understanding of their progress towards circularity (see Sect. 2.4.3).

Finally, spatial data analysis in GIS can create insights that aid decision-making in a circular built environment, often at a city or regional scale, for closing material loops. By analysing spatial parameters of locations such as accessibility and proximity to amenities, site selection and facility location analysis can be conducted to find suitable locations for circular activities, such as facilities for recycling or remanufacturing (see Sects. 2.4.2.1 and 2.4.3.1). Clustering analysis can be conducted on hotspots of circular industrial clusters, highlighting areas that could further scale up their circular activities in closing material loops (see Sect. 2.4.1). Network analysis can be conducted on (circular) supply chains, allowing policymakers to identify important players in a network of secondary material flows. Stakeholders in (circular) supply chains can be identified using a dataset that represents locations of material flows, such as waste statistics or shipping movement data (see Sect. 2.4.3.1).

2.4 Example Use Cases

The following subsections present use cases of GIS in the circular built environment, with examples from academia, industry, and governments.

2.4.1 Academia

Academia has played a role in the development of GIS in circular built environment research and in improving the accuracy of estimating future locations of secondary material availability in cities, identifying optimal locations for circular infrastructure, and developing circular city information infrastructures.

2.4.1.1 Estimating Locations of Future Secondary Material Availability: From a Top-Down to a Bottom-Up Approach

One of the most well-known mapping and accounting methods in circular economy research is material flow and stock analysis (MFSA). This method is rooted in the scholarship on societal metabolism and has a history of theoretical developments and methodological advancements since the nineteenth century (Fischer-Kowalski 1998). MFSA was mostly done with statistical data and in a top-down approach based on the accounting principle of mass balance, where a stock of materials is seen as the difference between inflows and outflows of a certain material in a certain area (often an administrative unit) for a time period (usually a year) (Lanau et al. 2019). Often, top-down approaches use statistical data of the amounts of the materials without specific building location instances, which makes it difficult for interested parties to locate where and when these materials could become available for future reuse.

In 2009, Tanikawa and Hashimoto published a seminal paper which proposed using GIS data for estimating material flows and stocks (Tanikawa and Hashimoto 2009). Integrating GIS provides information about the location and timing of the availability of reclaimable materials. By incorporating spatial data using GIS, the results of MFSA become useful for specialised deconstruction companies, urban miners, and other reuse actors at the local level, in contrast to nonspatial MFSA (Wuyts et al. 2022). If there is a time series of GIS layers (cadastral data), researchers can study patterns, such as the average life span of buildings, categorised according to building year or period. This can be used as input in estimating the future potential supply of (secondary) materials from demolition (Van den Berghe and Verhagen 2021).

While many practitioners and academics still use the top-down approach and the EUROSTAT compilation guidelines (European Commission 2018), Tanikawa and Hashimoto's work has inspired more researchers to use a bottom-up approach, shifting towards more local estimations of material stocks and flows (Wuyts et al. 2022). This approach entails the quantification of materials in a certain location (e.g. a building) by multiplying the area of the location (e.g. in square metres) by the material intensity coefficient typical for the building or location (e.g. tonnes of a certain material per square metre). The coefficient refers to the material intensity, which is derived as part of a multiplication with the volume of that material present in that building. The coefficients are often based on existing planning documentation retrieved from archival work, communications with the demolition or construction companies, on-site investigations through laser scanning (see Chap. 3 by Gordon et al. on scanning technologies), or, if available, BIM models (Sprecher et al. 2022; Honic et al. 2023). However, these material intensity coefficients are different for the period, the location, and the functional unit, as, for example, demonstrated through comparing material intensity databases in the Canadian city of Toronto, the Australian city of Perth, and the island of Luzon in the Philippines (Arceo et al. 2023). These databases of material intensity coefficients are often not organised in standard structures, which makes comparisons and data exchanges difficult.

In addition, this bottom-up approach, often called the 'coefficient-based approach', requires a lot of data and is labour-intensive but provides spatial information on the specific building location. These coefficients are multiplicated by gross volumes, often derived from cadastral data, and the outputs are maps showing where materials are located in a geographical area. When materials embedded in buildings are seen as future urban resources, mapping them is like developing an inventory of future available materials. Hence, associated researchers have been developing urban resource cadastres for a circular economy in European cities, such as in Odense, Denmark (Lanau and Liu 2020), Vienna, Austria (Kleeman et al. 2017), and Gothenburg, Sweden (via CREATE project, 2022–2025). In some cases, researchers can predict when these materials could become available for future reuse, via municipal demolition and construction agendas (cf. building permits), or if time-related data (such as building ages and average life spans of buildings) is available. While most cadastral data is digitally available, older cadastral data can be digitised using artificial intelligence and machine learning methods (e.g. see the Nested Phoenix in Melbourne and Brussels (Stephan et al. 2022)).

The outputs of coefficient-based approaches in MFSA can inform urban mining studies and plans, thus helping to estimate and visualise the location, availability, and reusability potential of secondary construction materials within cities and regions (Wuyts et al. 2022). An academic cluster of researchers associated with Tanikawa and Hashimoto is using their method to collect data for informed decisions for sustainable urban and regional development. For example, Guo et al. (2021) used this method for estimating the material stocks and flows as well as the lifetime of buildings over a chronicle in Tiexi district of the Chinese city Shenyang. This district is often seen as a microcosm of Chinese studies and representative of many Chinese urban neighbourhoods.

Building upon coefficient-based approaches in MFSA, researchers have started investigating how this method can help the circular economy transition in the built environment. One of the explorations is the combination of insights from historical studies, political economy, and innovation with spatially explicit material stock studies. GIS is used twice: first to map the material stocks, and then to make an estimation model of where vacant sites are (and, thus, where materials could be available for reuse) (Wuyts et al. 2020). However, this approach should not be used without critical thinking. First of all, predictive or speculative mapping of vacant sites for future mining can be seen as a colonial capitalist practice that erases the histories and presences of specific groups of people and other beings (Noterman 2022). Second, it is important to note that mining the materials and reusing construction materials is a short-term perspective, while a long-term perspective is renovating and repurposing these constructions, and a multispecies perspective is giving the land back to other species and letting it overgrow (Wuyts and Marjanović 2022; Marin and De Meulder 2021a).

All these methods often require specialised equipment and consume time and large file types and databases, especially if some data is retrieved through laser scanning (Uotila et al. 2021). A promising new approach for locating reclaimable materials is using data from street-view images (e.g. Google Street View) of facades to train machines to create classification maps that can assist in defining protocols and urban planning, as demonstrated in Zurich and Barcelona (Raghu et al. 2022). (See Chap. 4 by Armeni et al. to learn more about artificial intelligence and image recognition for reuse.)

2.4.1.2 Identifying Locations of Existing and Future Circular Facilities Using Spatial Analysis

GIS can be used in speculative mapping studies to understand the location of existing and future facilities and infrastructure associated with a circular built environment. Speculative mapping or cartography is a tool to make the future or frontier visible for extracting potential resources (Noterman 2022). In the circularity context, this frontier could be possible locations of reclaimable secondary materials or circular infrastructure (Tsui et al. 2023). Speculative mapping is often used in urban planning. In Belgian cities such as Brussels and Leuven, landscape architects are using GIS to map circular practices within an area or landscape. By doing so, they map and speculate how a facility such as a 'material bank' can facilitate circularity in a city (Marin and De Meulder 2021b). Verga and Khan (2022) created an urban circular practice atlas in Brussels, which is a combination of different GIS layers for different facilities and organisations for different sectors that shows the spatial configuration of logistic infrastructure such as collection points for different material flows (e.g. textile, construction materials).

Furthermore, GIS can be utilised to conduct spatial statistical analysis to identify optimal locations of present and future circular facilities – whether they are facilities for waste recycling or hubs for material exchange. In the Netherlands, spatial analysis is conducted to quantify the spatial clustering of waste reuse activities, as well as to find hotspot locations for waste reuse (Tsui et al. 2022). Further work was also conducted to estimate the optimal number and locations of concrete recycling plants in the Netherlands (Hodde 2021). In industrial symbiosis, proximity is key, which requires local optimisation calculations requiring GIS (e.g. see Yu et al. 2021).

Other researchers build further on spatially explicit material stock studies stored in GIS, where origin-destination calculations are conducted to criticise missing infrastructures for recycling concrete in a city, such as Den Hague (Van den Berghe and Verhagen 2021). In Singapore, spatially explicit material stock studies were performed to estimate the potential of building materials that could be transferred to the growing housing market in Indonesia, which is only a few kilometres away (Arora et al. 2019, 2020). Spatially explicit material stock and flow studies have been shown to benefit circular city implementation (Wuyts et al. 2022).

2.4.1.3 Developing Circular City Information Infrastructures

Different cities are developing circular city information infrastructure to monitor and support policy planning. Mostly the information is analytical: digital twins are developed, (top-down) indicators are refined, and material flows are mapped. In Flanders, Belgium, the Vlaamse Open City Architectuur (VLOCA 2023) hosts a knowledge hub for smart cities. Other similar initiatives include the circular economy monitor in Flanders (Vlaanderen Circulair 2021), Ganbatte World (Ganbatte

World 2023), and the Amsterdam circular economy monitor (Gemeente Amsterdam 2023). However, none of these initiatives integrates experiential knowledge, avoiding the fact that cities are also experiential information systems (De Franco and Moroni 2023).

In 2022, NTNU Sustainability at the Norwegian University of Science and Technology funded the Circular City Project (2022–2026). The researchers will apply a bottom-up technique to assist Trondheim, Norway, in catalysing circular material flows (NTNU 2022). The idea is to create digital twins of individual buildings within the larger city-scale digital twin of Trondheim, fusing macro-level data (GIS layers, with graph data) with micro-level data such as BIM objects. This application is similar to a research project on modelling and predicting building blocks in Vienna, where BIM models provided a material intensity database that could be multiplied by the gross volumes obtained from GIS (Honic et al. 2023).

A wide array of GIS applications in academia are working towards the circularity transition of the built environment industries. These GIS techniques are beginning to leave the academic sphere, leading to action in industry and government, as seen in the section below.

2.4.2 *Industry*

The following section provides an overview on how GIS is used in industry for a more circular built environment. Industry uses GIS to plan locations of reuse infrastructures, to track locations of components and materials via digital platforms, and to facilitate the efficiency of reverse logistics.

2.4.2.1 Planning Reuse Infrastructures

Companies such as reclaimable material brokers and manufacturing companies use GIS analysis to inform their spatial strategies for facility location of circular construction hubs and (reverse) logistics. In southern Norway, more than 30 partners, representing different actors of the forestry, timber construction, deconstruction, waste industry value network, and research institutes, started the SirkTRE consortium and received funding for research, development, and innovation projects in 2021–2024 (SirkTRE 2022). The first phase encompassed a stakeholder mapping process, including missing roles. One of the missing roles was circular hubs where wood waste, mostly from demolition projects, would get collected for quality check and pretreatment (drying, removing hazardous substances, cutting it ready for industrial sale) and assembly in new building elements and components. In Belgium and Norway, these circular hubs were more the result of the availability of land, often placed in restored brownfields (e.g. Materialenbank Leuven in Belgium; Omtre's Materialenbank in Hønefoss, Norway) or vacant public spaces that are in development (e.g. Sirkulær Ressurssentral in Oslo, Norway).

Because stakeholders wanted to deal with high material volumes and withstand higher investment risks, the planning went through a methodology cocreated by SirkTre consortium partners, external consultants, and seed funders. In early 2022, Omtre AS started the planning of a circular hub to be in operation before 2030 and collected insights from Norwegian experts but also looked at the existing and emerging circular hubs in Belgium and the Netherlands. Informed by theories from economic geography, investigations will be made in different locations, spatial configurations, and setups (e.g. temporary vs permanent) under different input parameters and future scenarios. One of the research-for-informed-planning tasks considers a forecasting GIS-material stock analysis to estimate the potential availability (when, where, and how much) of various wood waste fractions of demolition projects. Noteworthy is that this spatially explicit material and flow analysis will not follow standard MFSA guidelines by going beyond administration boundaries. Omtre uses the metaphor of the circumference. At a UNESCO site at the former mining mountain town of Røros, Norway, mining happened within a circumference because of the location of the copper and the economic costs related to the transport of the copper and the input resources (e.g. trees for fire). Since it is seen as reasonable to drive 2 hours to pick up materials in the Norwegian cultural context, this distance is the radius of the circumference for urban mining.

Omtre AS is also setting up GIS to map existing infrastructure (e.g. storage, transport, etc.), power relationships, technical lock-ins (risks), and required partnerships to enable the relocation of building materials. This data collection task has two objectives. First, it will inform a speculative mapping of how a material bank facilitates timber flows (inspired by Marin and De Meulder 2021b). Second, it will feed the setup of an optimal routing calculation of the collection of the selected wood waste fractions and distribution of the building materials and elements to construction sites or intermediary partners (e.g. prefabricated module builders).

2.4.2.2 Tracking and Tracing via Digital Platforms

Presently, digital platforms are emerging to enable circularity practices such as reusing building materials and components, selling tools or advice, calculating life cycle costs, or even providing a marketplace (Wuyts et al. 2023). In this chapter, we are interested in the functionality of tracking and tracing the locations of buildings, elements, or materials using GIS. In some circularity practices and strategies, geographical proximity matters. Industrial symbiosis platforms are taking the role of intermediary third parties that match supply and demand (Krom et al. 2022) – often by sharing not only data on available materials but also other resources (storage space, equipment, trucks). These platforms require GPS coordinates, so the logistics of the relocation can be arranged and optimised. Especially in industrial symbiosis, proximity is key and requires the integration of GIS (e.g. Yu et al. 2021).

Digital platforms can provide two services: tracking and tracing. Tracking is recording data on where the material is at the moment. Tracing means knowing where the material comes from, including its history and exposure to harmful events.

Tracing can help estimate risk and identify which application the material could be used for. Tracing is also key in the increased demand for transparency and social sustainability controls, especially upstream of the value chain. Tracking and tracing are presently already part of cyber logistic systems. Here, GIS is used to automatically calculate the optimal routing when moving materials or components from one location to another.

There have been speculations that material banks and other temporary storage spaces will become obsolete in the future and will be replaced by systems that track the required materials in planned demolitions and constructions and facilitate a direct relocation of materials from the deconstruction to the construction site. These digital solutions would substitute spatial requirements (e.g. land and infrastructures needed for storage). Nevertheless, due to the conservatism of the construction sector and the slow uptake of digital solutions in general, the rapid wide deployment of digital solutions replacing spatial requirements is not expected.

2.4.2.3 Tracking in Reverse Logistics and Remanufacturing

While digital markets match different building industry actors and can facilitate the optimisation of logistics of the reclaimed building material from one actor to another (Sect. 2.4.2.2), the relocation of materials can also happen via reverse logistic systems within the same company's value chain. Optimisation of logistics from a building to a remanufacturing plant and again integration in the old or new building project is again key to reducing the costs of remanufacturing. As part of the linear economy, these companies have normally established internal tracing and tracking systems which enables the logistic managers to follow a product (often only within the factory boundaries) until the ownership of the product is transferred to the next actor in the value chain.

One important step for enabling tracing and tracking is that products get tagged with a label or a unique identifier (e.g. barcode or QR code). Later, in the value chain or during the use or demolition phases, this label can disappear for various reasons, and the connection with a digital tracking and tracing system can get lost. Hence, setting up reverse logistic systems often means the creation of a new tag at the source of the collection of secondary materials (e.g. waste collection points, which would become the new point zero of the tracking system). If companies create tags which will not disappear, they can set up a signalling system when these materials should be reclaimed and transferred back for remanufacturing into the current or new building projects. (For more information about information needs for the complex process from deconstruction via reverse logistics to remanufacturing, see Chap. 11 by van den Berg.)

Different companies in the circular built environment transition look into different labelling systems that contribute to tracking and tracing, even over different life cycle phases of the material and when it is owned by other actors. Are barcodes or QR codes the right tags from a reuse perspective, knowing they can disappear in the use or deconstruction phase? For example, in the timber construction industry, the constellation of the knots in each wood element – or so-called wood fingerprint – is unique and can be recognised with cameras of scanners (e.g. Pahlberg 2017); it could be used as a natural QR code which can be scanned at any life cycle phase, from forest to second or third use cycles. These tags or unique identifiers are coupled with data from industry foundation classes (IFC), which is a data exchange schema describing architectural, building, and construction industry data. There are developments where the geolocation of products would be part of IFC specifications. If these geolocation requirements were part of IFC data in tracking, this would create information about where products end up in the first use cycle and also in the second, third, and next-use cycles, which would lead to insights about environmental impacts related to transport and on the service time of these products in that location and building. (See Chap. 5 by Honic et al. for the role of data templates and material passports in tracking assets over more life cycle phases.)

2.4.3 *Government*

Governments use GIS to monitor circularity in their areas of jurisdiction and to assess the circular spatial development potential of (industrial) land. This section will introduce two examples: Project Zuid-Holland and the RePair project.

2.4.3.1 Project Zuid-Holland: Prioritising Industrial Land for the Circular Economy

Project Zuid-Holland (South Holland) is a collaboration between the Delft University of Technology and the Province of Zuid-Holland in the Netherlands. The aim of the project is to evaluate the importance of water-bound industrial areas in the province in accordance with its existing and future needs, with a special emphasis on the transition to a circular economy. The project arose from the province's need to prioritise the preservation of its existing scarce industrial areas. In many municipalities within the province, existing water-bound industrial areas are being transformed into residential and commercial land use that do not take advantage of the spatial and logistical possibilities of industrial activity and water transport for circular activities such as locally reusing or recycling construction materials. Responding to a lack of understanding of circular economy from a spatial perspective, this project focuses on the spatial requirements for future transitions. This includes location conditions such as available firms and technologies, the presence and diversity of labour forces, environmental restraints, and logistical multimodal possibilities. This leads to the question: What spatial planning strategies are necessary for the current and future stock of water-bound industrial areas in Zuid-Holland in order to foster future transitions?

The project has three main work packages: (1) mapping existing water-bound industrial areas, (2) determining spatial requirements for future transition-related activities, and (3) offering policy recommendations. The first work package, mapping of water-bound industrial areas, uses GIS to generate insights for the subsequent deliverables. The mapping process includes two main steps: the topographical mapping of existing commercial and industrial activities and the topological movement of materials via water transportation infrastructure. By using spatial data on the locations of commercial companies, a map was created showing industrial land within the region that hosted circular economy-related industries. Additionally, shipping data was used to identify industrial lands that were visited by ships, indicating the utilisation of water transportation. The spatial analysis work of the project resulted in two maps. The first is a geographical map of industrial sites in the province, showing whether each site was water bound, if it utilises water transport, and if it contains circular industries. The second is a network map (or diagram) showing which industrial sites are connected to each other via water transportation, as well as highlighting industrial sites that are strongly connected within the network.

To summarise, this project addresses circularity by deepening our understanding of the circular economy transition from a spatial perspective (for the full report, see Van den Berghe et al. 2023). Spatial analysis allows key industrial sites to be selected and prioritised for future circular activities, contributing to narrowing, slowing, and closing material cycles.

2.4.3.2 The RePair Project: Geo-design Decision Support Environment for Circular Spatial Strategies

Funded by the European Commission, the RePair project (2016–2020) (RePair 2023) aimed to develop a methodology that allowed for the creation of integrated, place-based, and eco-innovative spatial development strategies to reduce waste flows in periurban areas. The methodology was implemented in six metropolitan areas, using a geo-design decision support environment (GDSE) in multiple workshop settings. This method extends the assessment of urban metabolism to include concepts related to urban drivers, urban patterns, environmental and spatial quality, and potential co-benefits of strategies.

GDSE is a digital platform based on the geo-design framework, which allows for a geographical study area to be described, evaluated, and (re-)designed according to a predetermined goal (Steinitz 2012; Arciniegas et al. 2019). The GDSE combined and visualised spatial data collected from local contexts, such as locations of waste production and processing, land use, and company location data. An example of this can be seen in Fig. 2.2, showing the movement of waste to and from Amsterdam. This information was presented within a series of workshops to key local stakeholders in the development of a circular economy, including planning authorities, public/private organisations involved in strategic environmental assessment, and industrial actors in waste and resource management. The workshops with key stakeholders were used to aid decision-making in the development of place-based spatial development strategies for each of the six metropolitan areas.

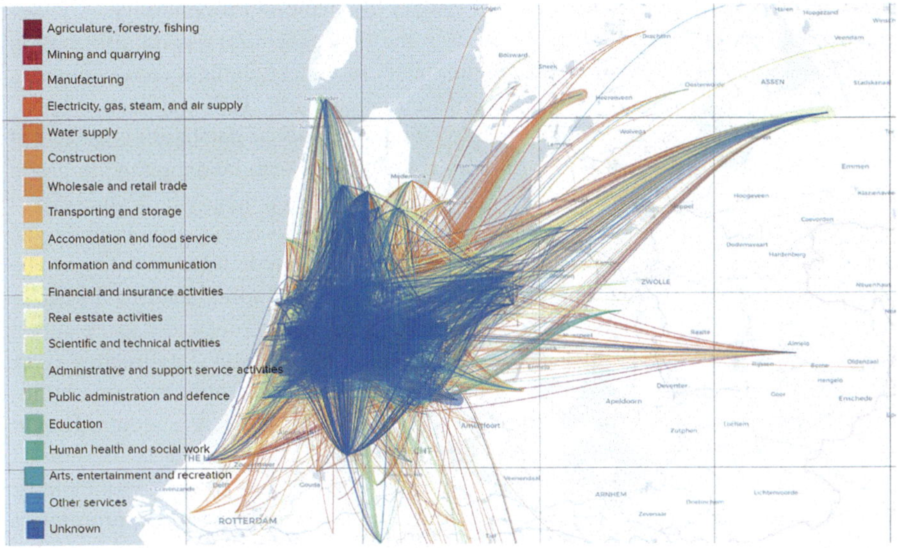

Fig. 2.2 Mapping waste flows in Amsterdam (Furlan et al. 2020)

The project resulted in the creation of a spin-off company, GeoFluxus (GeoFluxus 2023), which provides material flow monitoring services to governmental bodies and private companies. Using material and waste statistics, GeoFluxus provides insight into where waste streams are available for circular (business) opportunities and develops methods for monitoring those streams at various scales (municipal, provincial, national), thus allowing organisations to measure their progress towards the circular economy.

2.5 Discussion

2.5.1 Connecting to Other Technologies

Collecting spatial data is key in academic research for informing circular city and built environment projects. GIS is at a high technology readiness level (TRL 9), meaning that the technology and information systems are widely known and adopted, in most European and North American regions, and is often integrated with other digital technologies (e.g. BIM) to support circular systemic solutions. Recently, software companies ESRI and AUTOCAD have worked together to smooth data exchange between GIS and BIM graph data (ArcGIS 2023b), allowing higher data transfer speeds and more seamless integration. For better integration of software systems, data handlers need more sensitivity for the different data formats. Some scholars propose standardised structures for databases, such as for organising material intensity data (e.g. Guven et al. 2022) to enable interoperability. Another measure is clear-cut communication between the GIS executioners and information and communications technology (ICT) architects to foster seamless data exchange. For example, one risk is that these systems are designed by ICT architects who do not realise that GIS can be based on graph data and relational database management systems. The shapes used in GIS (polygons, lines, points) can be expressed in graph data, made of arcs and nodes. They are single attributes in a table which can be part of a relational database in the case of software programmes such as ArcGIS and qgis. However, the shapes are not necessarily explicitly related; tabular data can be exchanged instead. Tabular-centric architecture requires more effort to integrate with GIS expressed in graph data.

GIS can also be linked to technologies enabled by artificial intelligence (AI). Machine learning image recognition models can be used to identify reusable building components at an urban scale, using GIS and Google Street View data (Raghu et al. 2022). Additionally, as AI tools become increasingly available to the public, GIS technologies will become increasingly democratised and used by nonexperts. An example of this is ChatGPT, an AI chatbot that not only responds to prompts in text but in computer code as well, allowing users to potentially generate complex code for spatial analysis without prior knowledge (Tsui 2023). While these technologies can greatly empower the general public, it is important to take into account the dangers of releasing tools that are accessible to many but understood by few.

2.5.2 *Hurdles and Barriers*

In order to integrate GIS with circular economy solutions, a number of challenges need to be overcome. The availability and quality of spatial data, and the way insights are presented, need to be improved. GIS metadata – data about GIS data that provides information such as where the data was collected, who collected it, how it was processed, etc. – is crucial for trustworthiness and transparency, which is especially important to a circular built environment that strongly depends on information facilitation and sharing between different stakeholders (and in times of increased cybersecurity risks). Insufficient metadata often limits understanding of where the spatial data comes from or how it was created. Better metadata leads to more transparency and trustworthiness. There are already various ISO standards for trustworthiness and other frameworks for data and information facilitation (Naden 2019). In terms of data standards, industry actors should ensure that additional GIS data collection does not create more administrative hurdles, such as by necessitating additional agreements on who owns or stores spatial data. One important step forward would be to provide more process standards and data management plans that help define, for example, when to stop collecting and storing spatial data, finding a balance between cybersecurity and circularity.

The development of a circular built environment may also require the involvement of citizen communities. To ensure the participation of citizens, open and public access to GIS-based algorithms and methods would be ideal. Public access to GIS data and methods not only fosters collaboration and involves diverse groups of users (e.g. citizens and communities), but it also allows independent parties to detect possible data biases in algorithms that could discriminate against people of certain backgrounds (Lally 2022; McCall 2003).

2.5.3 *Future Trends*

The use of GIS for understanding and improving the current state of the circular economy will increase in importance, especially in terms of reducing hurdles in technology compatibility. The most promising future for the technology arguably lies in using GIS as a tool for making governance decisions, such as how to match the demand and supply of materials in time and space. Policymakers often lack the capability to use digital tools, while technology experts who use these tools often are uninformed of policy questions behind their application (Hollands 2020). We warn against the false promise that more quantitative measuring or digital tools will always lead to better results – a correlative but not causal relationship, debated already in the 1980s during the so-called quantitative revolution (Paasi et al. 2018). Additionally, circular city and regional initiatives and their information infrastructures focus mostly on analytical data, not lay or experiential knowledge. Significant progress is needed to bridge this gap.

In the end, GIS remains a tool that is operated by a designer or policymaker, starting with a question. That question must be given focused time and attention. We should thus be cautious about our enthusiasm and investing capacity (be it in financial, R&D, human, or other resources) in developing and applying digital tools such as GIS to solve circular economy problems. Ultimately, understanding why, when, where, and for whom we need a circular economy or circular built environment – and subsequently, why we need better tools – should be prioritised over investing further in the technology without a clear understanding of its utility for circularity. The past should serve as a warning. Although we have been developing tremendously sophisticated digital tools, material passports, and monitoring structures for about half a century, we have and are still losing the fight against climate change, more because of political reasons than a lack of data. If we view the circular economy as a strategy to cope with this climate change, then the same reasoning can be followed: are the digital tools we develop to foster a circular built environment really what we need? The correlation between policy and means needs more attention in research and practice to guarantee a more sustainable circular built environment.

2.6 Key Takeaways

- GIS can contribute to a circular built environment by creating, visualising, sharing, and analysing spatial data on the location of buildings, components, and materials.
- GIS can help identify where secondary building materials will be available in the future.
- GIS can use spatial statistical methods to identify optimal future locations for circular infrastructure, such as material banks, recycling facilities, or (reverse logistics) hubs for material exchange.
- In combination with digital platforms, GIS can facilitate the tracking and tracing of construction products, components, and materials.
- GIS can help governments prioritise spatial development strategies by highlighting future sites critical to the development of a circular economy.
- GIS data requires strong metadata in order to increase transparency and trustworthiness.

Acknowledgements The work of Tanya Tsui is funded by the European Union's Horizon 2020 Research and Innovation Programme under grant agreement No 821479. The work of Wendy Wuyts is funded by the Norwegian Research Council, as part of the circWOOD project (project no: 328698).

Declaration of Competing Interests The authors declare no conflicts of interest and no financial interests in the research presented in this paper.

References

Arceo A, MacLean HL, Saxe S (2023) Material intensity in single-family dwellings: variability between locations, functional unit and drivers of material use in Toronto, Perth, and Luzon. Resour Conserv Recycl 188:106683

ArcGIS (2023a) Extract 3D buildings using photogrammetry | Learn ArcGIS. https://learn.arcgis.com/en/projects/extract-3d-buildings-using-photogrammetry/. Accessed 14 Feb 2023

ArcGIS (2023b) ArcGIS for AutoCAD | Free plug-in for interoperability between AutoCAD & ArcGIS. https://www.esri.com/en-us/arcgis/products/arcgis-for-autocad. Accessed 14 Feb 2023

Arciniegas G, Šilerytė R, Dąbrowski M et al (2019) A geodesign decision support environment for integrating management of resource flows in spatial planning. Urban Plan 4:32–51. https://doi.org/10.17645/up.v4i3.2173

Arora M, Raspall F, Cheah L, Silva A (2019) Residential building material stocks and component-level circularity: the case of Singapore. J Clean Prod 216:239–248

Arora M, Raspall F, Cheah L, Silva A (2020) Buildings and the circular economy: estimating urban mining, recovery and reuse potential of building components. Resour Conserv Recycl 154:104581

Bahers JB, Athanassiadis A, Perrotti D, Kampelmann S (2022) The place of space in urban metabolism research: towards a spatial turn? A review and future agenda. Landsc Urban Plan 221:104376. https://doi.org/10.1016/j.landurbplan.2022.104376

Bucci Ancapi F, Bueren E, Van den Berghe K (2022) Circular cities. In: The Palgrave encyclopedia of urban and regional futures. Springer International, pp 1–12. https://doi.org/10.1007/978-3-030-51812-7_125-1

Chang KT (2018) Introduction to geographic information systems https://www.mheducation.com/highered/product/introduction-geographic-information-systems-chang/M9781259929649.html

De Franco A, Moroni S (2023) The city as an information system: urban agency, experiential inputs and planning measures. Cities 134:104183

ESRI (2020) Tracking and traceability 2020 – validations. In: ESRI community. https://community.esri.com/t5/gas-and-pipeline-blog/tracking-and-traceability-2020-validations/ba-p/887345. Accessed 14 Feb 2023

European Commission (2018) Economy-wide material flow accounts handbook. https://doi.org/10.2785/158567

Fischer-Kowalski M (1998) Society's metabolism: the intellectual history of materials flow analysis, Part I, 1860–1970. J Ind Ecol 2(1):61–78

Furlan C, Wandl A, Geldermans B, Sileryte R (2020) A refined waste flow mapping method.: addressing the material and spatial dimensions of waste flows in the urban territory through big data: the case of the Amsterdam metropolitan area. Contesti. Città, Territori, Progetti 1:74–89. https://doi.org/10.13128/contest-11909

Ganbatte World. (2023). https://www.ganbatte.world/cities/. Accessed 14 Feb 2023

Gemeente Amsterdam. (2023) Monitor Circulaire Economie | Website Onderzoek en Statistiek. https://onderzoek.amsterdam.nl/interactief/monitor-circulaire-economie. Accessed 14 Feb 2023

GeoFluxus. (2023). https://en.geofluxus.com/. Accessed 14 Feb 2023

Goodchild M (2010) Twenty years of progress: GIScience in 2010. J Spat Inf Sci 1(1):3–20. https://doi.org/10.5311/JOSIS.2010.1.2

Guo J, Fishman T, Wang Y, Miatto A, Wuyts W, Zheng L, Wang H, Tanikawa H (2021) Urban development and sustainability challenges chronicled by a century of construction material flows and stocks in Tiexi, China. J Ind Ecol 25(1):162–175

Guven G, Arceo A, Bennett A, Tham M, Olanrewaju B, McGrail M, Isin K, Olson AW, Saxe S (2022) A construction classification system database for understanding resource use in building construction. Sci Data 9(1):42

Hodde J (2021) Scoping the circular economy: finding the optimal spatial distribution of concrete recycling facilities. Master thesis dissertation. Wageningen University & Research.

Hollands RG (2020) Will the real smart city please stand up?: intelligent, progressive or entrepreneurial? Routledge companion to smart cities. Routledge, London, pp 179–199

Honic M, Ferschin P, Breitfuss D, Cencic O, Gourlis G, Kovacic I, De Wolf C (2023) Framework for the assessment of the existing building stock through BIM and GIS. Dev Built Environ 13:100110

Kleemann F, Lederer J, Rechberger H, Fellner J (2017) GIS-based analysis of Vienna's material stock in buildings. J Ind Ecol 21(2):368–380

Kontos TD, Komilis DP, Halvadakis CP (2005) Siting MSW landfills with a spatial multiple criteria analysis methodology. Waste Manag 25:818–832. https://doi.org/10.1016/j.wasman.2005.04.002

Krom P, Piscicelli L, Frenken K (2022) Digital platforms for industrial symbiosis. J Innov Econ Manage 39:I124-XXVI

Lally N (2022) What can GIS do? ACME Int J Crit Geogr 21(4):337–345

Lanau M, Liu G (2020) Developing an urban resource cadaster for circular economy: a case of Odense, Denmark. Environ Sci Technol 54(7):4675–4685

Lanau M, Liu G, Kral U, Wiedenhofer D, Keijzer E, Yu C, Ehlert C (2019) Taking stock of built environment stock studies: progress and prospects. Environ Sci Technol 53(15):8499–8515

Liu HK (2021) Crowdsourcing: citizens as coproducers of public services. Policy Internet 13:315–331. https://doi.org/10.1002/poi3.249

Marin J, De Meulder B (2021a) Time is life Ruimte 13(49):22–25

Marin J, De Meulder B (2021b) Wood(s): imagining how a materials bank can catalyze circular timber flows in Leuven, Belgium. Archit Theory Rev 25(1–2):117–135
McCall MK (2003) Seeking good governance in participatory-GIS: a review of processes and governance dimensions in applying GIS to participatory spatial planning. Habitat Int 27(4): 549–573
Naden C (2019) ISO – it's all about trust. https://www.iso.org/news/ref2461.html. Accessed 18 Apr 2023
Noterman E (2022) Speculating on vacancy. Trans Inst Br Geogr 47(1):123–138
NTNU (2022) Circular City Project. https://www.ntnu.edu/sustainability/circularcity. Accessed 23 Nov 2022
Paasi A, Harrison J, Jones M (2018) Handbook on the geographies of regions and territories. Edward Elgar
Pahlberg T (2017) Wood fingerprint recognition and detection of thin cracks (Doctoral dissertation, Luleå University of Technology)
Raghu D, Markopoulou A, Marengo M, Neri I, Chronis A, De Wolf C (2022) Enabling component reuse from existing buildings using machine learning-using Google Street View to enhance building databases. In: Lim C, Lee JH, Lee J, Kim M, Wessels L (eds) Proceedings of the 27th international conference on computer-aided architectural design research in Asia (CAADRIA 2022), vol 2. Association for Computer Aided Architectural Design Research in Asia (CAADRIA), 577–586
RePair (2023) https://h2020repair.eu/. Accessed 14 Feb 2023
Rotor (2023) Opalis. https://opalis.eu/nl. Accessed 14 Feb 2023
Santi P, Resta G, Szell M et al (2014) Quantifying the benefits of vehicle pooling with shareability networks. Proc Natl Acad Sci 111:13290–13294. https://doi.org/10.1073/pnas.1403657111
SirkTRE (2022). https://www.sirktre.no/. Accessed 23 Nov 2022
Spatial Morphology Group (2023) https://www.smog.chalmers.se. Accessed 14 Feb 2023
Spotr (2023) Spotr – inspect millions of buildings in seconds. https://www.spotr.ai/. Accessed 14 Feb 2023
Sprecher B, Verhagen TJ, Sauer ML, Baars M, Heintz J, Fishman T (2022) Material intensity database for the Dutch building stock: towards big data in material stock analysis. J Ind Ecol 26: 272–280
Steinitz C (2012) A framework for geodesign: changing geography by design. ESRI
Stephan A, Mulders G, Slavkovic K (2022) Nested Phoenix: a bottom-up Python model for the life cycle environmental performance of urban built stocks. In: IOP conference series: earth and environmental science, Vol. 1122(1): 012028. IOP Publishing
Superuse (2023) Oogskaart. https://www.oogstkaart.nl/. Accessed 14 Feb 2023
Tanikawa H, Hashimoto S (2009) Urban stock over time: spatial material stock analysis using 4d-GIS. Build Res Inf 37:483–502
Tsui T (2023) Coding with chatGPT. In: Medium https://medium.com/@tanyamarleytsui/coding-with-chatgpt-b50ab3fcb45f. Accessed 14 Feb 2023
Tsui T, Derumigny A, Peck D, Van Timmeren A, Wandl A (2022) Spatial clustering of waste reuse in a circular economy. Front Built Environ 8:154. https://doi.org/10.3389/fbuil.2022.954642
Tsui T, Furlan C, Wandl A, van Timmeren A. (2023) Spatial Parameters for Circular Construction Hubs: Location Criteria for a Circular Built Environment. Circ Econ Sust
TU Delft (2023) Detailed 3D building models automatically generated for very large areas (3D BAG). https://3d.bk.tudelft.nl/projects/3dbag/. Accessed 14 Feb 2023
Uotila U, Saari A, Junnonen JM (2021) Investigating the barriers to laser scanning implementation in building refurbishment. J Inf Technol Constr 26:249–262
Van den Berghe KB, Verhagen TJ (2021) Making it concrete: analysing the role of concrete plants' locations for circular city policy goals. Front Built Environ 7:136
Van den Berghe K, Tsui T, Iliopoulos G, Papadimitriou C, Bonte T, Fitzgeral T, Dillon T, Nefs M (2023) Watergebonden transport en bedrijventerreinen in Zuid-Holland. Fore- en backcasting van de circulaire economie om de implicaties voor ruimtelijke planning van vandaag te begrijpen. http://resolver.tudelft.nl/uuid:14c4ecb8-8cc0-4ea7-8a89-aa0e6ec9314f

van den Dobbelsteen A, Broersma S, Stremke S (2011) Energy potential mapping for energy-producing neighborhoods. Int J Sustain Build Technol Urban Dev 2:170–176. https://doi.org/10.5390/SUSB.2011.2.2.170

Verga GC, Khan AZ (2022) Space matters: barriers and enablers for embedding urban circularity practices in the Brussels capital region. Front Built Environ 8:810049

Vlaanderen Circulair (2021). https://cemonitor.be/. Accessed 14 Feb 2023

VLOCA (2023). https://vloca.vlaanderen.be/. Accessed 14 Feb 2023

Waag (2016) PUMA | Prospecting the urban mines of Amsterdam. http://code.waag.org/puma/index.html#52.366,4.8735,13. Accessed 14 Feb 2023

Wuyts W, Marjanović M (2022) The development of spatial circularity discourse in Japan: ecomodernist, territorialised, or both? The story of Onomichi's wastescapes. Circ Econ Sust:1–27. https://doi.org/10.1007/s43615-021-00146-6

Wuyts W, Sedlitzky R, Morita M, Tanikawa H (2020) Understanding and managing vacant houses in support of a material stock-type society—the case of Kitakyushu, Japan. Sustainability 12(13):5363

Wuyts W, Miatto A, Khumvongsa K, Guo J, Aalto P, Huang L (2022) How can material stock studies assist the implementation of the circular economy in cities? Environ Sci Technol 56:17523. https://doi.org/10.1021/acs.est.2c05275

Wuyts W, Liu Y, Huang X, Huang L (2023) Evaluating existing digital platforms enabling the reuse of reclaimed building materials and components for circularity. In: Hjelseth E, Sujan SF, Scherer RJ (eds) ECPPM 2022 – eWork and eBusiness in architecture, engineering and construction 2022: proceedings of the 14th European conference on product and process modelling (ECPPM 2022), September 14–16, 2022, Trondheim, Norway, 1st edn. CRC Press. https://doi.org/10.1201/9781003354222

Yu Y, Yazan DM, Bhochhibhoya S, Volker L (2021) Towards circular economy through industrial symbiosis in the Dutch construction industry: a case of recycled concrete aggregates. J Clean Prod 293:126083

Tanya Tsui is a PhD candidate at the Faculty of Architecture and the Built Environment in Delft University of Technology. Her research uses urban data science to analyse (waste) material flows to understand the spatial impacts of a circular economy in cities and regions. Her aim is to expand circular cities research beyond the study of governance and material accounting, towards a spatial analysis perspective.

Wendy Wuyts is a researcher at the Norwegian University of Science and Technology (NTNU). She also works as a project manager and consultant for the Norwegian purpose-driven start-up Omtre AS, which is building the first industrial plant for reclaimed timber for the construction sector. Her research focuses on circular economy, sustainable development, political-industrial ecology, and environmental justice. She is involved in the projects SirkTre, circWOOD, and 1-2-Tre-Steg, which aim to foster circular systemic solutions in the timber and architecture, engineering, construction, and deconstruction sector in Norway. She holds a PhD in Environmental Science from Nagoya University, Japan.

Karel Van den Berghe is an assistant professor in Spatial Planning and Urban Development at the Delft University of Technology. His research focuses on the (spatial) circular economy, strategic scenario planning, polycentric economic networks, planning of port cities, Dutch-Belgian cooperation (e.g. Westerschelde), and housing markets. He holds a BSc (2010) and MSc (2012) in geography, an MSc in spatial planning and urbanism (2013), and a PhD in urbanism and spatial planning (2018).

Chapter 3
Digitising Building Materials for Reuse with Reality Capture and Scan-to-BIM Technologies

Matthew Gordon, Luise von Zimmerman, Oushesh Haradhun, Dominik Campanella, Milena Bräutigam, and Catherine De Wolf

Abstract Effective building component reuse requires specific information about recoverable components. However, 85% of the European building stock predates the building information modelling (BIM) technology that stores and links such information. Digitisation technologies can be used to recover this information. Scanning and scan-to-BIM technologies such as LiDAR and photogrammetry enable us to capture and analyse large amounts of raw geometric data as point clouds to create digital records or BIM models of existing buildings. These digital representations can be used by building owners, inspectors, and deconstruction groups for deconstruction, new design, procurement, and new construction. They help implement closed circular resource strategies linking recovered materials to new projects. In this article, we look at a specific case study of these applications through the circularity consultant Concular. Digitisation technologies are compared based on their range and accuracy in conditions with noisy and cluttered data, as well as their cost and accessibility. Additional sensor technologies may integrate further compositional or structural details to ultimately produce insights beyond surface geometry that can be communicated through integrated digital platforms for data access and exchange. Further technological development will lower the time and labour costs during data collection, processing, and analysis.

Keywords Scanning technologies · LiDAR · Point cloud · Building stock digitisation · Reality capture · Scan-to-BIM

M. Gordon (✉) · C. De Wolf
Circular Engineering for Architecture, Institute of Construction & Infrastructure Management, Department of Civil, Environmental and Geomatic Engineering, ETH Zurich, Zurich, Switzerland
e-mail: gordon@ibi.baug.ethz.ch; dewolf@ibi.baug.ethz.ch

L. von Zimmerman · O. Haradhun · D. Campanella · M. Bräutigam
Concular GmbH, Berlin, Germany
e-mail: luise@concular.com; oushesh@concular.com; dominik.campanella@concular.com

C. De Wolf et al. (eds.), *A Circular Built Environment in the Digital Age*, Circular Economy and Sustainability, https://doi.org/10.1007/978-3-031-39675-5_3

3.1 Scanning Technologies: An Overview

Digitalised processes in the built environment require large amounts of data to describe an existing environment or new designs. To be useful, this data must be drawn from, or at least informed by, the physical world. This data is often encapsulated in building information modelling (BIM), which combines formal, compositional, and temporal information about a designed or completed building (Quirck 2012; see also Chap. 1 by Koutamanis on BIM and digital twins). The adoption of BIM, however, is incomplete: only 25% of EU countries mandated BIM for new projects as of 2019 (Charef et al. 2019), so many buildings lack models. Since BIM is mostly only used during design and construction, the models that do exist may become outdated (Heaton et al. 2019).

While *digitalisation* as a whole addresses the incorporation of digital processes into the industry, one central process is gathering data through *digitisation*. When working in a data-scarce environment, digitisation injects new information through procedures of measurement, description, and consolidation. One foundational type of digitisation in the field of architecture, engineering, and construction (AEC) is the capture of physical and spatial data, a set of processes often collectively referred to as reality capture. Data from reality capture technologies directly feed BIM and geographic information systems (GIS) (Waters 2018), which often deal with volumes of data for which manual data capture is impossible or inefficient. This technology is used at several stages throughout the building life cycle.

Spatial digitisation technologies – generally either laser-based light detection and ranging (LiDAR) or image-based photogrammetry – are utilised to capture data in large volumes. Both produce dense data in the form of point clouds: collections of individual points in space with associated data such as colour, reflectivity, or surface angle. Depending on the application and hardware, these clouds can cover scales and levels of detail from individual components to landscapes (Fig. 3.1). This data is then analysed using methods specific to its application to produce useful information. Geometrically processing point cloud data in semantically useful ways presents unique challenges compared to, for example, image analysis, because point clouds are inherently irregular, unstructured, and unordered (Bello et al. 2020).

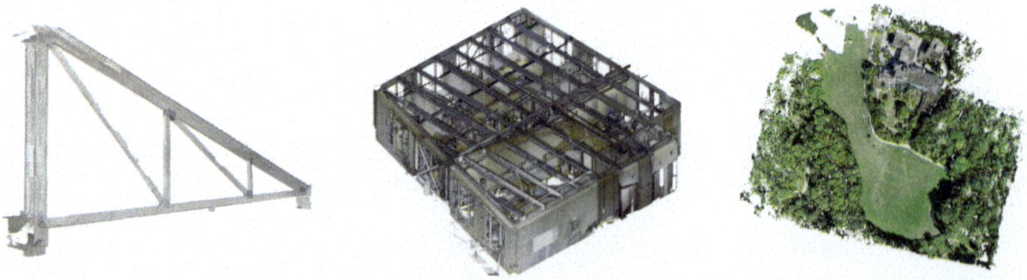

Fig. 3.1 Point clouds at the individual component scale (left: HIR building roof truss, ETH Zurich Hönggerberg Campus), building scale (middle: Kopfbau Halle 118, Winterthur), and landscape scale (right: rehabilitation clinic and grounds, Bettingen)

The relative efficiency, cost, and accuracy of reality capture methods distinguish their use for different situations, as the structure and depth of the data itself is generally standardised. For example, measuring deflection of structural components requires high accuracy, whereas capturing residential interiors for real estate requires communicating the visual experiences in high volume, without the need for millimetre accuracy. Within this range, LiDAR technology can produce very accurate measurements. LiDAR records distance by recording the time of flight (TOF) of a reflection of a laser from a relevant object in a scene. Millions of measurements are taken using different coverage strategies for moving through the site and combined with colour photography to create point clouds realistically coloured for the scene. These systems can be attached to unmanned aerial vehicles (UAVs), mounted on tripods, or operated as part of handheld or smartphone-integrated models. The TOF method can retain an accuracy of 5 mm or better at 100 m distance with contemporary hardware (Wu et al. 2022).

In scenarios with many captures over time, the efficiency of each capture becomes more important. This is a key scenario for the usage of mobile LiDAR systems, which continuously capture data over a path through a site, rather than from fixed positions. These systems may be handheld or attached to a variety of autonomous or driven vehicles. On average, these systems involve a trade-off of accuracy for speed (Di Stefano et al. 2021).

Alternatively, photogrammetry compares and triangulates distances from photographs to produce measurements. For the typical 'structure from motion' method of photogrammetry, photographs taken during a continuous walk through the site or from many sources over time are used to produce coloured point clouds. Photogrammetry is significantly cheaper than LiDAR, as it can be captured with standard smartphone models, and it can be more efficient, especially if capture is taken with video. Photogrammetry may, however, need additional manual work to correctly set the scale and orientation of a captured scene.

Some applications of reality capture may fuse different data sources depending on the data requirements and context. For instance, reality capture at the city scale may combine an overview of photogrammetry data taken from an aerial source with more detailed street-level LiDAR or photogrammetry, with the goal of maintaining complete coverage while allowing maximum detail in all areas.

3.2 Scanning Technology in the Built Environment

Although the construction industry has been slow to adopt comprehensive digitalisation, digitisation and reality capture have been adopted in applications throughout the building life cycle. As a first example of digitisation used in the industry, scanning is used on raw materials before construction even begins. Stockpiled materials such as sand for concrete production are scanned with extended tripods so they can be better managed and tracked. To avoid the accumulation of errors when only recording amounts of ingoing and outgoing materials, periodic scans of the entire stockpile can be taken, and volumes can be estimated from the resulting topology (Manish et al. 2022; Zhang and Yang 2019). These volume-based measurements may also be used to estimate excavation volumes for earthworks or the value of used material when working with sprayed concrete.

Many applications of scanning relate to specific BIM representations of buildings. Scans may either be used to compare data to a known existing BIM model, known as *scan*-vs-*BIM*, or to generate an entirely new model, known as *scan-to-BIM*. A second example of digitisation in construction makes use of scan-vs-BIM for combining and comparing dense spatial and often temporal data to track how buildings develop over time: large-scale progress tracking, for example, can be used in construction monitoring to find which components have already been installed at a particular point in time and to detect errors, such as if components have been installed or fabricated incorrectly. Both goals require a highly detailed BIM model as a goal state for the building. Overlap between the BIM model and the collected datapoints (or point cloud) can be checked to see if a particular component has been installed or to look for differences that reveal incorrect sizes or installation locations.

A third example of digitisation in construction can be found during a building's operational life span, when large-scale reality capture is also applied to monitor for defects or deformations that may affect their safety and usability (Guo et al. 2021). Monitoring deflection is similar to monitoring construction deviations: using an existing BIM model, individual element positions are compared to the local point cloud to detect deflection or rotation over time. The knowledge of the complete structural system allows for the system to detect which deviations represent the greatest immediate danger (Kaartinen et al. 2022). Other types of defects that can be monitored include cracks, spalling, and corrosion visible on the surface of timber, concrete, and steel structures. While these can be detected by comparing them to a BIM model, they can be more specifically identified by the unique geometric effects on the material surface, using techniques similar to 2D image analysis (Tzortzinis et al. 2022). For some materials, these types of digitisation may also involve a direct fusion with image-based techniques, where 3D data is used to broadly locate the critical areas in the component, while the specific defect detection is carried out using accompanying photographic information (see Chap. 4 by Armeni et al. on AI).

Digitisation is also used during the operational phase to maintain up-to-date, as-built modelling for facility management. During a building's life cycle, the digital twin of a building can be used to manage facilities, track changes, and optimise processes. This enables computer-aided facility management.

Monitoring defects by using reality capture has many applications in the fields of building heritage and preservation in particular. Similar techniques to those used for contemporary buildings can help detect and analyse defects in historic structures, though aimed at a wider array of damage and design scenarios in older structures. Assessments often lack existing documentation, so defects can be detected solely by their unique spatial characteristics. In scenarios with obvious damage or missing components, digitisation can also create highly detailed recordings for models used in the automated production of replacement and repair components (Weigert et al. 2019; see also Chap. 7 by Chadha et al. on additive manufacturing).

Scanning and digitisation can also rapidly capture historical sites as a whole for a complete view of their layout or design. Depending on the scale, ground-based mobile LiDAR systems, aerial systems, connected to low-flying aircraft, and UAVs or other drones might be useful (Adamopoulos and Rinaudo 2021; Rodriguez et al. 2019).

Different types and depths of digitisation are used throughout the building life cycle to manage construction, ensure its correctness, and check for problems over time. BIM information is often compared to the captured data, but the results often do not make it beyond the stage of capture to be communicated to other relevant parties, as the industry has not adopted comprehensive digitalisation.

3.3 Scanning Technology for a Circular Built Environment

The measurements and data that scanning technologies and digitisation collect are often not new in construction; human workers could collect (and historically have collected) similar data. The true value of digitisation technology comes from the large volume and efficiency of gathering digital data. Since the technology enables the rapid capture of entire buildings and structures, digitisation of the entire active built environment becomes a possibility. This technology could help close material loops in the built environment by efficiently tracking and working with the uniqueness of individual material components at a large scale after their initial life span.

Post-demolition material reuse is the primary method for closing material loops today. Historically, balancing supply and demand with uncertainty from the procurer regarding the history (and, thus, reliability) of a specific reused component has hindered post-demolition material reuse (Hobbs and Adams 2017). These challenges could be mitigated by thorough and data-rich inventories of components that are available or will soon become available (Chap. 5 by Honic et al. on material passports).

Scanning and digitisation can provide a source for these inventories in the form of automated inventory making or scan-to-BIM, primarily making use of techniques for identification and description, in contrast to the connection and comparison performed for scan-vs-BIM. BIM documentation of existing structures hinges on the connection between physical measurements and digital representations. Traditionally, human workers would draft from a series of manual measurements; more recently, they have drawn BIM elements over a point cloud, using their own judgement in response to noise and misalignments. Nowadays, contemporary research is increasingly enabling automated assistance in the generation and construction of BIM models.

Analysis techniques of construction verification and defect detection (as described in Sect. 3.2) require scan-to-BIM to combine noise handling, component detection, measurement, and material description concurrently. The complexity of techniques needed to identify building components and features depends on the components they handle. Some features may be identified through simple statistical analysis – for instance, the level of each floor may be identified by determining the highest concentrations of points from a series of horizontal slices through the building cloud. Individual elements with minimal surface detail (such as interior walls) may instead use a geometric approach – algorithms such as random sample consensus (RANSAC), which finds patterns in high-noise environments, are used to identify simple geometry primitives such as planes and cylinders within a dataset, which can then be further refined to locate doors or windows. Alternatively, the system may handle all interior components and objects simultaneously using techniques borrowed from 2D computer vision (which generally deal with image pixels). Here, each point is individually classified using deep learning techniques as one of several common classes for the type of site, which may be further instance-segmented into individual components (see also Chap. 4 by Armeni et al. on AI).

In all cases, the identified points are condensed into a single BIM element, at minimum containing information on the component's position and dimensions (such as the wall thickness or structural profile). From here, additional relevant information is extracted and integrated. Descriptions of the connections between elements are particularly relevant for making decisions about reuse. This may be at the fine-grain level – such as in determining details about a structural connection – or at a larger scale, such as for extracting the larger web of relationships between elements in a structural system. This BIM model can then be used as the basis of a component inventory for the building. Depending on the level of detail, these records can be used to estimate the economic value of or potential for reducing emissions by reusing the building's components, assisting deconstruction groups in planning removal operations, or helping designers and contractors estimate the amounts of local materials that will soon be available.

Scanning and digitisation may also be applied to material streams after demolition or deconstruction has occurred. In these cases, the system only has to consider a single component at a time, so analysis can be focused on capturing specific dimensional information or various types of defect detection, as described in Sect. 3.2. On an urban scale, scanning technologies such as airborne or satellite data can also be used to construct or enrich existing digital models. In addition to applications of 3D urban models ranging from solar potential analyses to a wide variety of environmental analyses (microclimate, flooding, etc.) to dynamic thermal simulation of buildings (Malhotra et al. 2022), they can be used to predict future trends for reuse. By combining geometric data with data on materials, material stocks can be identified on a larger scale and used as a basis for a material flow analysis.

By using scanning and digitisation techniques, scan-to-BIM can connect reality capture to reuse operations. While assistive modelling tools are increasingly being used in practice, fully automated methods have not yet seen commercial implementation. Active research is continuing for addressing finer details (Zhou et al. 2021; Yan and Hajjar 2022) and overcoming gaps and noise in data (Park et al. 2022).

3.4 Industrial Implementations of Scanning and Digitisation for a Circular Building Environment: Concular

While industry use of LiDAR and digitisation began with applications for detecting defects and manual drawings from scans, the importance of inventories and large-scale scan-to-BIM is prompting further practical implementation. The German start-up Concular is a primary example of a company that is applying scan-to-BIM to the circular built environment. To improve resource efficiency within the construction sector by closing material loops, Concular has developed a platform for matching the supply and demand of reusable materials (Concular 2021).

High reuse and recycling rates are crucial to achieve a circular construction sector. The reuse of materials and products must be considered in the design and building process, especially in anticipation of the end of the building life cycle. To this end, Concular is developing a platform storing material, component, and building passports to represent and reintroduce the materials used in a building throughout its life cycle. The dataset storage method for the building passports can be continuously updated during the material, component, and building life cycle. The resulting database of reusable material and components is utilised for estimating amounts of available resources and finding appropriate circular sources. An essential function of Concular's database is the ability to connect digitised 3D models to the platform. Existing and new buildings can be added through open-source interfaces (IFC or CSV file formats). However, there is often little or no digital information available on buildings built before 1978 (which account for 75% of all structures), meaning most of our built environment is only accessible via analogue plans if at all (Metzger et al. 2019).

An ideal dataset would include digital plans, an overview of the current renovation or planning status, and information on building materials and their suitability for reuse. Scanning technology offers an accurate and potentially efficient way to digitise the built environment and provide applications such as Concular with the necessary BIM data. Within this context, Concular focuses on the technical development of a unique database model that combines relevant data formats in the building industry with the capture as well as tracking of building elements. A scan/capture phase first evaluates 2D plans and records, including recognising the building envelope or envelope elements from floor plans. Then, 3D scans capture rooms to accurately locate building elements within a floor plan. This is primarily accomplished with photogrammetry, which leverages the image data to gain information on materials and qualities of single products and building elements during building assessments. The data generated by these capture methods is stored within a building life cycle passport on a platform provided by Concular.

Additional technical details relevant to further connections, such as GIS data and product-specific sustainability information, are also mapped to the building elements within the passport. To deal effectively with many elements at scale and verify the quality of the data input of the relevant file standards, an artificial intelligence (AI) application is trained through test projects to assign this sustainability information to the materials, products, and components.

Building life cycle passports store updatable information on trackable materials and products contained at the building level. As a result, they provide project managers and building owners with an overview of the materials and products used in their buildings, as well as the ability to evaluate buildings by carbon footprint, recyclability, or circularity. A 'living' building passport allows for changes during a building's life cycle, by updating information on products transferred between buildings or replacing and digitally renewing defective products. Linking building passports to external product passports enables detailed information (such as service cycle, replacement duration, circularity, and availability) about products and components to be retrieved, compared, or exchanged. Trackable technical details can be stored within a building product, material, and property table (Chap. 5 by Honic et al. on material passports).

Scanning and capture technologies help create the volume of data necessary to digitise the built environment. Subsequently generated BIM data is then available to analyse and evaluate the aspects of sustainability and reusability on an adequate scale. Data collection is supported by AI automation in relevant areas while supplemented by additional external data collected in traditional ways. While a fully automated scanning process is technically feasible, the costs of both manual and automation-assisted intensive building assessments, plus the need to transform and standardise data formats, are still prohibitive to fully automating the process. Nevertheless, providing clear use cases to a variety of stakeholders, such as building owners and planners, encourages further adoption.

Obtaining data about the built environment is essential for urban mining and reusing materials and building components in a circular manner. Concular's platform allows this data to be stored and made accessible in order to close material loops and reduce the amount of new resources needed for construction.

3.5 Business Models for Scanning in a Circular Built Environment

The circular economy retains and maintains the embedded value of products by creating continuous closed loops of materials or product parts and reclaiming value lost to waste. Today, the lack of available data through secure, quality assured, and automated methods is one of the main obstacles that industry actors point to when creating new circular value networks (Deloitte 2019). There is usually no digital information available on buildings predating the adoption of digital planning and tracking tools – information that is necessary for judging the suitability of reusing components for specific new uses. In addition, information is lacking on sustainability factors (e.g. embodied carbon emissions), and open interfaces to existing historical information often do not exist. Thus, evaluating a building's circularity potential and the necessary deconstruction according to ecological or sustainability measures remains a challenge, and doing so requires great manual effort by experts and reviewers. Data digitisation and capture via scanning can significantly reduce that effort and has relevance for stakeholders across the entire building sector.

The representations of existing buildings created by digitisation technology can be used to generate BIM models as a basis for renovations, retrofittings, or assessing reuse potential. For project developers as well as building owners, scanning can provide an overview of current building conditions as well as information on building elements. A detailed assessment of material and products is then possible through digital capture, and scan-to-BIM offers additional possibilities for analysis and evaluation of the building mass both in terms of economic calculations and sustainability and compliance. Calculating the material worth of materials and products for reuse or recycling increases the potential economic value of buildings and offers a more productive use of leftover materials.

BIM data is used as the basis for life cycle analyses, which enable the assessment of a building's compliance with sustainability standards and reveal areas for improvement. This also has implications for investors and property owners. Project developers and building owners who want to access 'green finance', for example, must ensure that their projects are resilient to climate change and resource scarcity. These requirements impact the financial market of sustainable investment but also the work of planners. As the demand for taxonomy-compliant properties increases in the future, sustainable investing will become relevant for private investors and governmental funding alike. Capture and scan-to-BIM technology contributes to demand for resource efficiency and creates measurable sustainability standards relevant for investors.

Companies offer digital BIM models for different project purposes such as renovations or retrofittings. Companies such as Plan3D, for example, offer as-built surveys of listed buildings and technical installations, deformation studies, and visualisations for marketing purposes (Plan3D 2021). Concular's services are an example of a complementary platform model that stores the resulting data and provides analysis tools in a subscription-based software-as-a-service model.

Scanning and capture technology can also provide a connection between digital planning and physical reality during the construction process. By monitoring the progress and placement of building elements, scan-vs-BIM supports planners and construction companies in planning new building projects and identifying potential construction issues. Components may also be tracked individually, for example, with embedded radiofrequency identification (RFID) chips or codes from production to gate to on-site application, where data can be updated in real time (Strabag 2023). This also allows manufacturers to track their products over an entire life cycle, thereby establishing takeback or refurbishment processes and diversifying material resource flows.

Scan-to-BIM can also capture necessary information as a basis to create inventories of reuse components for the planning process, including information on their availability or material makeup and on the geometry of building elements. In order to provide predictive information on material availability and flows, building-specific capture has the potential to be supplemented with large-scale datasets. For example, large-scale 3D building models are made freely available by surveying agencies as open data or are available for a fee as a file or as a web service (Geiger et al. 2022).

During the process of deconstruction, similar advantages can be gained: digital capture makes it possible to collect the necessary structural information as well as more detailed assessments on the location of pollutants. Building surveys and audits can quickly and accurately portray existing building conditions, including dimensions, structures, and finishes to generate BIM models for assessing reuse potential. As a basis for urban mining, this can be used for informing the planning process and resource availability.

These use cases provide an outlook on the possibilities to offset the current costs of labour- and time-intensive digitalised processes by establishing new applications, especially within the framework of urban mining and circular construction. Key factors for the application of scan-to-BIM within the built environment are high-quality, accurate results with concrete economic benefits. Adoption of circular principles is further fuelled by current political developments to reduce CO_2 emissions and diversify resource flows. For example, the European Commission has stated its intention to make Europe the first climate-neutral continent by 2050 and to create a nearly climate-neutral building stock (European Commission 2018). The necessary analysis of our built environment as a basis for reuse is only possible with large-scale digitisation – which can be reached by using large-scale methods such as capture and scan-to-BIM technology.

3.6 Discussion

Digitisation technology addresses an immediate gap in material recovery and reuse today – it provides an efficient, effective, and available way to fill a lack of knowledge about the location and quality of relevant materials. The density and coverage of models created by digitisation technologies contribute to the geometric side of digital twins (Chap. 1 by Koutamanis discussing BIM and digital twins), more complete and up-to-date building records, and building an initial record of pre-digital buildings. These technologies are increasingly being applied in different stages throughout the standard building life cycle and by reuse actors, such as in the case of Concular outlined here.

Going forward, automation and ease of access will be major factors in the further adoption of these technologies. Effective automation will be dependent on the development of focused models and machine learning systems for the reuse context. Existing machine learning work using 3D scans often focuses on natural features, geographic applications, or furniture and fit-out features in the building domain. Identification and analysis of construction components require the compilation of entirely new datasets. The specific needs of these models create unique challenges for analysis. The geometry of products and components may vary heavily, especially by region, leading to possible unexpected difficulties in generalising information about products. Additionally, the ability to recognise and process components from a variety of vernaculars and styles speaks to the design effects of circularity, where, for instance, a region may want to maintain a historic style while utilising components from nearby areas or alternatively create new combinations.

The contemporary adoption of scanning and scan-to-BIM technology still faces some existing challenges. Immediately relevant is the high entry cost for the capture hardware, with building-scale terrestrial models often costing more than 50,000 Euros. While accessible models are becoming widely available, the best resolution and accuracy may still be inaccessible for smaller groups interested in utilising digitisation, especially in scenarios requiring many captures over a period of time. Additionally, safe and efficient long-term storage of the huge volume of data required for a large portfolio of sites presents new operational challenges. These concerns may be mitigated through collaboration with specialised groups providing scanning services and with expertise in large-scale data processing and data storage. Finally, contemporary scanning technology can only gather a limited amount of detail about each component. Information about possible interior damage, the composition of multilayer parts, or even the materiality of painted surfaces must still be gathered through other technology or through manual means.

Going forward, technological advancements in scanning and capture technology will directly allow for a continuous record of buildings. While higher-end models will always be necessary in high-accuracy scenarios, the form factor and cost of LiDAR technology are already becoming accessible at the smartphone scale. Furthermore, image-based reconstruction is increasingly merging with deep learning techniques, fostering more accurate reproductions from smartphone hardware capture (Heipke and Rottensteiner 2020). As these tools become appropriate for full building reconstruction, they will become better integrated in the ongoing operations of building managers and owners, thus extending life cycles through target repair and ultimately more informed reuse of building components.

There are also paths for adapting the role of scanning technology in the building life cycle. Scanning and scan-to-BIM are often employed currently as a single explicit step, as a static representation of the building state. These same increases in accessibility will allow for regular or even continuous recording and updating of associated BIM models as part of the digital twin concept (Chap. 1 by Koutamanis on BIM and digital twins). In the context of circular construction and urban mining, it is also possible to provide up-to-date information on large-scale inventories of material and components, which are instrumental to closing material loops. These continuous and integrated models will be further enhanced with varied data sources. Alternative hardware, such as thermal imaging, has been studied to overcome noisy and uncertain lighting conditions in building sites, as well as its potential for locating hidden mechanical, electrical, and plumbing engineering components (Pazhoohesh et al. 2021; Penzel et al. 2019). Understanding the interior of components may be addressed by several tomography technologies, such as those using electrical resistance or ultrasonic audio for studying structural conditions in wood and concrete (Karhunen et al. 2010; Zielińska and Rucka 2021).

Digitisation via scanning technologies and scan-to-BIM is a key facilitator in a digitalised system for circularity in the built environment and is crucial for achieving the target of zero-carbon buildings by 2050. Presently, practitioners often lack precise pre-demolition information to understand the existing building stock. Scanning, geometry assignment, and material analysis together address this need. Centrally, they allow for the large-scale creation of BIM models and geometry for digital twins. This information can also aid in creating sharable material passports that compile the linked data describing a component's characteristics, location, history, and ownership status in previously non-digitised contexts. Automated techniques for capture and analysis are also a central application for AI and computer vision in reuse operations. Together, they are a key source of information that powers the subsequent systems for material tracking and design in a circular built environment as well as improving accuracy, efficiency, and collaboration in the built environment.

3.7 Key Takeaways

- Effective material reuse requires data about many available sites and components to close the loop into new projects.
- Reality capture technologies record large amounts of spatial data as point clouds, collecting useful information about as many relevant components as possible and contributing to the digital representation of the built environment ultimately needed for circular construction.
- Point cloud data can inform material passports and qualitative checks throughout the life span of a building, facilitating repair operations to slow the life span of individual buildings and enabling the analysis and tracking of materials and components throughout their life cycle.

Acknowledgements The associated research includes funding by Vlaanderen Circulair – Call Circulaire Bouweconomie – Vlaams Agentschap Innoveren and Ondernemen (VLAIO).

Declaration of Competing Interests Authors von Zimmerman, Haradhun, Campanella, and Bräutigam are or were employees of Concular, a Berlin-based start-up using digitalisation to further circular construction.

References

Adamopoulos E, Rinaudo F (2021) Close-range sensing and data fusion for built heritage inspection and monitoring—a review. Remote Sens 13(19):3936. https://doi.org/10.3390/rs13193936

Bello SA, Yu S, Wang C, Adam JM, Li J (2020) Review: deep learning on 3D point clouds. Remote Sens 12(11):1729. https://doi.org/10.3390/rs12111729

Charef R, Emmitt S, Alaka H, Fouchal F (2019) Building information modelling adoption in the european union: an overview. J Build Eng 25:100777. https://doi.org/10.1016/j.jobe.2019.100777

Concular (2021) Ausgewählte Projekte – Concular - Zirkuläres Bauen : Concular – Zirkuläres Bauen. https://concular.de/projekte/. Accessed 6 Feb 2023

Deloitte (2019) Complexity: overcoming obstacles and seizing opportunities. The Deloitte global chief procurement officer survey 2019. In: Deloitte Insights. https://www2.deloitte.com/si/en/pages/strategy-operations/articles/global-cpo-survey.html. Accessed 7 Feb 2023

Di Stefano F, Chiappini S, Gorreja A, Balestra M, Pierdicca R (2021) Mobile 3D scan lidar: a literature review. Geomat Nat Hazards Risk 12(1):2387–2429. https://doi.org/10.1080/19475705.2021.1964617

European Commission, Directorate-General for Climate Action (2018) Communication from the commission: a clean planet for all; a European strategic long-term vision for a prosperous, modern, competitive and climate neutral economy. Brussels. https://eur-lex.europa.eu/legal-content/EN/ALL/?uri=CELEX:52018DC0773

Geiger A, Nichersu A, Häfele K-H, Hagenmeyer V (2022) Usage profile enrichment of CityGML models for urban building energy modeling. In: BauSim 2022

Guo M, Sun M, Pan D, Huang M, Yan B, Zhou Y, Nie P, Zhou T, Zhao Y (2021) High-precision detection method for large and complex steel structures based on global registration algorithm and automatic point cloud generation. Measurement 172:108765. https://doi.org/10.1016/j.measurement.2020.108765

Heaton J, Ak P, Schooling J (2019) Design and development of BIM models to support operations and maintenance. Comput Ind 111:172–186. https://doi.org/10.1016/j.compind.2019.08.001

Heipke C, Rottensteiner F (2020) Deep learning for geometric and semantic tasks in photogrammetry and remote sensing. Geospat Inf Sci 23(1):10–19. https://doi.org/10.1080/10095020.2020.1718003

Hobbs G, Adams K (2017) Reuse of building products and materials – barriers and opportunities. In: International HISER conference on advances in recycling and management of construction and demolition waste. Delft

Kaartinen E, Dunphy K, Sadhu A (2022) Lidar-based structural health monitoring: applications in civil infrastructure systems. Sensors 22(12):4610. https://doi.org/10.3390/s22124610

Karhunen K, Seppänen A, Lehikoinen A, Monteiro PJM, Kaipio JP (2010) Electrical resistance tomography imaging of concrete. Cem Conc Res 40(1):137–145. https://doi.org/10.1016/j.cemconres.2009.08.023

Malhotra A, Bischof J, Nichersu A, Häfele K-H, Exenberger J, Sood D, Allan J, Frisch J, van Treeck C, O'Donnell J, Schweiger G (2022) Information modelling for urban building energy simulation—a taxonomic review. Build Environ 208:108552. https://doi.org/10.1016/j.buildenv.2021.108552

Manish R, Hasheminasab SM, Liu J, Koshan Y, Mahlberg JA, Lin Y-C, Ravi R, Zhou T, McGuffey J, Wells T, Bullock D, Habib A (2022) Image-aided LiDAR mapping platform and data processing strategy for stockpile volume estimation. Remote Sens 14(1):231. https://doi.org/10.3390/rs14010231

Metzger S, Jahnke K, Walikewitz N, Otto M, Grondev A, Fritz S (2019) Wohnen und Sanieren. Schuberth J, ed. Empirische Wohngebäudedaten seit 2002. Dessau-Rosslau: Umweltbundesamt Fachgebiet V 1.4

Park J, Kim J, Lee D, Jeong K, Lee J, Kim H, Hong T (2022) Deep learning–based automation of scan-to-BIM with modeling objects from occluded point clouds. J Manag Eng 38(4):04022025. https://doi.org/10.1061/(ASCE)ME.1943-5479.0001055

Pazhoohesh M, Zhang C, Hammad A, Taromi Z, Razmjoo A (2021) Infrared thermography for a quick construction progress monitoring approach in concrete structures. Archit Struct Constr 1(2):91–106. https://doi.org/10.1007/s44150-021-00008-7

Penzel M, Müller F, Zinecker M, Schubert A, Wuttke O, Gläser M (2019) Simulation of IR-heating and IR-scanning for the prediction of internal structures of walls. Comsol, Cambridge. https://www.comsol.com/paper/download/680891/Paper_Simulation_of_IR-heating_and_IR-scanning_for_the_prediction_of_internal_structures_of_walls.pdf
Plan3D (2021) https://www.plan3d-berlin.de/. Accessed 7 Feb 2023
Quirck V (2012) A brief history of BIM. ArchDaily https://www.archdaily.com/302490/a-brief-history-of-bim. Accessed 6 Feb 2023
Rodriguez BX, Simonen K, Huang M, De Wolf C (2019) A taxonomy for whole building life cycle assessment (WBLCA). Smart Sustain Built Environ 8(3):190–205. https://doi.org/10.1108/SASBE-06-2018-0034
Strabag (2023) RFID im Fenster. https://innovation.strabag.com/projekt/rfid-im-fenster/. Accessed 7 Feb 2023
Tzortzinis G, Ai C, Breña SF, Gerasimidis S (2022) Using 3D laser scanning for estimating the capacity of corroded steel bridge girders: experiments, computations and analytical solutions. Eng Struct 265:114407. https://doi.org/10.1016/j.engstruct.2022.114407
Waters N (2018) GIS: history. In: International encyclopedia of geography. Wiley, pp 1–13
Weigert A, Dhanda A, Cano J, Bayod C, Fai S, Santana Quintero M (2019) A review of recording technologies for digital fabrication in heritage conservation. Int Arch Photogramm Remote Sens Spatial Inf Sci XLII-2-W9:773–778. https://doi.org/10.5194/isprs-archives-XLII-2-W9-773-2019
Wu C, Yuan Y, Tang Y, Tian B (2022) Application of terrestrial laser scanning (TLS) in the architecture, engineering and construction (AEC) industry. Sensors 22(1):265. https://doi.org/10.3390/s22010265
Yan Y, Hajjar JF (2022) Geometric models from laser scanning data for superstructure components of steel girder bridges. Autom Constr 142:104484. https://doi.org/10.1016/j.autcon.2022.104484
Zhang W, Yang D (2019) Lidar-based fast 3d stockpile modeling. In: In: 2019 international conference on intelligent computing, automation and systems (ICICAS). IEEE, Chongqing, pp 703–707. https://doi.org/10.1109/ICICAS48597.2019.00152
Zhou X, Liu J, Cheng G, Li D, Chen YF (2021) Automated locating of replaceable coupling steel beam using terrestrial laser scanning. Autom Constr 122:103468. https://doi.org/10.1016/j.autcon.2020.103468
Zielińska M, Rucka M (2021) Using the ultrasonic tomography method to study the condition of wooden beams from historical building. 12th international conference on structural analysis of historical constructions (SAHC). https://doi.org/10.23967/sahc.2021.315

Matthew Gordon is an assistant with the Chair of Circular Engineering for Architecture at the ETH Zurich Department of Civil, Environmental and Geomatic Engineering. He studied architecture at the University of Virginia and obtained a master's degree in robotics and advanced construction from the Institute for Advanced Architecture of Catalonia (IaaC). His research focuses on accessible digitisation strategies for deconstruction planning and efficient assistive tools for designing with reused materials.

Luise von Zimmerman studied architecture at the Universität der Künste, Berlin, and the University of New South Wales, Sydney. She has worked in several architecture offices and in academia. In both her education and her work, she deals with the intersection of digital toolmaking, sustainability, and conceptual as well as computational design. After obtaining her master's in architecture, she is now a product owner at Concular. Additionally, she takes part in the ABC network, which advocates for a more sustainable construction industry.

Oushesh Haradhun earned a master of science degree in robotics, cognition, and artificial intelligence from the Technical University of Munich. He holds both academic and professional experiences in 3D computer vision, deep learning, and natural language processing. Prior to Concular, he worked in multiple start-ups involving AI products ranging from AI-based retail checkout system at Skidata, natural language processing at Munich RE, 2D to 3D movie conversion based on deep learning and neural radiance fields for BitAnimate Inc., and Virtual Try-On for shoes at Vyking. He is also an alumnus of Entrepreneur First: an international talent investor that supports individuals building their own tech companies.

Dominik Campanella is one of the co-founders of Concular, the digital ecosystem for circular construction. Concular allows to digitise materials in new and existing buildings and closes the material loop by getting them into new buildings before the buildings are demolished. Before founding Concular, he studied computer science and business administration and worked for several years for Google.

Milena Bräutigam majored in economics and mathematics from Maastricht University and studied econometrics and operations research at the Erasmus University Rotterdam. Before beginning her master's degree in data science and econometrics, Milena interned in the artificial intelligence team at Concular. Besides circular economy, her research interests lie in labour market and gender economics.

Catherine De Wolf is an Assistant Professor of Circular Engineering for Architecture at the Swiss Federal Institute of Technology of Zurich (ETH Zurich), where she conducts research on digital innovation towards a circular built environment. She has a dual background in civil engineering and architecture and completed her PhD at the Massachusetts Institute of Technology (MIT). She is a faculty member in the ETH Centre for Augmented Computational Design in Architecture, Engineering and Construction (Design++), the ETH AI Center, the Swiss Federal Laboratories for Materials Science and Technology (Empa), the Circular Future Cities group of the Future Cities Lab, and the National Centre of Competence in Research on Digital Fabrication (DFAB). She has worked in academia (e.g., University of Cambridge, University of Technology Delft, Nanjing University), in industry (e.g., Arup, Elioth, Ney & Partners), and in governmental institutions (e.g., Joint Research Centre of the European Commission). She also co-founded the circular construction company Anku, the Structural Engineers 2050 Commitment, and the Digital Circular Economy (DiCE) Lab.

Chapter 4
Artificial Intelligence for Predicting Reuse Patterns

Iro Armeni, Deepika Raghu, and Catherine De Wolf

Abstract Artificial intelligence, and specifically the subfields of computer vision and machine learning, has become a topic with great potential for predicting reuse patterns in the built environment. With sensors that collect visual data becoming more readily available, new opportunities are created to digitalise the built environment by applying technologies from these fields. Applications include exploring the design space, monitoring construction progress, and improving building performance during operation. Using these applications to increase circularity in the built environment requires information about in-use building products and their attributes (e.g. type, material, size, geometry, condition, etc.). This information is a starting point for many downstream circular processes and a core component of circular databases, which can enable designers, constructors, and facility managers to follow a circular paradigm. Many advancements have been made in academia and industry towards extracting such information from visual and other building data, e.g. for the downstream processes of predicting material reusability or automating the maintenance of building facades. This chapter presents efforts on this front and highlights the gaps in adopting and utilising these technologies for the circular built environment, including challenges in developing comprehensive systems for their deployment and in robustly evaluating them. It also discusses business and organisational considerations with respect to adoption, utilisation, and development of the technologies in the circular context.

Keywords Artificial intelligence · Computer vision · Machine learning · As-is building status · Buildings as material banks · Building inventory

I. Armeni (✉)
Department of Civil and Environmental Engineering, Stanford University, Stanford, CA, USA
e-mail: iarmeni@stanford.edu

D. Raghu · C. De Wolf
Circular Engineering for Architecture, Institute of Construction & Infrastructure Management, Department of Civil, Environmental and Geomatic Engineering, ETH Zurich, Zurich, Switzerland
e-mail: draghu@ethz.ch; dewolf@ibi.baug.ethz.ch

C. De Wolf et al. (eds.), *A Circular Built Environment in the Digital Age*, Circular Economy and Sustainability, https://doi.org/10.1007/978-3-031-39675-5_4

4.1 Introduction

'Can machines think?' asked Alan Turing, a British polymath who explored the mathematical possibility of artificial intelligence (AI). In 1950, Turing discussed how to build intelligent machines and test their intelligence (Turing A., 1950). But what is AI? John McCarthy, a prominent computer and cognitive scientist, coined the term in 1956 when he held the first academic conference on the subject. Although many definitions span different disciplines, from philosophical to very applied, AI can be defined as the computer science field that attempts to develop machines (i.e. algorithms and robots) with human-level intelligence, hence creating machines that can 'think'.

Two of the most well-known subfields of AI are machine learning (ML) and computer vision (CV). Both subfields enable algorithms to automatically analyse, learn from, and/or derive meaningful information from patterns in data and take actions or make recommendations based on that information. The field of ML includes deep learning, where a large amount of data is used to teach a neural network, consisting of multiple layers of increasing complexity and abstraction, how to perform a task. It also includes reinforcement learning, a type of ML that learns by doing to identify the best solutions based on rewards and penalties. Figure 4.1 illustrates a diagram of AI and subfields discussed in this chapter.

The field of CV focuses on digital images, 3D point clouds, and other visual inputs. Although CV can exist without ML to solve tasks related to photogrammetry (such as measuring distances, area, or volume), when combined with ML, it aims to

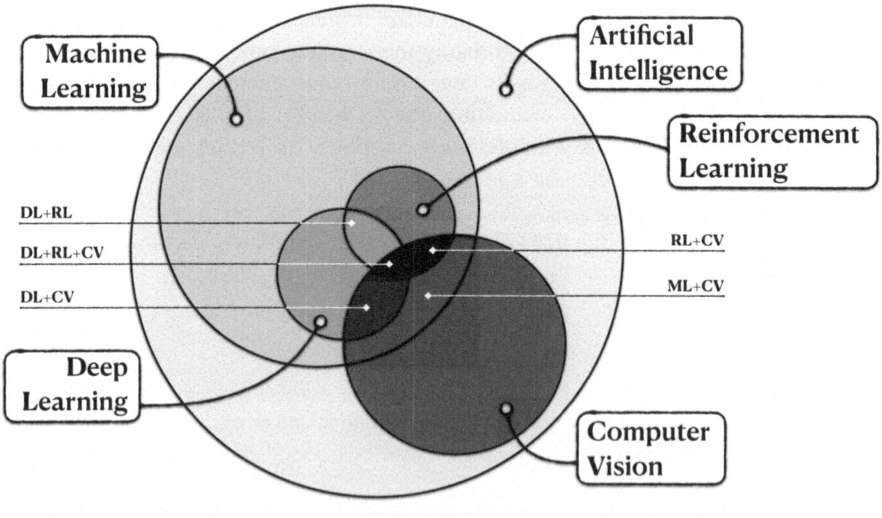

Fig. 4.1 Artificial intelligence and subfields. Machine learning (ML), computer vision (CV), deep learning (DL), and reinforcement learning (RL) are all subdomains of artificial intelligence that can be employed together for solving perceptual tasks. The decision to combine them depends on the nature of the task

develop algorithms with human-like visual perception and reasoning. In 1966, the Summer Vision Project at the Massachusetts Institute of Technology attempted to reach human-like visual perception (Papert 1966). However, the task proved to be harder than expected. Decades later, we are still far away from reaching this goal even though substantial progress has been achieved with the curation of large datasets (e.g. Deng et al. 2009), the development of more powerful computation, and the advancements in deep learning algorithms that began around 2010 (e.g. Krizhevsky et al. 2017). Regardless, CV and ML research have matured significantly and entered the daily lives of consumers (e.g. fingerprint identification, face recognition, virtual reality games, and more), as well as many industries (e.g. retail, agriculture, manufacturing, medicine, and more). In this chapter, we examine the impact, potential, and present limitations of CV and ML on a circular built environment.

Throughout the chapter, we refer to using CV algorithms for recognising information on visual data. (Hereafter and for simplification, algorithms that address visual perception will be considered as belonging to the CV field, even if they contain an ML component.) But how do these algorithms work? Commonly they are supervised learning techniques, i.e. during the training process, they are given paired data points of inputs (raw data) and correct outputs (the answer that needs to be predicted, e.g. the type of an object or a material) to learn the task at hand. Given a random unseen input, the goal is for the algorithm to predict the output as accurately as possible. The most commonly used algorithms for the above tasks are convolutional neural networks, which is a type of deep learning network that can operate on image data. This means that datasets with numerous and diverse (paired) data points need to be curated that allow the algorithms to gain generalised (i.e. broadly applicable) understandings of varying real-world scenarios. However, this curation is a difficult process, and, depending on the task, one must carefully design the collection and annotation process that creates the paired data points.

We specifically discuss, among other topics, recognising defects or materials and their attributes in the context of the construction industry. This corresponds to the following CV tasks: image classification, object detection, and/or semantic segmentation. Image classification is the task of predicting a single object label for an entire image in which the object of interest is highly represented in it. Object detection is the task of finding object instances in an image and localising them on it in the form of bounding boxes (2D boxes that tightly include all pixels of each identified object instance). Semantic segmentation is similar to the previous task, but instead of localising an object with a bounding box, an object label is given to each pixel that corresponds to it. It should be noted that, in all cases, an object can also be a material or any other type of semantic information. In their basic formulation, all three tasks follow the close-world assumption. This means that the algorithm has knowledge of a fixed set of object labels during training, and any object outside of that set is considered unknown or irrelevant. Hence, the algorithm is only able to recognise objects that it has been explicitly trained on and cannot recognise new or unexpected ones.

4.2 Computer Vision and Machine Learning in the Built Environment

Present use of CV and ML at the different life cycle stages of the buildings (whether *operation, construction,* or *design*) shows the potential that such algorithms can have in the domain of architecture, engineering, construction, and operations (AECO) towards multiple building lifetimes (Hong et al. 2020).

The *operation* stage is one of the first life cycle stages for which CV and ML algorithms were developed. In this life cycle stage, algorithms contribute to high-performance building operation. The term 'high-performance building' means a building that integrates and optimises all major high-performance building attributes, including energy efficiency, durability, life cycle performance, and occupant productivity (Act, E. P., 2005). The way such algorithms contribute is by measuring, modelling, and predicting building performance for existing buildings based on metrics such as energy consumption and loss (Macher et al. 2020; Barahona et al. 2021; Mirzabeigi et al. 2022; Rakha et al. 2022; Mayer et al. 2023; Motayyeb et al. 2023) and occupant trajectory and density (Sonta and Jain 2019; Tien et al. 2020; Sonta et al. 2021) to develop better strategies and processes for building operation. Temperature, stress, and ventilation levels, among other indicators, are similarly employed to achieve better user comfort (Acquaah et al. 2021; Ashrafi et al. 2022).

Digital twins of buildings are being developed to provide more insights into and control over building operation processes. (A digital twin is defined as the geometric, semantic, operational, and organisational twin of a real-world building in the digital world. See more in Chap. 3 by Gordon et al.) They connect metrics, such as the aforementioned, with spatial and building system information for more accurate modelling and prediction. Presently, a digital twin is a manually updated building model that reflects the as-is state of the building, and in most cases, it is an adapted, as-designed building information model (BIM) of the building. Research has made steps to automate this process with the use of CV and ML for both new and existing constructions (Bassier and Vergauwen 2020; Chen et al. 2022; Ma et al. 2022; Pan et al. 2022; Kim et al. 2023); however, as mentioned in Chap. 3, the capability to fully replicate a building automatically has not been entirely reached.

During the *construction* life cycle stage, CV and ML are used to monitor and quantify in detail construction progress (Kim 2020). This level of monitoring is not achievable through traditional practices which lack automation. These technologies can improve construction scheduling by connecting, optimising, and spatially distributing sequences of tasks given prior knowledge (Fischer et al. 2018; Amer and Golparvar-Fard 2021; Awada et al. 2021; Fitzsimmons et al. 2022). They can, for example, automatically count material loaded in trucks to verify delivery to the construction site without the use of a human (Yella and Dougherty 2013; Sun et al. 2021; Li and Chen 2022; Li and Zheng 2023), or machinery (e.g. dump trucks and excavators) can be autonomously navigated and operated to relieve humans of dangerous tasks and to provide labour despite an ageing workforce or pandemic situation (Guo et al. 2020; Ali 2021; Eraliev et al. 2022). They also help ensure the safety of workers by predicting hazardous settings when given visual data with respect to their surroundings and activities (Liu et al. 2021; Pham et al. 2021; Tang and Golparvar-Fard 2021; Koc et al. 2022; Alkaissy et al. 2023; Xu et al. 2023). Here again, the most advanced solutions are still in the research stage, with many industry players relying largely on manual and time-consuming work and with a few start-ups leading new efforts towards automation or integrating AI into construction work. Developing robust, safe, and generalisable solutions that automate entire processes remains a challenge. Several technical questions are unsolved, and additional challenges include the need for hardware located in strategic positions to track on-site activities of humans and machines; problems with illumination conditions, clutter, and occlusions that prevent from using CV on site; the absence of clear regulations and guidelines with respect to capturing and monitoring human activity on sites; and the resistance of construction workers and contractors to introducing these technologies on site (especially given the absence of regulations and guidelines).

The *design* stage is one of the most recent stages to benefit from CV and ML technology. Although expert-based AI algorithms have been developed since many years for the design stage (Maher and Fenves 1985; Fischer 1993), data-driven methods have mainly started to appear since the development of generative adversarial networks (GANs). GANs are a type of deep neural network that allow the generation of new data points by synthesising learned attributes and patterns given an underlying dataset. These networks are mainly used to generate 2D facades or small-scale simplified floor plans (Chaillou 2020; Chang et al. 2021; Nauata et al. 2021; Bisht et al. 2022; Sun et al. 2022; Yin et al. 2022; Zhao et al. 2023). These approaches demonstrate promising results, but there are certain shortcomings. GANs are well known for being hard to train and for suffering from mode collapse where, no matter the input, the algorithm will continually produce the same exact result.

The use of CV and ML goes beyond the spatial scale of the building: to the smaller scale of materials to the larger scales of neighbourhoods and cities. In addition to counting materials, CV algorithms are employed to recognise the type and spatial extent of material in images (Bell et al. 2015; Upchurch and Niu 2022). Most applications recognise the material of objects in daily indoor environments with the goal of enabling robotic agents to perform tasks. For example, the knowledge of the material on a traversable surface can inform the agent of how much friction can be expected (Suryamurthy et al. 2019; Hosseinpoor et al. 2021; Guan et al. 2022). Also on the material scale, CV algorithms have been used to understand the condition of materials, to detect defects (e.g. cracks on concrete), and to monitor the effect of time on them. This is very commonly applied in infrastructural settings, e.g. in tunnels, roads, and bridges (Dung et al. 2019; Fan et al. 2019; König et al. 2019; Xu et al. 2019; Li et al. 2020; Liu et al. 2020; Lei et al. 2021; Yu et al. 2021). In the scale of neighbourhoods and cities, CV is used to create 3D maps with the use of drones and satellite images, respectively, as well as to identify the land use type and material of a patch of land (e.g. vegetation, house, road) (Albert et al. 2017; Diakogiannis et al. 2020; Rousset et al. 2021) and to detect changes that take place over time (Shi et al. 2021; Zheng et al. 2021). Reinforcement learning is often used in the scale of cities to define new or renewed models for different types of transportation (e.g. private vehicles, buses, trams, trains, etc.) (Haydari and Yılmaz 2020).

Although the aforementioned applications progress the state of practice in several tasks that relate to creating and maintaining the built environment, they do not create a cohesive ensemble. In almost all cases, applications of CV and ML are viewed in isolation, without considering potential information exchange or the creation of standards for the collection, sharing, and processing of this information with different stakeholders. The AECO domain typically approaches projects in this way – in a siloed and non-centralised manner where knowledge learned from one project is limited to the people involved and not in processes and data.

4.3 Computer Vision and Machine Learning for a Circular Economy

CV and ML methods can be used to foster a transition towards a circular economy (CE). We will be approaching this from three levels: *micro*, *meso*, and *macro* (Ghisellini et al. 2016). The *micro* level focuses on a particular company, consumer, or product; the *meso* level on eco-industrial networks or supply chains; and the *macro* level on regions, cities, and municipalities. We will also approach this from four main resource loop strategies: narrowing, slowing, closing, and regenerating the loop (Bocken et al. 2016; Çetin et al. 2021).

4.3.1 Narrowing the Loop

Researchers have been investigating how to narrow the loop by minimising resources via improving efficiency (Bocken et al. 2016; Çetin et al. 2021) on several aspects of building performance during the design process. This commonly involves developing techniques that generate designs by jointly optimising several variables, such as those related to structural integrity (Kraus et al. 2022; Málaga-Chuquitaype 2022), fabrication (Ramsgaard Thomsen et al. 2020), waste (Akanbi et al. 2020), energy (Płoszaj-Mazurek et al. 2020; Di Natale et al. 2022; Tien et al. 2022), wind (Kareem 2020; Wang and Wu 2020; Li and Zheng 2023; Li and Yi 2023; Zuo et al. 2023), and more. The goal is to generate a design that serves exactly what is needed by the project definition. This line of work integrates domain expertise and proven mathematical or physics theorems to guide the design process. Such prior knowledge is essential for ML execution; although data-driven methods have the advantage of learning from real-world data distributions (i.e. learning from how the real-world looks and behaves), if left unconstrained, there is a high probability that they will make associations from data that do not lead to plausible solutions. Thus, approaches have employed physics-based reasoning, reinforcement learning, and other domain knowledge and constraints to steer the learning process.

Other researchers are exploring new methods of design and construction by using waste materials from the demolition of structures with the help of material-informed reinforcement learning (Huang 2021). However, the process includes time-consuming steps, such as a full high-resolution scan of each material fragment in isolation. In the future, easier pipelines should be explored that do not require high-end and expensive capturing systems in lab settings. Instead, pipelines and systems should be highly automated, operate in different conditions, and employ commodity devices. Fragkia and Foged (2020) proposed a new integrated circular design workflow that employs ML to predict fabrication files of material components for robotic production that follow specific material performance and design requirements. Such an approach optimises material distribution and promotes material economy. Akanbi et al. (2020) developed a deep learning method to predict the demolition waste of a building given basic building features, such as gross floor area, volume, number of floors, building architype, and usage. The authors demonstrate the use of the method to predict waste from four different design alternatives of a building. The above can be categorised as being at the *micro* level.

At the *macro* level, Cha et al. (2023) developed an ML algorithm to predict the demolition waste generation rate for cities in South Korea under redevelopment. They gathered data from hundreds of buildings before and after demolition. For the prediction, they employ as input similar information as Akanbi et al. (2020), on location, structure, usage, wall type, roof type, gross floor area, and number of floors. Such an algorithm can provide valuable information on the generated waste in case of demolition (or deconstruction), hence enabling to design regions that produce less waste and as such more sustainable.

4.3.2 Slowing the Loop

Achieving longevity of the built environment requires adopting appropriate measures for preventative maintenance (Çetin et al. 2021). In turn, a well-maintained product – be that an entire building, an element within, or a material – results in higher reusability potential in the future. Of particular interest for preventative maintenance is the understanding of a product's condition, early detection of any defects, monitoring of their progression to ensure appropriate intervention, and predicting remaining useful life (Khuc and Catbas 2018; Srikanth and Arockiasamy 2020; Dong and Catbas 2021; Li et al. 2021; Berghout and Benbouzid 2022). Researchers employ CV algorithms to inspect infrastructure projects whose failure mode can have catastrophic consequences, such as bridges (Galdelli et al. 2022), tunnels (Tichý et al. 2021), and dams (Khaloo et al. 2018; Feng et al. 2020). A common pipeline consists of the following steps: (1) capturing video sequences and point cloud data with scanners or cameras, (2) processing this data to identify and analyse defects and localise the defects on the 3D model, and (3) comparing the data with the previous state of any defects. Not all steps are fully automated, and user participation is required to complete them. In the case of ML, critical data, such as year of construction, traffic load, and temperature range, are used to predict the remaining life of a product. To achieve this, historical data from inspections of similar products are extracted from – public – databases.

Although the ultimate goal of many such pipelines would be to reach full automation, it is unrealistic to expect that we can fully achieve this goal while guaranteeing robust algorithmic performance, especially in high-stakes settings (e.g. dam collapse). The inclusion of a user can bridge the gap between automation and accuracy requirements (Muin and Mosalam 2021). However, user actions should be integrated into helping the algorithm increase learning capacity over time with human-in-the-loop techniques (e.g. reinforcement learning-based). Such techniques involve human and machine collaboration, where input from the (human) user serves as a learning signal for the machine to correct or further expand previously learned states.

One might ask: why do we need both image *and* point cloud data? To answer this, we need to make the distinction between images and point clouds: images capture high-resolution information related to colour, edges, and texture, whereas point clouds capture geometry, shape, and general 3D composition. Hence, detecting defects that relate to cracks, discolorations, texture changes, etc., requires images, since point cloud data is not able to contain this visual information. On the other hand, detecting general structural failures, such as bending, dislocation, etc., requires geometric cues, which are contained in point clouds and are not easy to extract from images. With photogrammetry, one can create point clouds with a sequence of images (video), but the final point density is lower and geometric error is higher, which makes this technique less appropriate for this type of analysis. (For more details on photogrammetry, see Chap. 3 on scanning technologies by Gordon et al.)

4.3.3 Closing the Loop

With respect to closing the loop, and essentially reusing or recycling materials (Bocken et al. 2016; Çetin et al. 2021), CV can play an instrumental role in managing materials and waste by identifying useful recovery pathways at various scales during a project's life cycle. On the *micro* level of recycling or reusing materials, CV algorithms can be used to reduce labour costs by automating the classification of recyclable materials (Mao et al. 2021; Zhang et al. 2021) at the end of a building's life in sorting centres. The classification of scrap metals and estimation of their masses (Díaz-Romero et al. 2022) can also be carried out to obtain a better understanding of the physical properties of the objects being sorted. Multilabel waste detection models are additionally being explored to identify glass, fabric, metal, plastic, and paper waste (Zhang et al. 2022), as well as earth-based components such as concrete, clinker, and bricks (Kuritcyn et al. 2015). For effective construction and demolition waste management, inspectors at disposal facilities must determine the amount of different waste components loaded in incoming trucks. The large quantities of waste and their mixed nature make it difficult to access. Chen et al. (2022) developed an algorithm to automatically quantify specific materials from a single image obtained from monocular cameras installed on site, while Díaz-Romero et al. (2021) built a real-time deep learning system to separate cast and wrought-aluminium scrap on conveyor belts.

To evaluate the reusability of materials at the *meso* scale, tools are being investigated that employ ML on data acquired from networks of organisations and companies. At this level, efforts are made to optimise resource use in businesses and supply chains. This is achieved by designing products and processes that are modular and adaptable for a closed-loop system. An example is the predictive model in Rakhshanbabanari (2021) that was built to evaluate using ML the reusability of load-bearing building elements given data from a network of stakeholders. The results of this study indicate that the most important economic factor is cash flow implications associated with the need to purchase reused elements early in a project. This highlights the need for supply chain innovation to enable the timely sharing of resources and collaboration in reuse projects, which can in turn lead to significant resource and cost savings.

In recent years, the use of CV to recognise construction materials (such as concrete, brick, metal, wood, and stone) on existing buildings is being increasingly explored. This is of particular significance to creating a database of building materials in use on a global scale – information that is lacking for the present building stock. Material identification from imagery has been previously studied in constrained settings, where the imagery contains close-up pictures of the material (Dimitrov and Golparvar-Fard 2014), because curating datasets with this type of information in the context of a scene, a building, or a neighbourhood is difficult. As a result, CV algorithms trained on such data fail to recognise materials under real-world conditions and scenarios.

To overcome this limitation, Sun and Gu (2022) created a dataset of construction material found on building facades. The images contain the full building structure (facades), but only the prominent building material label is given (i.e. the building material which occupies the most pixels). This simplification can confuse the algorithm when processing data from regions in the image that correspond to other materials. Raghu et al. (2022) generated an image dataset of materials where street-view imagery of urban settings is annotated with material labels (for more details, see Sect. 4.4). In this case, since a building's facade is only partially captured within one image, the strategy of the one prominent label provides a more accurate description of the image content. This is due to the nature of the data; there is usually not enough front-facing line of sight from the bottom to the top of a building in normal city streets.

Most existing CV algorithms developed for the circular analysis of material stocks focus on identifying the material in a certain image but still lack finer granularity with respect to the location, geometry, and boundaries of the materials both in the images and 3D space (semantic segmentation of or object detection in images and/or 3D data) (Dimitrov and Golparvar-Fard 2014; DeGol et al. 2016;

Raghu et al. 2022). This is highly relevant information to create in-use material stocks since it can define quantities and usability. Such information can be further used for design decisions on reuse in new constructions and would allow the development of ML-based approaches that can assist or automatically identify these decisions.

There is vast literature in CV on semantic segmentation and object detection algorithms that address the problem of what objects exist in the data on a finer granularity. Although the underlying algorithms would be the same when addressing the problem of what materials exist in the data, the lack of comprehensive datasets that are tailored for CE tasks does not allow the algorithms' easy use. An example is the OpenSurfaces dataset (Bell et al. 2013), commonly used in CV for material segmentation. It contains images from occupied real-world indoor environments (i.e. they contain furniture and other objects) along with pixel-wise material annotations for most contained objects. However, these annotations do not include building and structural elements that are important in a circularity setting, and as such they cannot be leveraged in a CE setting.

4.3.4 Regenerating the Loop

Regeneration is about leaving the environment in a better state than before (Bocken et al. 2016; Çetin et al. 2021). Zhang et al. (2022) developed a CV algorithm that operates on the *macro* level and detects construction and demolition waste in remotely sensed images in China. Specifically, the algorithm localises abandoned soil and other waste piled in open environments and targets a more efficient and prompt waste management. Konietzko et al. (2020) introduced the 'regenerate' strategy to place the focus on minimising the use of toxic substances and renewable materials. Studies with respect to such regeneration strategies on the *macro* level include calculating the maximum volume of long-term storage depots to optimise the flow of mineral resources (Globa et al. 2021) and predicting the presence of hazardous materials in urban building stocks (Wu et al. 2021, 2022). Such inventories can reduce the risk of unexpected costs and delays during the demolition process but, most importantly, enable the well-being of users and their surroundings. The removal of hazardous materials and their replacement with natural, non-harming, potentially reused ones enable a safer and regenerated environment for future generations. Çetin et al. (2022) report an AI-based system operating on the *micro* and *meso* levels with the capability to detect toxic or hazardous contents on building facades. However, it is important to carefully perform the disposal of hazardous materials; otherwise, the regeneration loop cannot be achieved. As discussed in Chap. 15, algorithms can help AECO stakeholders make informed decisions for regenerative design and architecture.

A summary of the above is offered in Table 4.1. Specifically, it contains examples of CV and ML use per resource loop strategy explored in this section.

Table 4.1 Examples of prominent tasks that CV and ML methods focus on with respect to CE. They are categorised per their contribution to each main resource loop strategy

Narrowing the loop (through design)	**Slowing the loop (prolonging life span)**	**Closing the loop (recycling/reuse)**	**Regenerating the loop (improving living)**
Lean structure	Material identification	Recyclable material identification (end of life)	Abandoned waste localisation
Fabrication rules	Structural condition assessment	Building material identification (lifetime)	Hazardous material prediction
Energy performance	Material condition assessment	Material quantification (end of life)	Storage depot quantification
Reused material	Material deterioration progress		
Waste prediction	Remaining useful life		

4.4 An Example of Using Computer Vision for Mining Materials in Urban Settings

In this section, we describe a research effort of using CV for a CE and particularly of creating a material inventory of urban environments using street-view imagery towards urban mining. Urban mining is the process of recovering and reusing a city's materials when buildings, infrastructure, or products become obsolete. For more information on building demolition valuation and decision-making, we point the reader to Chap. 11 on digital deconstruction by van den Berg. This concept treats the entire city as a storage depot of prospective materials that can be recovered. Information on which materials are available and where they are located in the city is key to forecasting their end-of-life destinations. In this context, the discourse on buildings as 'material banks' is emerging, where high-value materials are retained for future use instead of discarding them as waste (BAMB 2022). To achieve a CE, it is necessary to adopt long-term and innovative thinking, considering that up to 25% of material in a traditional residential structure can be easily reused and up to 70% of material can be recycled (Bohne and Wærner 2014).

The construction sector still faces significant bottlenecks in scaling up urban mining. Given the lacking as-is documentation of the existing building stock, efficient reverse-engineering tools are required to appropriately quantify waste in the built environment. The increase of open access repositories documenting the world around us contains untapped potential. Using a street-view image API (i.e. application programming interface), ocular observations were conducted to analyse building-specific characteristics that can enable reuse using CV by Raghu et al. (2022). Images of buildings marked for demolition were retrieved, following which material and component detection algorithms were implemented to observe the various facade materials and window counts in two European cities (Barcelona

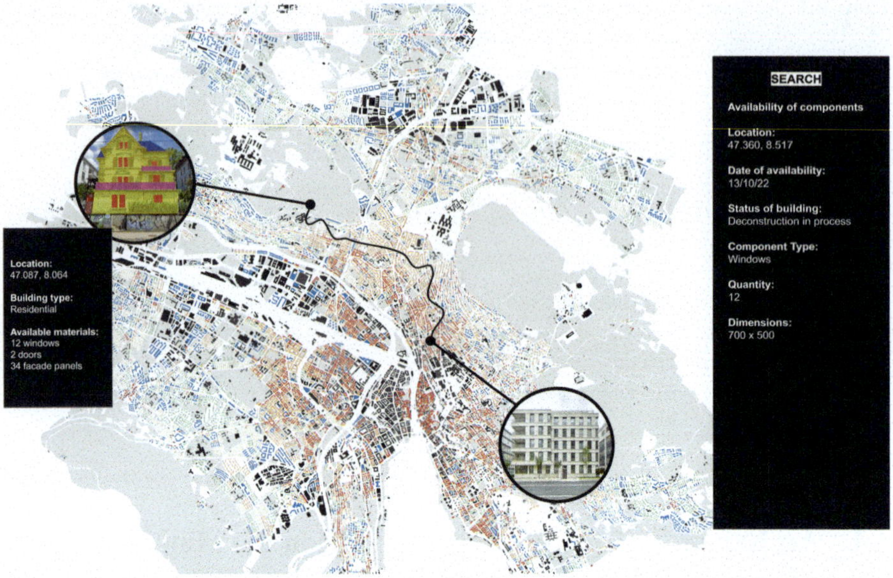

Fig. 4.2 Visualisation of database with insights on materials and components in supply project (or donor project) derived using CV that can be used for new construction projects (image from Raghu, D., adapted from De Wolf et al., 2023)

and Zurich). The algorithm was tested on these two cities to understand the replicability of the model despite variation in architectural building styles.

Such forecasting techniques help create a material inventory that designers can check to verify the forthcoming availability of components and assess their suitability for reuse in new projects. In contrast to other data gathering techniques that are carried out at the building end of life, which require complex and sophisticated software (Uotila et al. 2021), the methodology developed in this project can help in expeditiously procuring information about reusable components earlier on in the project life cycle. Figure 4.2 illustrates the potential use of CV, specifically semantic segmentation of materials and components on building facades. This allows for automated recognition and information retrieval of availability of materials for reuse prior to building demolition. The planning of new construction projects with a circular agenda requires such information about reusable materials, along certainty about their quality and quantity, that is provided to designers or architects well in advance.

4.5 Business Models for Computer Vision and Machine Learning in a Circular Built Environment

Services related to CV and ML in CE are the identification and characterisation of building elements; their materials, types, quantities, and conditions; ways to repair them; patterns for reuse; and more. These can be tailored to the interiors and exteriors of buildings, and one can consider whether these can be provided as separate services. An example is that of Spotr (Spotr.ai 2022), a real estate inspection company based in the Netherlands, which has built a tool to conduct digital building surveys, thus enabling stakeholders to make well-founded decisions on property maintenance. As such, Spotr supports slowing the loop by targeting early identification of damages, hence extending the life of products through timely repairs. This information is identified on images collected from various sources, including human-captured images, drones, and satellite imagery. Using CV, defects are assigned to an exact location which helps easily plan follow-up actions. Spotr then enables large-scale analyses by projecting the results on a map to gain insights into specific areas. Primary information about the buildings is brought together, including images, information on their constituent elements, and present condition.

As more and more services are being developed, other aspects need consideration. While CV and ML are being marketed as a universal solution for many problems faced by the industry, the reality is that CV and ML readiness and adoption in organisations is slow and uncertain. Stakeholders that can benefit financially from these technologies in the frame of CE include real estate agents, asset managers, building owners, and construction companies. Although large organisations tend to prefer in-house solutions so that they may maintain a high degree of integration with and customisation to their existing processes, several considerations should be taken into account before deciding on an in-house or external solution.

First, these technologies require specialised personnel that will be able to develop the algorithms and systems from both research and an engineering perspective. Second, there is a requirement for certain infrastructure, such as computation and annotation services (when lacking access to curated datasets) or patent licencing (for CV and ML methods, systems, and hardware for data collection) if building an internal database and more. Third, apart from the inherent cost, it is important to make the following distinction with respect to in-house and outside solutions: the former provides more control, but the latter allows users access to better-performing algorithms. Dedicated providers of such services have access to more and diverse data that, in turn, can create better generalisable and more robust algorithms. Last, one should also consider the benefits and downfalls of performing data collection in addition to data processing: this allows companies to keep the data and build databases and datasets to train better algorithms, expand services, and sell or share the data. However, there is an associated cost with hardware and personnel for data collection, especially considering how fast hardware evolves and the bottleneck it can create due to the maximum amount of devices one can acquire with respect to cost and return on investment.

4.6 Discussion

Throughout the chapter, we highlighted limitations related to AI approaches and technologies. CV and ML can be considered the eyes and brain of machines, respectively, and can play a fundamental role in mapping and understanding what we have at hand now for reuse, maintenance, or demolition, as well as in identifying ways to design, fabricate, and construct for a CE. In our opinion, there are three fundamental barriers to creating such methods, systems, and tools:

Breaking Down Silos The AECO industry is known for operating in a very isolated manner both in terms of projects and involved stakeholders. However, here we would like to point out shortcomings regarding data and processes. CV and ML technology, especially in a holistic approach such as CE, cannot be developed in isolation without a comprehensive understanding of how to connect the dots between different sub-problems. Efforts that take place in isolation result in duplicating data, processing, and information. Essentially, we should consider how we can apply circularity not only in buildings but also in the technology we are creating (i.e. recycle data, share data across processes, and more).

In addition, a non-targeted and convulsive way of trying existing AI methods – initially developed for other domains – for circularity will not help solve these problems. Practitioners and researchers would benefit from going beyond trying to see what existing AI approaches can do for circularity and instead first consider what should be achieved for the transition to a CE followed by how AI can help towards this goal. Only then should they consider what existing AI methods can solve now and what is still needed to be developed.

Certainly, using algorithms to better understand the *micro*, *meso*, and *macro* levels of the world around us is still an open question in the field of CV and ML, one that will take time to answer. However, advances in CE should not take place after technological developments in the field of AI but instead should proceed hand in hand, given the importance of this domain in sustainability and future generations.

Regulations and Standards What happens when an algorithm provides a wrong prediction? What safety measures can we establish – both in terms of technology and organisation – to minimise or address such cases? What are the evaluation procedures, metrics, and thresholds that we will use to assess CV and ML algorithms for CE and that go beyond technical accuracy? How can we ensure privacy and security of data storage, processing, and sharing to allow operation on different scales and achieve CE in the built environment? What information should be in global public databases and what in private ones? What should be funded and how should it be regulated by the government and what is in the hands of private companies? How do we control and ensure the privacy of information in both public and private settings? These are important questions to answer going forward since they can drastically alter open access and trust of information on a problem that concerns the entirety of humanity.

Ethical Considerations Using algorithms to create and sustain the built environment can offer benefits as long as we have ethical, legal, and environmental processes in place to keep the use of such technology in check. CV and ML are not only here to automate what humans can already do; they can enable us to explore aspects of the data that humans cannot associate or imagine ourselves. In certain cases, these technologies will help achieve superhuman understanding of the world around us. However, it is not only in the hands of CV and ML scientists to achieve this. Expertise from researchers that address different facets of realising CE needs to come together and collaborate with the CV and ML scientists.

What are fundamental steps going forward? In our opinion, it is imperative to formalise the minimum and maximum viable information to achieve a circular built environment. With this information defined, decisions can be made by CE and AI experts on which technologies can play a role and what developments are required to create truly robust systems that operate in real-world settings.

4.7 Key Takeaways

- Machine learning (ML) and computer vision (CV) can play a pivotal role in circularity, by providing stakeholders with actionable information on the state of the building stock.
- A variety of CV and ML methods can be helpful in each of the four strategies of narrowing, slowing, closing, and regenerating the loop.
- Described methods focus on understanding the present state of the building stock. The next step is to develop methods that go beyond this and can propose best actions to take, even going as far as implementing actions directly.
- Challenges in creating global and robust circular CV and ML circular systems for the built environment require multidisciplinary collaboration between experts in artificial intelligence, circular economy, and the architecture, engineering, construction, and operations (AECO) industry.

Acknowledgements This research was partially conducted at the Future Cities Lab Global at ETH Zurich. Future Cities Lab Global is supported and funded by the National Research Foundation, Prime Minister's Office, Singapore, under its Campus for Research Excellence and Technological Enterprise (CREATE) programme, and ETH Zurich, with additional contributions from the National University of Singapore, Nanyang Technological University Singapore, and the Singapore University of Technology and Design. This research also received funding from the Swiss National Science Foundation (SNSF) under grant n° 200021E_215311 for the Lead Agency MINT project: "Re-engineering informal construction through circular practices and models".

References

Acquaah YT, Gokaraju B, Tesiero RC III et al (2021) Thermal imagery feature extraction techniques and the effects on machine learning models for smart HVAC efficiency in building energy. Remote Sens 13(19):3847. https://doi.org/10.3390/rs13193847

Act EP (2005) Building standards – energy policy act of 2005. US Congr 914:1–27

Akanbi LA, Oyedele AO, Oyedele LO et al (2020) Deep learning model for demolition waste prediction in a circular economy. J Clean Prod 274:122843. https://doi.org/10.1016/j.jclepro.2020.122843

Albert A, Kaur J, Gonzalez MC (2017) Using convolutional networks and satellite imagery to identify patterns in urban environments at a large scale. In: 23rd ACM SIGKDD international conference on knowledge discovery and data mining. https://doi.org/10.1145/3097983.3098070

Ali D, Frimpong (2021) DeepHaul: a deep learning and reinforcement learning-based smart automation framework for dump trucks. Prog Artif Intell 10:157–180. https://doi.org/10.1007/s13748-021-00233-7

Alkaissy M, Arashpour M, Golafshani EM et al (2023) Enhancing construction safety: machine learning-based classification of injury types. Saf Sci 162. https://doi.org/10.1016/j.ssci.2023.106102

Amer F, Golparvar-Fard M (2021) Modeling dynamic construction work template from existing scheduling records via sequential machine learning. Adv Eng Inform 47. https://doi.org/10.1016/j.aei.2020.101198

Ashrafi R, Azarbayjani M, Tabkhi H (2022) Machine learning-based automated thermal comfort prediction: integration of low-cost thermal and visual cameras for higher accuracy. In Architectural research centers consortium – European association for architectural educators international conference

Awada M, Srour FJ, Srour IM (2021) Data-driven machine learning approach to integrate field submittals in project scheduling. J Manag Eng 37(1). https://doi.org/10.1061/(ASCE)ME.1943-5479.0000873

BAMB (2022) Enabling a circular building industry. https://www.bamb2020.eu. Accessed 26 Mar 2023

Barahona B, Buck R, Okaya O et al (2021) Detection of thermal anomalies on building façades using infrared thermography and supervised learning. J Phys Conf Ser 2042(1):012013. https://doi.org/10.1088/1742-6596/2042/1/012013

Bassier M, Vergauwen M (2020) Topology reconstruction of BIM wall objects from point cloud data. Remote Sens 12(11):1800. https://doi.org/10.3390/rs12111800

Bell S, Upchurch P, Snavely N et al (2013) OpenSurfaces: a richly annotated catalog of surface appearance. ACM Trans Graph (TOG) 32(4):1–17. https://doi.org/10.1145/2461912.2462002

Bell S, Upchurch P, Snavely N et al (2015) Material recognition in the wild with the materials in context database. In: IEEE conference on computer vision and pattern recognition, pp 3479–3487. https://doi.org/10.1109/CVPR.2015.7298970

Berghout T, Benbouzid M (2022) A systematic guide for predicting remaining useful life with machine learning. Electron 11(7):1125. https://doi.org/10.3390/electronics11071125

Bisht S, Shekhawat K, Upasani N et al (2022) Transforming an adjacency graph into dimensioned floorplan layouts. Comput Graph Forum 41(6):5–22. https://doi.org/10.1111/cgf.14451

Bocken NM, De Pauw I, Bakker C et al (2016) Product design and business model strategies for a circular economy. J Ind Prod Eng 33(5):308–320

Bohne RA, Wærner E (2014) Barriers for deconstruction and reuse/recycling of construction materials in Norway. CIB Publication 397. Working Commission W115. Construction Materials Stewardship, pp 89–107

Çetin S, De Wolf C, Bocken N (2021) Circular digital built environment: an emerging framework. Sustain 13(11):6348

Çetin S, Gruis V, Straub A (2022) Digitalization for a circular economy in the building industry: multiple-case study of Dutch social housing organizations. Resour Conserv Recycl Adv 15: 200110. https://doi.org/10.1016/j.rcradv.2022.200110

Cha GW, Choi SH, Hong WH et al (2023) Developing a prediction model of demolition-waste generation-rate via principal component analysis. Int J Environ Res Public Health 20(4):3159. https://doi.org/10.3390/ijerph20043159

Chaillou S (2020) Archigan: artificial intelligence x architecture. In Architectural intelligence selected papers 1st international conference on computational design and robotic fabrication. Springer Nature, Singapore, pp 117–127. https://doi.org/10.1007/978-981-15-6568-7_8

Chang KH, Cheng CY, Luo J et al (2021) Building-GAN: Graph-conditioned architectural volumetric design generation. In IEEE/CVF international conference on computer vision, pp 11956–11965. https://doi.org/10.1109/ICCV48922.2021.01174

Chen J, Lu W, Yuan L et al (2022) Estimating construction waste truck payload volume using monocular vision. Resour Conserv Recycl 177:106013. https://doi.org/10.1016/j.resconrec.2021.106013

DeGol J, Golparvar-Fard M, Hoiem D (2016) Geometry-informed material recognition. In: IEEE conference on computer vision and pattern recognition, pp 1554–1562. https://doi.org/10.1109/CVPR.2016.172

De Wolf C, Raghu D, Sentic A, Fivet C (2023) Innovations practitioners need for circularity in the Swiss architecture, engineering, and construction sector. Cirkla. https://doi.org/10.5281/zenodo.7520351

Deng J, Dong W, Socher R et al (2009) Imagenet: a large-scale hierarchical image database. In: IEEE conference on computer vision and pattern recognition, pp 248–255. https://doi.org/10.1109/CVPR.2009.5206848

Di Natale L, Svetozarevic B, Heer P et al (2022) Physically consistent neural networks for building thermal modeling: theory and analysis. Appl Energy 325:119806. https://doi.org/10.1016/j.apenergy.2022.119806

Diakogiannis FI, Waldner F, Caccetta P et al (2020) ResUNet-a: A deep learning framework for semantic segmentation of remotely sensed data. ISPRS J Photogramm Remote Sens 162:94–114. https://doi.org/10.1016/j.isprsjprs.2020.01.013

Díaz-Romero DJ, Sterkens W, Van den Eynde S et al (2021) Deep learning computer vision for the separation of cast- and wrought-aluminum scrap. Resour Conserv Recycl 172:105685. https://doi.org/10.1016/j.resconrec.2021.105685

Díaz-Romero DJ, Van den Eynde S, Sterkens W et al (2022) Simultaneous mass estimation and class classification of scrap metals using deep learning. Resour Conserv Recycl 181:106272. https://doi.org/10.1016/j.resconrec.2022.106272

Dimitrov A, Golparvar-Fard M (2014) Vision-based material recognition for automated monitoring of construction progress and generating building information modeling from unordered site image collections. Adv Eng Inform 28(1):37–49. https://doi.org/10.1016/j.aei.2013.11.002

Dong CZ, Catbas FN (2021) A review of computer vision–based structural health monitoring at local and global levels. Struct Health Monit 20(2):692–743. https://doi.org/10.1177/1475921720935585

Dung CV, Sekiya H, Hirano S et al (2019) A vision-based method for crack detection in gusset plate welded joints of steel bridges using deep convolutional neural networks. Automation Constr 102:217–229. https://doi.org/10.1016/j.autcon.2019.02.013

Eraliev OMU, Lee KH, Shin DY et al (2022) Sensing, perception, decision, planning and action of autonomous excavators. Automation Constr 141:104428. https://doi.org/10.1016/j.autcon.2022.104428

Fan R, Bocus MJ, Zhu Y et al (2019) Road crack detection using deep convolutional neural network and adaptive thresholding. In: 2019 IEEE symposium on intelligent vehicle IV, pp 474–479. https://doi.org/10.1109/IVS.2019.8814000

Feng C, Zhang H, Wang H et al (2020) Automatic pixel-level crack detection on dam surface using deep convolutional network. Sens 20(7):2069. https://doi.org/10.3390/s20072069

Fischer M (1993) Linking CAD and expert systems for constructibility reasoning. In: Computing in civil and building engineering, pp 1563–1570. ASCE

Fischer M, Garcia-Lopez NP, Morkos R (2018) Making each workhour count: improving the prediction of construction durations and resource allocations. In: Advanced computing strategies for engineering: 25th EG-ICE international workshop 2018 Part I 25, pp 273–295. Springer International. https://doi.org/10.1007/978-3-319-91635-4_15

Fitzsimmons JP, Lu R, Hong Y et al (2022) Construction schedule risk analysis–a hybrid machine learning approach. J Inform Technol Constr. https://doi.org/10.36680/j.itcon.2022.004

Fragkia V, Foged I (2020) Methods for the prediction and specification of functionally graded multi-grain responsive timber composites. In: 38th eCAADe conference. TU Berlin

Galdelli A, D'Imperio M, Marchello G et al (2022) A novel remote visual inspection system for bridge predictive maintenance. Remote Sens 14(9):2248. https://doi.org/10.3390/rs14092248

Ghisellini P, Cialani C, Ulgiati S (2016) A review on circular economy: the expected transition to a balanced interplay of environmental and economic systems. J Clean Prod 114:11–32. https://doi.org/10.1016/j.jclepro.2015.09.007

Globa SB, Ashkerov M, Arnold V et al (2021) Optimisation of material flows in the concept urban mining based on the use of long-term storage depot. In: Popkova EG, Ostrovskaya VN, Bogoviz AV (eds) Socio-economic systems: paradigms for the future. Springer International, pp 491–498. https://doi.org/10.1007/978-3-030-56433-9_51

Guan T, He Z, Song R et al (2022) TNS: Terrain traversability mapping and navigation system for autonomous excavators. In: Robotics: science systems

Guo Y, Xu Y, Li S (2020) Dense construction vehicle detection based on orientation-aware feature fusion convolutional neural network. Automation Constr 112:103124. https://doi.org/10.1016/j.autcon.2020.103124

Haydari A, Yılmaz Y (2020) Deep reinforcement learning for intelligent transportation systems: a survey. IEEE Trans Intell Transp Syst 23(1):11–32. https://doi.org/10.1109/TITS.2020.3008612

Hong T, Wang Z, Luo X et al (2020) State-of-the-art on research and applications of machine learning in the building life cycle. Energ Buildings 212:109831. https://doi.org/10.1016/j.enbuild.2020.109831

Hosseinpoor S, Torresen J, Mantelli M et al (2021) Traversability analysis by semantic terrain segmentation for mobile robots. In: IEEE 17th international conference on automation science and engineering, pp 1407–1413. https://doi.org/10.1109/CASE49439.2021.9551629

Huang C (2021) Reinforcement learning for architectural design-build – opportunity of machine learning in a material-informed circular design strategy. In: 26th association for computer-aided architectural design research in Asia international conference, pp 171–180. https://doi.org/10.52842/conf.caadria.2021.1.171

Kareem A (2020) Emerging frontiers in wind engineering: computing, stochastics, machine learning and beyond. J Wind Eng Ind Aerodynamics 206:104320. https://doi.org/10.1016/j.jweia.2020.104320

Khaloo A, Lattanzi D, Jachimowicz A et al (2018) Utilizing UAV and 3D computer vision for visual inspection of a large gravity dam. Front Built Environ 31. https://doi.org/10.3389/fbuil.2018.00031

Khuc T, Catbas FN (2018) Structural identification using computer vision–based bridge health monitoring. J Struct Eng 144(2):04017202. https://doi.org/10.1061/JMENEA.MEENG-5143

Kim J (2020) Visual analytics for operation-level construction monitoring and documentation: state-of-the-art technologies, research challenges, and future directions. Front Built Environ 6: 575738. https://doi.org/10.3389/fbuil.2020.575738

Kim S, Jeong K, Hong T et al (2023) Deep learning–based automated generation of material data with object–space relationships for scan to BIM. J Manag Eng 39(3):04023004

Koc K, Ekmekcioğlu Ö, Gurgun AP (2022) Accident prediction in construction using hybrid wavelet-machine learning. Automation Constr 133:103987. https://doi.org/10.1016/j.autcon.2021.103987

Konietzko J, Bocken N, Hultink EJ (2020) A tool to analyze, ideate and develop circular innovation ecosystems. Sustain 12(1):417

König J, Jenkins MD, Barrie P et al (2019) A convolutional neural network for pavement surface crack segmentation using residual connections and attention gating. In IEEE international conference on image processing, pp 1460–1464. https://doi.org/10.1109/ICIP.2019.8803060

Kraus MA, Bischof R, Kaufmann W et al (2022) Artificial intelligence-finite element method-hybrids for efficient nonlinear analysis of concrete structures. Acta Polytechnica CTU 36:99–108. https://doi.org/10.14311/APP.2022.36.0099

Krizhevsky A, Sutskever I, Hinton GE (2017) Imagenet classification with deep convolutional neural networks. Commun ACM 60(6):84–90. https://doi.org/10.1145/3065386

Kuritcyn P, Anding K, Linß E et al (2015) Increasing the safety in recycling of construction and demolition waste by using supervised machine learning. J Phys Conf Ser 588(1):012035. https://doi.org/10.1088/1742-6596/588/1/012035

Lei M, Liu L, Shi C et al (2021) A novel tunnel-lining crack recognition system based on digital image technology. Tunn Undergr Sp Technol 108:103724. https://doi.org/10.1016/j.tust.2020.103724

Li Y, Chen J (2022) Computer vision–based counting model for dense steel pipe on construction sites. J Constr Eng Manag 148(1):04021178. https://doi.org/10.1061/10.1061/(asce)co.1943-7862.0002217

Li Y, Yi YK (2023) Optimal shape design using machine learning for wind energy and pressure. J Build Eng. https://doi.org/10.1016/j.jobe.2023.106337

Li S, Zheng J (2023) Design of vehicle log image acquisition system based on deep learning and laser sensor. In: 3rd international symposium on computer engineering and intelligent communications, 12462:37–46. SPIE. https://doi.org/10.1117/12.2660793

Li G, Ma B, He S et al (2020) Automatic tunnel crack detection based on u-net and a convolutional neural network with alternately updated clique. Sens 20(3):717. https://doi.org/10.3390/s20030717

Li Z, Jin Z, Wang P et al (2021) Corrosion mechanism of reinforced bars inside concrete and relevant monitoring or detection apparatus: a review. Constr Build Mater 279:122432. https://doi.org/10.1016/j.conbuildmat.2021.122432

Liu YF, Nie X, Fan JS et al (2020) Image-based crack assessment of bridge piers using unmanned aerial vehicles and three-dimensional scene reconstruction. Comput Aided Civ Infrastruct Eng 35(5):511–529. https://doi.org/10.1111/mice.12501

Liu W, Meng Q, Li Z et al (2021) Applications of computer vision in monitoring the unsafe behavior of construction workers: current status and challenges. Build 11(9):409. https://doi.org/10.3390/buildings11090409

Ma JW, Czerniawski T, Leite F (2022) Automated scan-to-building information modeling. In: Weishing L, Chimay A (eds) Research companion to building information modeling, Edward Elgar. Edward Elgar, pp 169–189. https://doi.org/10.4337/9781839105524.00018

Macher H, Landes T, Grussenmeyer P (2020) Automation of thermal point clouds analysis for the extraction of windows and thermal bridges of building facades. Int Archives Photogramm Remote Sens Spat Inform Sci 43:287–292. https://doi.org/10.5194/isprs-archives-XLIII-B2-2020-287-2020

Maher ML, Fenves SJ (1985) HI-RISE: an expert system for the preliminary structural design of high rise buildings. In: Gero JS (ed) Knowledge engineering in computer-aided design. North-Holland, Amsterdam, pp 125–135

Málaga-Chuquitaype C (2022) Machine learning in structural design: an opinionated review. Front Built Environ 8. https://doi.org/10.3389/fbuil.2022.815717

Mao WL, Chen WC, Wang CT et al (2021) Recycling waste classification using optimized convolutional neural network. Resour Conserv Recycl 164:105132. https://doi.org/10.1016/j.resconrec.2020.105132

Mayer Z, Kahn J, Hou Y et al (2023) Deep learning approaches to building rooftop thermal bridge detection from aerial images. Automation Constr 146:104690. https://doi.org/10.1016/j.autcon.2022.104690

Mirzabeigi S, Razkenari M (2022) Design optimization of urban typologies: a framework for evaluating building energy performance and outdoor thermal comfort. Sustain Cities Soc 76:103515

Motayyeb S, Samadzedegan F, Javan FD et al (2023) Fusion of UAV-based infrared and visible images for thermal leakage map generation of building facades. Heliyon. https://doi.org/10.1016/j.heliyon.2023.e14551

Muin S, Mosalam KM (2021) Human-machine collaboration framework for structural health monitoring and resiliency. Eng Struct 235:112084. https://doi.org/10.1016/j.engstruct.2021.112084

Nauata N, Hosseini S, Chang KH et al (2021) House-gan++: generative adversarial layout refinement network towards intelligent computational agent for professional architects. In: IEEE conference on computer vision and pattern recognition, pp 13632–13641. https://doi.org/10.1109/CVPR46437.2021.01342

Pan Y, Braun A, Brilakis I et al (2022) Enriching geometric digital twins of buildings with small objects by fusing laser scanning and AI-based image recognition. Automation Constr 140: 104375. https://doi.org/10.1016/j.autcon.2022.104375

Papert SA (1966) The summer vision project. MIT AI Memo 100

Pham HT, Rafieizonooz M, Han S et al (2021) Current status and future directions of deep learning applications for safety management in construction. Sustain 13(24):13579. https://doi.org/10.3390/su132413579

Płoszaj-Mazurek M, Ryńska E, Grochulska-Salak M (2020) Methods to optimize carbon footprint of buildings in regenerative architectural design with the use of machine learning, convolutional neural network, and parametric design. Energies 13(20):5289. https://doi.org/10.3390/en13205289

Raghu D, Markopoulou A, Marengo M et al (2022) Enabling component reuse from existing buildings using machine learning-using Google Street View to enhance building databases. In: 27th association for computer-aided architectural design research in Asia international conference, vol 2, pp 577–586. https://doi.org/10.52842/conf.caadria.2022.2.577

Rakha T, El Masri Y, Chen K et al (2022) Building envelope anomaly characterization and simulation using drone time-lapse thermography. Energ Buildings 259:111754. https://doi.org/10.1016/j.enbuild.2021.111754

Rakhshanbabanari K (2021) A predictive model for assessing the reuse potential of structural elements at the end-of-life of a building. Doctoral dissertation, Coventry University

Ramsgaard Thomsen M, Nicholas P, Tamke M et al (2020) Towards machine learning for architectural fabrication in the age of industry 4.0. Int J Archit Comput 18(4):335–352. https://doi.org/10.1177/1478077120948000

Rousset G, Despinoy M, Schindler K et al (2021) Assessment of deep learning techniques for land use land cover classification in southern New Caledonia. Remote Sens 13(12):2257. https://doi.org/10.3390/rs13122257

Shi Q, Liu M, Li S et al (2021) A deeply supervised attention metric-based network and an open aerial image dataset for remote sensing change detection. IEEE Trans Geosci Remote Sens 60: 1–16. https://doi.org/10.1109/TGRS.2021.3085870

Sonta AJ, Jain RK (2019) Building relationships: using embedded plug load sensors for occupant network inference. IEEE Embed Syst Lett 12(2):41–44. https://doi.org/10.1109/LES.2019.2937316

Sonta A, Dougherty TR, Jain RK (2021) Data-driven optimization of building layouts for energy efficiency. Energ Buildings 238:110815. https://doi.org/10.1016/j.enbuild.2021.110815

Spotr.ai (2022) Inspect millions of buildings in seconds. https://www.spotr.ai/. Accessed 26 Mar 2023

Srikanth I, Arockiasamy M (2020) Deterioration models for prediction of remaining useful life of timber and concrete bridges: a review. J Traffic Transp Eng 7(2):152–173. https://doi.org/10.1016/j.jtte.2019.09.005

Sun Y, Gu Z (2022) Using computer vision to recognize construction material: a trustworthy dataset perspective. Resour Conserv Recycl 183:106362. https://doi.org/10.1016/j.resconrec.2022.106362

Sun X, Li X, Xiao D et al (2021) A method of mining truck loading volume detection based on deep learning and image recognition. Sens 21(2):635. https://doi.org/10.3390/s21020635

Sun J, Wu W, Liu L et al (2022) WallPlan: synthesizing floorplans by learning to generate wall graphs. ACM Trans Graph 41(4):1–14. https://doi.org/10.1145/3528223.3530135

Suryamurthy V, Raghavan VS, Laurenzi A et al (2019) Terrain segmentation and roughness estimation using Rgb data: Path planning application on the centauro robot. In: IEEE-RAS 19th international conference on humanoid robots, pp 1–8. https://doi.org/10.1109/Humanoids43949.2019.9035009

Tang S, Golparvar-Fard M (2021) Machine learning-based risk analysis for construction worker safety from ubiquitous site photos and videos. J Comput Civ Eng 35(6):04021020. https://doi.org/10.1061/(ASCE)CP.1943-5487.0000979

Tichý T, Brož J, Bělinová Z et al (2021) Analysis of predictive maintenance for tunnel systems. Sustain 13(7):3977. https://doi.org/10.3390/su13073977

Tien PW, Wei S, Calautit J (2020) A computer vision-based occupancy and equipment usage detection approach for reducing building energy demand. Energies 14(1):156. https://doi.org/10.3390/en14010156

Tien PW, Wei S, Darkwa J et al (2022) Machine learning and deep learning methods for enhancing building energy efficiency and indoor environmental quality–a review. Energy AI. https://doi.org/10.1016/j.egyai.2022.100198

Uotila U, Saari A, Junnonen J (2021) Investigating the barriers to laser scanning implementation in building refurbishment. J Inform Technol Constr 26:249–262. https://doi.org/10.36680/j.itcon.2021.014

Upchurch P, Niu R (2022) A dense material segmentation dataset for indoor and outdoor scene parsing. In: European conference on computer vision, pp 450–466. https://doi.org/10.1007/978-3-031-20074-8_26

Wang H, Wu T (2020) Knowledge-enhanced deep learning for wind-induced nonlinear structural dynamic analysis. J Struct Eng 146(11):04020235. https://doi.org/10.1061/(ASCE)ST.1943-541X.0002802

Wu PY, Mjörnell K, Mangold M et al (2021) A data-driven approach to assess the risk of encountering hazardous materials in the building stock based on environmental inventories. Sustain 13(14):7836. https://doi.org/10.3390/su13147836

Wu PY, Sandels C, Mjörnell K et al (2022) Predicting the presence of hazardous materials in buildings using machine learning. Build Environ 213:108894. https://doi.org/10.1016/j.buildenv.2022.108894

Xu H, Su X, Wang Y et al (2019) Automatic bridge crack detection using a convolutional neural network. Appl Sci 9(14):2867. https://doi.org/10.3390/app9142867

Xu Z, Huang J, Huang K (2023) A novel computer vision-based approach for monitoring safety harness use in construction. IET Image Process 17(4):1071–1085. https://doi.org/10.1049/ipr2.12696

Yella S, Dougherty M (2013) Automatically detecting the number of logs on a timber truck. J Intell Syst 22(4):417–435. https://doi.org/10.1515/jisys-2013-0026

Yin Y, Hutchcroft W, Khosravan N et al (2022) Generating topological structure of floorplans from room attributes. In: International conference on multimedia retrieval, pp. 295–303. https://doi.org/10.1145/3512527.3531384

Yu Z, Shen Y, Shen C (2021) A real-time detection approach for bridge cracks based on YOLOv4-FPM. Automation Constr 122:103514. https://doi.org/10.1016/j.autcon.2020.103514

Zhang Q, Zhang X, Mu X et al (2021) Recyclable waste image recognition based on deep learning. Resour Conserv Recycl 171:105636. https://doi.org/10.1016/j.resconrec.2021.105636

Zhang C, Zhou L, Du M et al (2022) A cross-channel multi-scale gated fusion network for recognizing construction and demolition waste from high-resolution remote sensing images. Int J Remote Sens 43(12):4541–4568. https://doi.org/10.1080/01431161.2022.2115864

Zhao P, Liao W, Huang Y et al (2023) Intelligent beam layout design for frame structure based on graph neural networks. J Build Energy 63:105499. https://doi.org/10.1016/j.jobe.2022.105499

Zheng Z, Ma A, Zhang L et al (2021) Change is everywhere: Single-temporal supervised object change detection in remote sensing imagery. In: IEEE/CVF international conference on computer vision, pp 15193–15202. https://doi.org/10.13140/RG.2.2.36048.02562

Zuo C, Liang C, Chen J et al (2023) Machine learning-based urban renovation design for improving wind environment: a case study in Xi'an. China Land 12(4):739. https://doi.org/10.3390/land12040739

Iro Armeni is an assistant professor at Stanford University, conducting interdisciplinary research between architecture, civil engineering, and visual machine perception. She conducted her post-doctoral research at the Swiss Federal Institute of Technology of Zurich (ETH Zurich) with Prof. Marc Pollefeys (computer science) and Prof. Catherine De Wolf (circular engineering for architecture). She completed her PhD at Stanford University in civil and environmental engineering, with a PhD minor in computer science. Prior to the PhD, she received a master of science in computer science (Ionian University), an MEng in architecture and digital design (University of Tokyo), and a diploma in architectural engineering (National Technical University of Athens). She is the recipient of the ETH Zurich Postdoctoral Fellowship, the Google PhD Fellowship on Machine Perception, and the Japanese Government (MEXT) scholarship.

Deepika Raghu is a PhD student at the Chair for Circular Engineering for Architecture at the Swiss Federal Institute of Technology of Zurich (ETH Zurich), where she is conducting research on how the obsolescence of the existing building stock can be prevented using emerging digital technologies. Deepika Raghu works with big data and machine learning methods to analyse existing buildings to arrive at solutions for retaining high-value materials for future use. She studied architecture at Siddaganga Institute of Technology (SIT), after which she obtained a master's in advanced architecture at the Institute for Advanced Architecture of Catalonia (IAAC).

Catherine De Wolf is an Assistant Professor of Circular Engineering for Architecture at the Swiss Federal Institute of Technology of Zurich (ETH Zurich), where she conducts research on digital innovation towards a circular built environment. She has a dual background in civil engineering and architecture and completed her PhD at the Massachusetts Institute of Technology (MIT). She is a faculty member in the ETH Centre for Augmented Computational Design in Architecture, Engineering and Construction (Design++), the ETH AI Center, the Swiss Federal Laboratories for Materials Science and Technology (Empa), the Circular Future Cities group of the Future Cities Lab, and the National Centre of Competence in Research on Digital Fabrication (DFAB). She has worked in academia (e.g., University of Cambridge, University of Technology Delft, Nanjing University), in industry (e.g., Arup, Elioth, Ney & Partners), and in governmental institutions (e.g., Joint Research Centre of the European Commission). She also co-founded the circular construction company Anku, the Structural Engineers 2050 Commitment, and the Digital Circular Economy (DiCE) Lab.

Chapter 5
From Data Templates to Material Passports and Digital Product Passports

Meliha Honic, Pedro Meda Magalhães, and Pablo Van den Bosch

Abstract Lack of data and difficulty in tracking materials and elements are two major obstacles in the construction industry that hinder the realisation of a circular economy. Data templates, material passports (MPs), and digital product passports (DPPs) are passport instruments that provide valuable information about buildings. Data templates deliver digital standardised data structures for MPs (digital data sets describing building characteristics of, e.g. elements) and DPPs (cross-sectoral passports developed by the European Union to collect product data for sustainability).

MPs, which are associated with the built environment, help urban miners and building owners assess the value and reuse potential of building materials and elements. Several initiatives, such as Madaster, Concular, and Platform CB'23, have produced data templates and MPs for new and existing buildings. Challenges to their use include the lack of standardisation of data templates and MPs and difficulties in collecting and tracing data needed to create and maintain MPs through a building's life cycle. Standardisation would foster the implementation of passports, but aligning existing concepts and identifying overlaps remains a present challenge. Future research and practice suggest that using geographic information systems, laser scanning, and computer vision will help deploy MPs more effectively in practice.

Keywords Building passport · Circular building · Data traceability · Standardisation · Reuse

---

M. Honic
Circular Engineering for Architecture, Institute of Construction & Infrastructure Management, Department of Civil, Environmental and Geomatic Engineering, ETH Zurich, Zurich, Switzerland
e-mail: mhonic@ethz.ch

P. M. Magalhães (✉)
CONSTRUCT/Gequaltec, Instituto da Construção, Faculty of Engineering, University of Porto, Porto, Portugal
e-mail: pmeda@fe.up.pt

P. Van den Bosch
Madaster, Utrecht, The Netherlands
e-mail: pablo.vandenbosch@madaster.com

C. De Wolf et al. (eds.), *A Circular Built Environment in the Digital Age*, Circular Economy and Sustainability, https://doi.org/10.1007/978-3-031-39675-5_5

5.1 Data Templates, Material Passports, and Digital Product Passports

The European Union Circular Economy Action Plan (EU 2020) identified the digitisation of material and product information as a key driver for enhancing the transition to a circular economy. Several existing concepts are gaining attention in the construction industry: data templates, material passports (MPs), digital product passports (DPPs), circularity passports, product circularity data sheets, building renovation passports, and digital building logbooks. This chapter provides an overview of existing passport approaches. It discusses their common and distinct aspects and demonstrates their application to real-world examples and their role as enablers of a circular economy. It further presents the business models of Madaster, Concular, and Platform CB'23.

5.1.1 Differences Between MPs and DPPs

MPs and DPPs are valuable concepts that can enable a circular economy. Both types of digital data sets contain valuable information such as material properties and potentials for reuse and disassembly: MPs about buildings (BAMB 2020) and DPPs about any product (European Commission 2022a). MPs and DPPs make it easier to assess the value of materials and elements incorporated in existing buildings or products and can prevent them from being demolished or disposed of and enable reuse. While MPs, circularity passports, and building passports are more common in the built environment, DPPs can be used in any industry (including the built environment) and have a focus on products. (In this chapter, MP is used to refer to passports in the built environment unless the term DPP is stated explicitly in literature or practice.)

Although the differences between MPs and DPPs are not clearly stated in the academic literature, MPs have mostly been applied in the built environment. Many MPs exist in academia and practice, yet no consensus on the content, the data formats, and the main requirements are given. DPPs are a cross-sectoral, relatively new concept by the EU (European Commission 2020a), and they pursue the same aim as MPs: a circular economy. MPs lack a regulative framework to standardise and align various MP approaches, while for DPPs the backbone (EU regulatory framework) is established.

MPs have been identified as the main enablers of a circular economy in the built environment next to, e.g. blockchain, artificial intelligence, and building information modelling (BIM) (Çetin et al. 2021). The EU Horizon 2020 project Buildings as Material Banks (BAMB) defines MPs as "digital sets of data describing defined characteristics of materials and components in products and systems that give them value for present use, recovery, and reuse" (Mulhall et al. 2017). BAMB describes how MPs can be used by various stakeholders across the value chain for different purposes (Luscuere and Mulhall 2018). They can also be generated at the material, element, or building scales. An MP can, e.g. indicate the type of timber (material scale), display the material composition of a slab (element scale), or show the amount of timber used in the entire building (building scale) (Honic et al. 2019b). The lack of information on the material composition of buildings is a major obstacle in the construction industry (Rose and Stegemann 2019), and implementing MPs would allow circular material flows in the built environment.

In the European strategy for data, DPPs are described as passports that "provide information on a product's origin, durability, composition, reuse, repair and dismantling possibilities, and end-of-life handling" (European Commission 2020a). According to the same European strategy, DPPs are "a structured collection of product-related data with a predefined scope and agreed data ownership and access rights conveyed through a unique identifier", set on a "decentralised system with a central registry" with "information related to sustainability, circularity, value retention for reuse/remanufacturing/recycling". One practical example is circularise (Circularise 2023), which ensures supply chain tracking through DPPs. Even though the EU regulative framework for DPPs exists, the specific content, data formats, and data structures of DPPs are not defined. To enable an implementation of both, MPs and DPPs, in the daily practice of companies, their standardisation is needed, wherefore data templates can play a crucial role.

5.1.2 Data Templates for MPs and DPPs

Data templates are digital data structures used to generate MPs, DPPs, or any other digital passport. To establish standardised and structured MPs and DPPs, data templates are of utmost importance. Data templates provide data structures or "skeletons" that can support all types of characteristics from material to building scale of such passports and can be used to generate structured and standardised MPs and DPPs. The International Organization for Standardization (ISO) provides data templates for the built environment, but they are also used in other sectors, albeit often with different terminology. Data templates can also be referred to as "metadata structures" or "digital templates" and apply to all industries and areas of activity where digitalisation is a trend. The ISO 23387 standard established the information and digital requirements for data templates to become digital, traceable, and interoperable (ISO 2020). Data templates enable construction project stakeholders to exchange information about construction objects through an asset life cycle, using the same data structure, terminology, and globally unique identifiers to ensure machine-readability. Data templates set the framework for MPs by providing common data structures and are key to evaluating the value of materials and products over time (Mêda et al. 2020). In this respect, they constitute a key background to enable the realisation of a digital twin (see Chap. 1 by Koutamanis on BIM to digital twins) at the building scale (Mêda et al. 2021).

5.1.3 The Development of Passport Instruments

The automotive industry was one of the first sectors to adopt passport instruments. The International Material Data System, which was introduced in the early 2000s (Frühbuss et al. 2000), is used to collect and transfer information about all materials in a product across the whole supply chain, thereby supporting the requirements of the European End-of-Life Vehicles Directive 2000/53/EC (EU 2000). This EU legal document states that a reuse and recovery rate of a minimum of 95% by an average weight per vehicle and year should be achieved (Walden et al. 2021). As part of the clean energy transition, the electrification of vehicles and, accordingly, batteries plays a crucial role.

A battery passport was introduced in 2019 by the World Economic Forum and the Global Battery Alliance, aiming for a sustainable battery value chain by 2030 (World Economic Forum 2019). In 2020, the EU mandated battery passports for new industrial and electric-vehicle batteries by 2026. Each passport should have a unique identifier, be linked to the information about the basic characteristics, be accessible online, and be allowed access to information (European Commission 2020b).

Similar to the battery passport, the EU proposed a regulation in which DPPs would be used in “a framework for setting eco-design requirements for sustainable products” in March 2022 (European Commission 2022b), which applies to all sectors except food, animal fodder, and medicinal products (European Commission 2022b). The regulation builds on the European Green Deal, the Circular Economy Action Plan, and the Ecodesign Directive. Under this regulation, DPPs should ensure access to product information, improve the traceability of products along the value chain, foster the verification of product compliance, and include relevant data attributes to enable tracking substances. The DPP should provide information on materials’ origin and composition, on opportunities for repair and disassembly, and on possibilities for recycling and disposal at the end of life (Götz et al. 2022) This regulation shall apply to any physical good which is placed on the market or put into service. It includes components (a product intended to be incorporated into another product) and intermediate products (a product which requires further manufacturing or transformation such as mixing, coating, or assembling to make it suitable for end-users).

Other industries that used passports in a very early stage include the shipping industry, which introduced a “resource passport” in 2007 (de Brito et al. 2017), a “Cradle to Cradle Passport” in 2011 (Maersk 2011), and a “Circularity Passport” (EPEA Netherlands 2023). The electrical and electronic equipment industry introduced a “recycling passport” in 2000 (Hesselbach et al. 2001).

5.2 Passport Instruments in the Built Environment

MPs are not new to the built environment. A passport for buildings was first introduced by Eichstädt in 1982 and was foreseen as a document that records changes and enables a qualitative evaluation of factories (Eichstädt 1982). In the report of the United Nations Environment Programme, the MP "guides materials through industrial cycles, routing them from production through reuse, defining optimum uses and intelligent practices" (McDonough & Braungart, 2003, p. 15). The Horizon 2020 project BAMB has extensively explored MPs and their use in the built environment.

5.2.1 MPs and Life Cycles

MPs play a crucial role across the entire life cycle of buildings and materials. A building's life cycle includes every phase from the conceptualisation until the end of life. A material's life cycle starts with the extraction of the raw materials and proceeds with the manufacturing process, the use phase, and the end-of-life stage. Similar to a material's life cycle, a construction element's or product's life cycle begins outside a project context. Most elements and products are produced without knowing exactly where or for what they will be used. During their life cycle, elements or products might have the same lifetime as the built object for which they are used. They might be replaced or even last beyond the life of that specific built object. The reuse, recyclability, and disposal of all of those materials, elements, and products should be considered in the life cycles.

Using MPs throughout the life cycle of buildings would facilitate the planning of renovations and retrofit (Çetin et al. 2022), thus slowing the resource loops (Bocken et al. 2016). MPs could also enable managing sustainable end-of-life material flows, such as reusing and recycling materials and elements (Çetin et al. 2022), thereby closing the resource loops (Bocken et al. 2016). Environmental impact of all life cycle stages can be recorded in the MP to assess the impacts of the entire building (Honic et al. 2019b, further described in Sect. 5.4).

5.2.2 MPs and Digital Platforms

MPs can be generated to provide information that spans different life cycles and scales – from construction materials to elements, buildings, and cities – though a standard structure and scale for MPs do not exist. At a city scale (see Chap. 2 by Tsui et al. on geographic information systems), using MPs for the building stock and embedding them in digital platforms could provide several benefits for municipal authorities, urban miners, architects, and waste auditors. Municipalities could, for example, plan retrofits and renovations, predict upcoming waste streams, and implement sustainable end-of-life streams (e.g. reuse and recycling) in their existing building stock. Urban miners would be able to detect where valuable elements and materials are located within a city and be informed about when these will be available. Architects could design new buildings with materials and elements provided in the platform, thus facilitating reuse, and waste auditors would have an easier job while investigating existing buildings since most of the information needed for a waste audit would be available in the platform. Such a digital platform could also be used as an ecosystem for trading materials and elements, where MPs constitute the backend information provider.

To capture the value of materials, elements, and buildings, digital platforms will play an increasingly important role in the future (Chan et al. 2020). One example of a digital platform has been developed by Honic et al. (2023), who established a framework for a digital urban mining platform for the city of Vienna. BIM (see Chap. 1 by Koutamanis on BIM and digital twins), laser scanning, ground-penetrating radar, and GIS technologies were used to compile MPs and assess material intensities (tonnes/m^3 gross volume of a building) of single buildings. These material intensities were extrapolated to calculate the material composition of similar building types, which enabled a prediction of a large number of buildings in the city. The predicted material compositions were integrated into MPs, which were made available in the digital urban mining platform (Honic et al. 2023).

5.3 Passport Instruments for a Circular Economy

Considering "buildings as a material depot" (Rau and Oberhuber 2022) helps view the building stock as a potential provider of materials for new buildings. Reusing materials and elements from existing buildings, thereby avoiding extraction of raw materials and associated carbon emissions, serves the circular economy principle of "closing the loop" (Bocken et al. 2016). In existing buildings, valuable materials can only be reused for new construction if the necessary information on the materials, such as their quality, remaining lifetime, allocation within the building, accessibility, disassembly potential, etc., is available. However, the scarcity of information on materials and elements embedded in existing buildings (Arora et al. 2019) is a major obstacle in reaching high reuse rates.

5.3.1 MP-Related Concepts

To support and provide guidance on best practices for performing the assessment of demolition waste streams prior to demolition, the EU Commission developed the "Guidelines for the Waste Audits before Demolition and Renovation Works of Buildings" (European Commission 2018), which specify information to be collected during audits on existing buildings, such as the type of materials embedded in buildings and if they consist of harmful substances. However, the audits are conducted mainly to assess the amount of waste materials and plan how much of what type of material will be incinerated or disposed of at which landfill type (e.g. specific landfills for harmful substances). The waste audit documents are structured by waste categories (e.g. metallic, plastic, wood) established by the EU.

As the name implies, waste audits are made to assess the amount of waste and not to assess its potential reuse or provide information on disassembly. This is where MPs come into play. MPs provide more information than a waste audit: they store information about the disassembly, reuse, and recycling potential, as well as the allocation and amount of materials and elements (CB'23 2023; Madaster 2023). If generated in the design stage of a building and updated during its life span, at the end-of-life stage of a building, an MP can prevent building materials from being demolished, incinerated, or disposed of since information on the incorporated materials and elements exists which can be used to design a new building with the existing stock.

5.3.2 MPs for New and Existing Buildings

The use of MPs can be beneficial for both new and existing buildings. For new buildings, MPs could help implement all principles of a circular economy, from narrowing to slowing and closing the loop as well as regenerating nature. In the conceptualisation and design stage of buildings, an MP could serve as an optimisation tool to assess and reduce the amount of materials used for the building (thus narrowing the loop) and to choose materials with a long life span (slowing the loop), elements and products with a high reuse potential (closing the loop), and bio-based materials (regenerating nature). Creating an MP for new buildings is feasible due to existing 2D plans, 3D models, BIM models, environmental product declarations, declarations of performance, life cycle assessments, and energy certificates, all of which provide important information that could be stored in MPs.

Compiling MPs for existing buildings is a challenging task due to the lack of information about the existing building stock (Rose and Stegemann 2019). Several digital technologies can be applied to gather information on existing buildings at the city, building, and element scales. At the city scale, these are computer vision (see Chap. 4 by Armeni et al. on AI) and geographic information systems (see Chap. 2 by Tsui et al. on GIS). Laser scanning can be applied at the city and building scales (see Chap. 3 by Gordon et al. on Reality Capture). At the element scale, ground-penetrating radar is a useful technology. Some examples of these technologies are described in the next paragraph. More examples can be found in the associated chapters.

To acquire information at city, building, and element scale, several approaches have been developed. Raghu and De Wolf (2022) applied computer vision technology to detect facade materials and elements such as windows and doors and developed machine learning algorithms to detect cracks and evaluate the quality of materials and elements in the city of Zurich. Wu et al. (2022) explored the prediction of asbestos-containing materials in residential buildings in Gothenburg and Stockholm through artificial neural networks. Similar work using computer vision has been conducted by Koch et al. (2018), Mahami et al. (2020), and Nordmark and Ayenew (2021). Geographic information system models are provided from various

European cities which can be applied to assess material stocks and to investigate where specific materials are allocated (Bradshaw et al. 2020). Laser scanning can be used to gather geometrical information, e.g. the height of a building, and to generate a BIM model of a building (Mill et al. 2013). To acquire information on the materials, Honic et al. (2021a, b) used a ground-penetrating radar at element scale. They automatically identified building elements' material compositions through machine learning algorithms.

Combining the information gathered at city, building, and element scales can help generate new MPs for existing buildings or, if available, feed existing MPs with further information. The availability of MPs at city scale could enable the assessment of expected waste, planning of sustainable end-of-life streams, and generation of a digital platform for cities, as described in Sect. 5.2.2. Some examples of MPs for new and existing buildings are presented in Sect. 5.4.

5.4 Examples from Research and Practice

In the last decade, several MPs and data templates were developed in research and practice. Some of these examples are presented in this section: first, three academic examples, followed by one practical example and the initiatives CB'23 (CB'23 2023) from the Netherlands, Product Circularity Data Sheet from Luxembourg (PCDS 2023), and GrowingCircle project from Portugal (GrowingCircle 2023).

5.4.1 Examples from Research

Within the EU Horizon 2020 project BAMB (BAMB 2020), 300 MPs for various products, components, and materials, as well as an MP platform, were created. The platform was developed to facilitate the appropriate accessibility of information for different stakeholders at specific stages in the process (Fig. 5.1). The aim of the MP in BAMB was to keep or increase the value of materials, products, and components across their life cycles; to support developers, managers, and renovators in their selection of circular materials; to create incentives for suppliers to produce healthy and sustainable materials; and to facilitate the return of products, materials, and components. BAMB provided an MP platform in which various characteristics of elements and products could be uploaded by manufacturers or others. These elements and products were given an identification number, which helped identify them in a BIM model. However, changes in the BIM model could not be automatically integrated in the platform (BAMB 2020).

An automated process to generate MPs was developed by Honic et al. (2019a). BIM was used to compile an MP that optimises the design with regard to the reduction of waste and environmental impacts. This automated process has the advantage that different design variants can be compared with each other.

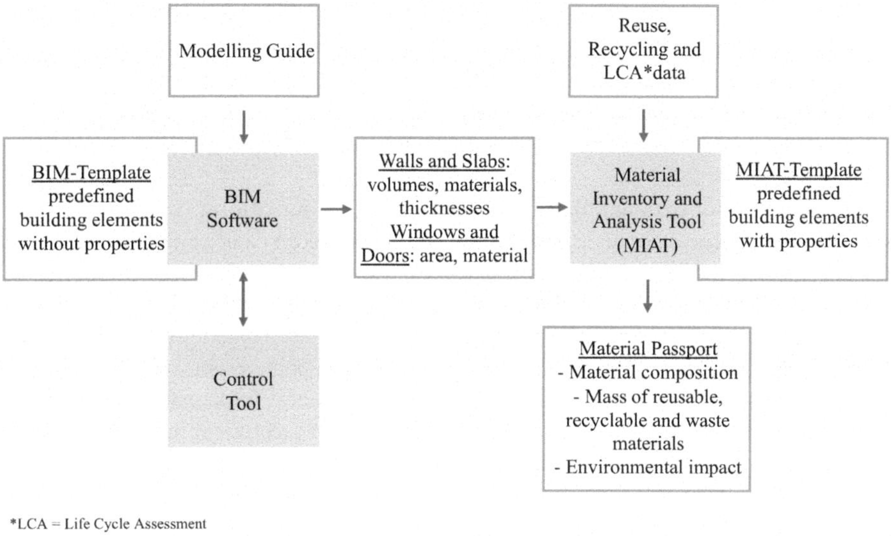

Fig. 5.1 BIM-based workflow for the generation of an MP. (Adapted from Honic et al. 2019a)

Automating the generation of the MP requires proper modelling in BIM (the use of proper BIM objects, geometry and materials, etc.). The researchers provide the framework, templates, and a modelling guideline to help users apply the MP method (Fig. 5.1). Since the basis for an MP and an environmental impact assessment is the same, namely, a bill of quantities on the element scale, the environmental impact assessment was also integrated into the MP. Three indicators were used for the environmental impact assessment: the global warming potential (CO_2 emissions), acidification potential (SO_2 emissions), and primary energy intensity. Insights and optimisations can be conducted on the material, element, and component (the aggregation of elements of the same type) to the building scale.

The researchers also explored MPs for existing buildings (Honic et al. 2021b). Digital technologies such as laser scanning and ground-penetrating radar were used to acquire the geometry and materials of existing buildings. The results showed that a compilation of MPs for existing buildings is possible and delivers valuable information on incorporated materials and elements. However, the cost and time needed to apply the digital technologies and process the data were very high, and optimisation is needed in terms of user-friendliness to be applicable in practice (Honic et al. 2021a).

5.4.2 *Examples from Practice*

An MP example from practice is provided by Concular, a digital platform that collects information on new and existing buildings. Concular aims to facilitate a circular and resource-efficient built environment. Concular uses BIM models and CSV files to generate MPs in the form of data sheets. The MPs enable the tracing of materials and elements throughout the whole life cycle. Next to MPs, CO_2 emissions and circularity assessments can be conducted for new and existing buildings. For existing buildings, Concular uses 3D scans and computer vision algorithms to detect objects (Concular 2023).

An example for data templates and MPs is CB'23, a platform developed by the joint efforts of professionals (market parties, policymakers, and scientists) from the Dutch construction sector. The platform offers guides for measuring circularity in and creating passports for the construction sector as well as a web tool for creating MPs for the whole life cycle of buildings. CB'23 provides a data template in form of a spreadsheet, which contains many parameters and characteristics needed for the MP. (To mention a few, these parameters include general information such as the object number, length, width, and origin of the data and technical information such as adaptability, existing certificates, and units in which this information should be provided.) The template is also aligned with the life cycle stages of buildings, thus giving information on which data is required at a certain stage.

Another example for data templates, namely, the Product Circularity Data Sheet, is provided by the Circularity Dataset Initiative. The initiative aims to provide an industry standard for product circular data using the Product Circularity Data Sheet (Mulhall et al. 2022). The data sheet works at the product level and consists of specific sections devoted to product identification, product composition, designs for better use, disassembly, and reuse.

A case study using data templates and MPs was conducted by the GrowingCircle project (GrowingCircle 2023), which was funded by the EEA grants (2014–2021). GrowingCircle focuses on the awareness and provides a proof of concept of how circular data is key to enabling circular economy processes. One of their case studies is a residential building renovation process where the digitalisation of meaningful elements was realised to generate an element catalogue (Mêda et al. 2022) that could feed a 3D model of the building with data. Using a data template framework, the element data from the renovation design, which was delivered by manufacturers, was integrated in MPs next to other performance-related data. Further, the data from the MP was attached to the 3D model to enable several analyses and deliverables associated with sustainability.

The examples from research and practice show the variety of existing MP and DPP concepts. Several approaches exist for generating MPs. Although not yet standardised, BIM is often used for generating MP. Some solutions offer the upload of a BIM model to create insights in terms of circularity or environmental impact assessment. One of these solutions is Madaster, which is further presented in the next section. The use of data templates is becoming more common and is expected gain importance in the future in order to create a common basis for MPs.

5.5 Business Models for Passport Instruments in a Circular Built Environment

A digital service for registering, storing, and exchanging data – in this case, in the form of MPs – is like a utility service, for which the user wants to be sure that the service is trustworthy, efficient, and effective. The business model of the service provider is based on providing the service, not on the value or relevance of the content of the service – similar to the business model of transportation companies where the price is based on transport, not on the content of the goods transported.

Madaster (Madaster, 2023), which started in 2017, is an example of an independent, international digital service provider that provides MPs. Technically Madaster works by using a BIM environment as a data source for the generation of MPs (Fig. 5.2). The interoperable exchange format within BIM – the industry foundation class (IFC) – and an XLS file can also be uploaded on the Madaster platform. As output, the platform delivers an MP next to the Madaster Circularity Index and embodied carbon and costs assessments.

The Madaster business model is based on the delivery of data registration, storage, exchange, processing, reporting, and analysis services, where the MP delivery is one of the offerings provided. Its business model consists of the following components:

- Initiation and market rollout: The Madaster platform development is funded by private investors (50%), grants and subsidies (25%), and early adopters (25%).
- Service provision and functional development: Usage of the Madaster platform requires a yearly subscription, where pricing depends on the country and type of

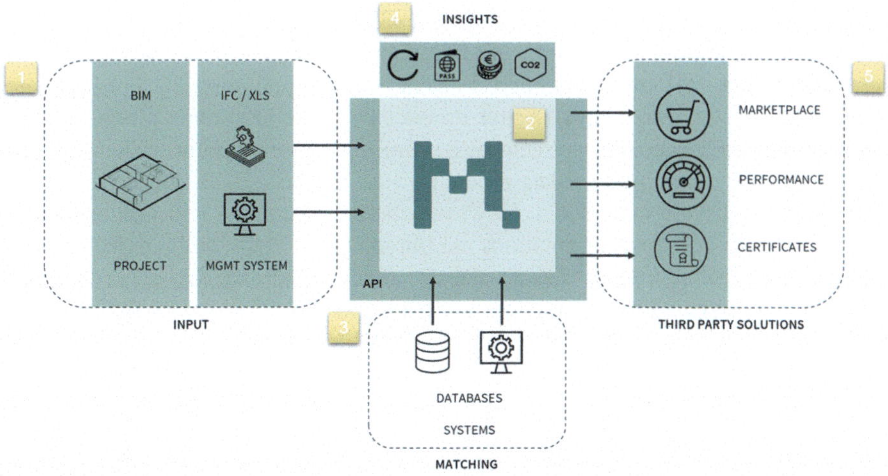

Fig. 5.2 Framework of the Madaster platform showing data inputs, outputs, and insights on MPs, cost, circularity, and CO_2 assessments (Madaster, 2023)

client. The subscription includes MPs, embodied carbon calculations, environmental and circularity assessments, and residual value reporting.
- Supervision and R&D: Madaster is supervised by the independent, not-for-profit Madaster Foundation. This foundation owns the brand IP and assures the availability of the (meta) data for public usage. A partnership agreement between Madaster Foundation and the service company describes the supervisory activities, including the possibility to select an alternative service company in case of noncompliance with the supervision criteria.

5.6 Discussion

A circular economy relies on digital data that can be tracked throughout the entire life cycle of buildings. Data templates, MPs, and DPPs all rely on digital data and aim to enable a circular economy. MPs, which rely on data templates and differ from DPPs only in terms of the industries and scales they are applied, can be generated for new and existing buildings whereby digital technologies are used to feed them with relevant information. The academic and practical examples in this chapter showed different approaches to generating an MP (e.g. BIM-based or with data templates from CB'23) and that the insights created from MPs can vary from circularity assessments including the reuse potential of buildings to environmental impact assessments.

Finding common technical language and standardising key data have been major challenges in construction since the publication of the first Directive on Construction Products in the late 1980s (Council of the European communities 1988). Presently, there is still no consensus on the data requirements, structures, and formats of MPs or DPPs. The initiatives associated with data templates, MPs, and DPPs mentioned in this chapter consider the need for the digital representation of the built environment. The integration of product data into an MP remains an obstacle if the data formats of the DPP and the MP are not aligned. Present efforts in aligning the concepts are being made by the Ecosystem Digital Product Passport Initiative, which aims to present an unambiguous cross-sectoral definition and description of a DPP and define a cross-sectoral product data model for it (CIRPASS 2023). In the initiative, MPs are presently not considered. However, the developments will influence how knowledge, standards, and orientations are implemented in the built environment.

A further challenge to increasing the use of digital passports lies in how the required data can be gathered, tracked, and made available throughout the life cycle of buildings. To overcome these obstacles, digital technologies can play a crucial role. Computer vision, geographic information systems, laser scanning, and ground-penetrating radar show potential for gathering data on existing buildings for MPs. For new buildings, BIM plays a crucial role, since, if modelled properly, models contain detailed information on each material and element used in a building. Blockchain technology (see Chap. 13 by Shojaei et al. on blockchain technology) promises greater transparency, improved traceability, and increased efficiency (IBM

2018). The integration of blockchain technology and the internet of things could enable tracking of the present state of materials and elements throughout their life cycle, thereby keeping information up-to-date and enabling predictions for reuse of materials and elements used in the built environment (Esmaeilian et al. 2020). The integration of the information delivered by different technologies into an MP needs further manual steps since the data formats and structures can vary significantly. Thus, aligning data formats and structures for data templates will be an important step in enabling the compilation of standardised MPs.

5.7 Key Takeaways

- Material passports (MPs) are a main enabler of a circular economy in the built environment, because they consist of relevant information that facilitates maintenance, repair, reuse, etc.
- MPs give building materials and elements value for present use, recovery, and reuse.
- Several passport instruments exist in academia and practice.
- Format, structure, and terminology for MPs are not standardised.
- Several digital technologies can support the gathering, storing, and maintaining information of new and existing buildings.

Acknowledgements Meliha Honic would like to acknowledge the funding of the ETH Zurich Postdoctoral Fellowship (21-2 FEL-06). Pedro Mêda would like to acknowledge the European Economic Area (EEA) Financial Mechanism 2014–2021; the Environment, Climate Change and Low Carbon Economy Programme: 13_Call#2 GrowingCircle; and the Base Funding of CONSTRUCT – Instituto de I&D em Estruturas e Construções – funded by national funds through the FCT/MCTES (PIDDAC): UIDB/04708/2020. The authors declare that no potential conflicts or financial interests exist.

Declaration of Competing Interests Author Van den Bosch is a co-founder and director of Madaster.

References

Arora M, Raspall F, Cheah L, Silva A (2019) Residential building material stocks and component-level circularity: the case of Singapore. J Clean Prod 216:239–248. https://doi.org/10.1016/j.jclepro.2019.01.199

BAMB (2020) Buildings as material banks. https://www.bamb2020.eu/. Accessed 15 Feb 2023

Bocken NMP, de Pauw I, Bakker C, van der Grinten B (2016) Product design and business model strategies for a circular economy. J Ind Prod Eng 33(5):308–320. https://doi.org/10.1080/21681015.2016.1172124

Bradshaw J, Jit Singh S, Tan S-Y, Fishman T, Pott K (2020) GIS-based material stock analysis (MSA) of climate vulnerabilities to the tourism industry in Antigua and Barbuda. Sustainability 12(19):8090. https://doi.org/10.3390/su12198090

CB'23 (2023) Platform CB'23. https://platformcb23.nl/english. Accessed 15 Feb 2023
Çetin S, De Wolf C, Bocken N (2021) Circular digital built environment: an emerging framework. Sustainability 13(11):6348. https://doi.org/10.3390/su13116348
Çetin S, Gruis V, Straub A (2022) Digitalization for a circular economy in the building industry: multiple-case study of Dutch social housing organizations. Resour Conserv Recycl Adv 15: 200110. https://doi.org/10.1016/j.rcradv.2022.200110
Chan PW, De Wolf C, Koutamanis A (2020) The digital potential in creating a circular construction economy. Technische Universiteit Delft
Circularise (2023) Circularise. In: circularise. https://www.circularise.com/dpp. Accessed 13 Apr 2023
CIRPASS (2023) Cirpass project EU. https://cirpassproject.eu/. Accessed 15 Feb 2023
Concular (2023) Concular. https://concular.de/. Accessed 17 Feb 2023
Council of the European Communities (1988) Council directive of 21 December 1988 on the approximation of laws, regulations and administrative provisions of the member states relating to construction products. Off J Eur Communities L 40(12)
de Brito MP, van der Laan EA, Irion BD (2017) Extended producer responsibility in the aviation sector
Eichstädt J (1982) Modernisation rationalisée des usines. Batiment Int Build Res Pract 10(3): 177–181. https://doi.org/10.1080/09613218208551081
EPEA Netherlands (2023) EPEA circularity passport. https://epea.com/nl/en/. Accessed 14 Feb 2023
Esmaeilian B, Sarkis J, Lewis K, Behdad S (2020) Blockchain for the future of sustainable supply chain management in industry 4.0. Resour Conserv Recycl 163:105064. https://doi.org/10.1016/j.resconrec.2020.105064
EU (2000) Directive 2000/53/EC of the European parliament and of the council of 18 September2000 on end-of life vehicles. Off J Eur Communities 21:34–42
EU (2020) European Union Action Plan
European Commission (2018) Guidelines for waste audits before demolition and renovation works of buildings
European Commission (2020a) A European strategy for data
European Commission (2020b) Regulation of the European parliament and of the council concerning batteries and waste batteries, repealing Directive 2006/66/EC and amending Regulation (EU) No 2019/1020
European Commission (2022a) Regulation of the European parliament and of the council establishing a framework for setting Ecodesign requirements for sustainable products and repealing Directive 2009/125/EC
European Commission (2022b) Regulation of the European parliament and of the council laying down harmonised conditions for the marketing of construction products, amending Regulation (EU) 2019/1020 and repealing Regulation (EU) 305/2011
Frühbuss T, Nottebom F, Seibert R (2000) 20 years of International Material Data System (IMDS)
Götz T, Berg H, Jansen M, Adisorn T, Cembrero D, Markkanen S, Chowdhury T (2022) Digital Product Passport: the ticket to achieving a climate neutral and circular European economy?
GrowingCircle (2023) GrowingCircle data driven construction to circular economy. https://growingcircle.netlify.app/. Accessed 17 Feb 2023
Hesselbach J, Herrmann C, Ohlendorf M, Graf R (2001) Approach of substance flow oriented closed loop supply chain management in the electrical and electronic equipment industry. In: Proceedings of second international symposium on environmentally conscious design and inverse manufacturing. IEEE, pp 725–728
Honic M, Kovacic I, Rechberger H (2019a) Concept for a BIM-based material passport for buildings. IOP Conf Ser Earth Environ Sci 225:012073. https://doi.org/10.1088/1755-1315/225/1/012073
Honic M, Kovacic I, Rechberger H (2019b) Improving the recycling potential of buildings through material passports (MP): an Austrian case study. J Clean Prod 217:787–797. https://doi.org/10.1016/j.jclepro.2019.01.212

Honic M, Hinterleitner A, Schloegel I, Kovacic I, Sreckovic M (2021a) Application of GPR-technology for identifying the material composition of building components, pp 366–372
Honic M, Kovacic I, Aschenbrenner P, Ragossnig A (2021b) Material passports for the end-of-life stage of buildings: challenges and potentials. J Clean Prod 319:128702. https://doi.org/10.1016/j.jclepro.2021.128702
Honic M, Ferschin P, Breitfuss D, Cencic O, Gourlis G, Kovacic I, De Wolf C (2023) Framework for the assessment of the existing building stock through BIM and GIS. Dev Built Environ 13: 100110. https://doi.org/10.1016/j.dibe.2022.100110
IBM (2018) Top five blockchain benefits transforming your industry. https://www.ibm.com/blogs/blockchain/2018/02/top-five-blockchain-benefits-transforming-your-industry/. Accessed 15 Feb 2023
ISO (2020) ISO 23387:2020 Building information modelling (BIM) – data templates for construction objects used in the life cycle of built assets – concepts and principles. https://www.iso.org/standard/75403.html. Accessed 14 Feb 2023
Koch D, Despotovic M, Sakeena M, Döller M, Zeppelzauer M (2018) Visual estimation of building condition with patch-level ConvNets. In: Proceedings of the 2018 ACM workshop on multimedia for real estate tech. ACM, Yokohama Japan, pp 12–17
Luscuere L, Mulhall D (2018) Circularity information management for buildings. In: Charter M (ed) Designing for the circular economy, 1st edn. Routledge, pp 369–380
Madaster (2023) Madaster Platform. https://madaster.com/. Accessed 15 Feb 2023
Maersk (2011) Maersk Cradle to Cradle® Passport. In: C2C-Cent. http://www.c2c-centre.com/library-item/maersk-cradle-cradle%C2%AE-passport. Accessed 14 Feb 2023
Mahami H, Ghassemi N, Darbandy MT, Shoeibi A, Hussain S, Nasirzadeh F, Alizadehsani R, Nahavandi D, Khosravi A, Nahavandi S (2020) Material recognition for automated progress monitoring using deep learning methods. arXiv preprint arXiv:2006.16344
McDonough W, Braungart M (2003) Towards a sustaining architecture for the 21st century: the promise of cradle-to-cradle design. Indu Environ 26(2):13–16
Mêda P, Sousa H, Hjelseth E (2020) Data templates-traceability and digital record through project life-cycle. J Appl Bus Econ 22(12):274–291
Mêda P, Calvetti D, Hjelseth E, Sousa H (2021) Incremental digital twin Conceptualisations targeting data-driven circular construction. Buildings 11(11):554. https://doi.org/10.3390/buildings11110554
Mêda P, Calvetti D, Sousa H, Hjelseth E (2022) What comes first when implementing data templates? Refurbishment case study. In: Proceedings of the 14th European conference on product and process modelling. p 8
Mill T, Alt A, Liias R (2013) Combined 3D building surveying techniques–terrestrial laser scanning (TLS) and total station surveying for BIM data management purposes. J Civ Eng Manag 19 (sup1):23–32
Mulhall, D, Hansen K, Luscuere L, Zanatta R, Willems R, Boström J, Elfström L, Heinrich M, Lang W (2017) Framework for materials passports. BAMB consortium, EPEA, SundaHus
Mulhall D, Ayed A-C, Schroeder J, Hansen K, Wautelet T (2022) The product circularity data sheet – a standardized digital fingerprint for circular economy data about products. Energies 15(9):3397. https://doi.org/10.3390/en15093397
Nordmark N, Ayenew M (2021) Window detection in facade imagery: a deep learning approach using mask R-CNN. arXiv preprint arXiv:2107.10006
PCDS (2023) Product circularity data sheet. Luxembourg. https://pcds.lu/. Accessed 17 Feb 2023
Raghu D, De Wolf C (2022) Digitalizing existing buildings: moving towards circularity using big data and machine learning. In: In artificial intelligence in architecture, engineering and construction conference 2022. online
Rau T, Oberhuber S (2022) Material matters: developing business for a circular economy. Taylor & Francis
Rose CM, Stegemann JA (2019) Characterising existing buildings as material banks (E-BAMB) to enable component reuse. Proc Inst Civ Eng: Eng Sustain 172(3):129–140. https://doi.org/10.1680/jensu.17.00074

Walden J, Steinbrecher A, Marinkovic M (2021) Digital product passports as enabler of the circular economy. Chem Ing Tech 93(11):1717–1727. https://doi.org/10.1002/cite.202100121

World Economic Forum (2019) A Vision for a Sustainable Battery Value Chain in 2030. 91–93 route de la Capite CH-1223 Cologny/Geneva Switzerland

Wu P-Y, Mangold M, Sandels C, Johansson T, Mjörnell K (2022) Modeling artificial neural networks to predict asbestos-containing materials in residential buildings. IOP Conf Ser Earth Environ Sci 1122(1):012050. https://doi.org/10.1088/1755-1315/1122/1/012050

Meliha Honic obtained her bachelor's and master's degree in architecture at TU Wien. Her master's thesis was about urban densification whereby BIM and LCA tools were used to assess the environmental impact of a designed building. Within her PhD at the Faculty of Civil Engineering, she developed a BIM-based material passport which serves as resources and LCA optimisation tool in the design stage of buildings. She initiated and led several research projects about digitalisation of existing building stocks and BIM-based material passports for new design and the end-of-life stage of buildings, where she was working with partners from the industry and academia. She is presently an ETH Zurich postdoctoral fellow at the Circular Engineering for Architecture (CEA) lab, directed by Prof. De Wolf.

Pedro Meda Magalhães graduated in civil engineering at Porto University, Faculty of Engineering, in 2005. He later got a master's degree in the same university with a thesis entitled Integrated Construction Organization: Contributions to the Portuguese Framework. He has worked since 2009 at Instituto da Construção with the role of project manager and researcher. He participated in several national and international projects scoped in construction data management, namely, the standardisation of bill of quantities and construction products properties. In terms of training, he took part in several actions focused on public procurement for construction-related services and works. He participated in several commissions and committees related to the construction industry. Presently, he is a PhD candidate and invited professor at Porto University.

Pablo Van den Bosch is co-founder and director of Madaster and focuses on public affairs and internationalisation. He studied public affairs at the Erasmus University Rotterdam and was partner in a global consultancy that he co-founded in 2003. Since 2016, Pablo has invested his time and experience in new ventures that aim to deliver a positive social impact. He teamed with Thomas Rau, a renowned architect who has created energy-neutral buildings since the 1990s. Based on Thomas' vision that "Waste is material without an identity", they and four others created Madaster, the register for products and materials in the built environment.

Part II
Design and Fabrication

Chapter 6
Enabling Design for Circularity with Computational Tools

Felix Heisel and Joseph McGranahan

Abstract Circular construction is a design task that requires new datasets and computational tools for matching supply and demand within an urban circular system. Material passports (MPs) contain detailed inventories of materials and products, as well as their specifications, location, and connection details. Circularity indicators (CIs) allow an assessment of a design's environmental impacts with respect to circularity: the degree to which solutions minimise extraction and waste in favour of reusable, renewable, or recyclable resources both in construction and at end-of-use. Often implemented as an extension to detailed BIM models, MPs and CIs are presently applied in the permit and documentation phases. However, these metrics also establish parameters in early design phases, where circular design thinking and evaluation are most impactful. Circular construction consequently calls for a new suite of design tools that can be integrated into existing workflows, are applicable within the uncertain context of the early design phase, and ideally offer immediate feedback related to formal deliberations, structural considerations, material selection, and detailing. This chapter describes the importance of CIs as design parameters across phases with a special focus on recent early design developments such as the software application RhinoCircular.

Keywords Design for circularity · Early design · Circularity indicator · Circularity assessment · Material passports

6.1 Computational Tools Enabling Design for Circularity

A crucial step in building the capacity of a circular economy is assessing the quantities and impacts of material inputs, assets, and outputs in the built environment to ensure that building materials remain at their highest utility and value through use cycles. This serves two main purposes: they help understand the carbon impact of

---

F. Heisel (✉) · J. McGranahan
Circular Construction Lab, Cornell University, Ithaca, NY, USA
e-mail: felix.heisel@cornell.edu

C. De Wolf et al. (eds.), *A Circular Built Environment in the Digital Age*, Circular Economy and Sustainability, https://doi.org/10.1007/978-3-031-39675-5_6

construction through the quantification and measurement of a building's component materials, and they help create a material inventory which allows for their future activation in the form of a material depot. Both tasks require computational tools which can generate data on material quantities and qualities, connections, environmental impacts, recyclability, and reusability (Heisel et al. 2022a).

A solution can be found in the creation of material passports (MPs), digital inventories of material assemblies, and their associated metrics (Heisel and Rau-Oberhuber 2020). MPs document material quantities as well as weights, volumes, dimensions, specifications, and locations of materials within a structure (van Capelleveen et al. 2023). Efforts from companies and organisations such as Madaster, BAMB, DGNB, and the European Union offer platforms for creating, storing, and sharing MPs, often integrated with building information models (BIM) to determine the quantities of components automatically (BAMB 2020; DGNB 2023; Madaster 2023; WBCSD 2023).

Such passports may also report the overall circularity of a construction represented by a circularity indicator (CI), a metric which evaluates both reuse and recyclability of the input material stock and their end-of-use pathways (Heisel and Nelson 2020). The metric itself is a number between 0% and 100%, determined via a set of equations that are dependent on parameters such as life span, efficiency of recycling, and the fraction of feedstock taken from renewable, recycled, or reused sources. Typically, equal weight is given to reused, recycled, or biogenic/compostable material pathways in both the production and end-of-use phase. An initial CI version was developed by the Ellen MacArthur Foundation in 2016 and then further advanced for use in the built environment by Dutch company Madaster in 2018 (Ellen MacArthur Foundation & Granta Design 2019). Building on this work, a variety of definitions and equations on material circularity have been published since. Recent examples are the Urban Mining Index developed at the University of Wuppertal, which takes deconstruction times and difficulties into consideration (Rosen 2022), or the Gebäuderessourcenpass (building resource passport), which is now officially integrated into the certification system of the German Sustainable Building Council (DGNB) (Concular 2023).

CIs are an important reference not only in the design phases of buildings. Since these metrics can inform stakeholders of possible or anticipated pathways for various materials and components, i.e. whether they are structurally sound and/or designed for disassembly, they are also relevant at a building's end-of-use; in cases of adaptive reuse; when tracking renovations, retrofits, and repairs; or in evaluating recycling or reuse scenarios. Both CIs and MPs are applicable across scales, ranging from construction details to the scale of large urban areas. The tools function in similar ways across these scales; however, the feedback they provide informs different actions and stages of the design process. This makes CIs useful across the entire life span of a building and applicable from early design-to-construction documentation.

Despite the versatility of circularity metrics, present software tools are typically applied in the later stages of building development, when most building parameters are known, leaving little flexibility for informed design. The reason for this lies in the availability of data: at this stage, designs are highly resolved, and information which is needed for calculations, such as detailed product specifications and material quantities, can be computed and stored in MPs. Although they can help evaluate the circularity of a building, CIs generated at this stage rarely actuate significant changes in design or specifications because many critical design decisions have already been determined by the time they are used. Instead, they are often used for compliance to meet technical standards or legislative requirements.

This approach to circularity is effective at creating detailed accounts of a building's material content for future reference but does not give designers flexibility within the early design phases in making major changes based on design evaluation and feedback. This creates an opportunity for new design tools which can address this phase. We argue that greater impact can be achieved by implementing circularity analysis in the early, pre-BIM design phase. Design is an iterative process, and a building's design is more prone to change in early phases. More importantly, major changes in early design phases are easier and cheaper to implement (Lawson 2005). As a design develops, decisions are more finely tuned as architects synthesise feedback from clients, engineers, contractors, stakeholders, and various consultants until the design is finalised. Because of this, the earlier circularity analysis occurs within the design process, the greater impact it will have in later design phases. Additionally, it can help reduce the uncertainty of decisions in the early design phase by providing information on the impact of individual decisions through an (ideally rapid) information feedback loop.

Before CIs can be incorporated in early design phases, the strengths and weaknesses of existing computational tools in addressing varying scales should be assessed to better understand how they can achieve the greatest impact in designing for circularity. These tools need to be adaptable and be able to reduce the uncertainty of decisions regarding design and material salvage. This chapter will discuss the scales, stages, and metrics needed to apply these tools. It will evaluate some of the existing technologies and establish guideposts for future tools. By improving flexibility in their inputs, and introducing user-adjustable parameters and immediate feedback within the CAD environment, new design tools will hopefully steer the hand of designers towards increased circularity.

6.2 Computational Tools Across Scales of the Built Environment

Computational tools play a key role in today's practice of architecture. Since their introduction, they have greatly influenced a wide range of areas relevant to the design and construction of buildings, ranging from the design and modelling of buildings to the simulation and analysis of their performance (or the performance of their component parts), their visualisation and presentation of their design strategies to the construction, and off-site fabrication of parts or on-site fabrication of full buildings. In the future, the influence of computational tools is bound to increase even further, whereby a building's whole-life perspective in both evaluation and implementation should be a focus.

Computational tools that assess circularity apply to a wide range of scales. These tools could be employed to assess the circularity of items as small in scale as a product and construction detail or as massive as extended urban areas and material mines (Heisel and Hebel 2021). This section will discuss some of the applications CIs and computation tools for circularity can have at these varying scales.

Computational tools for circularity are extremely relevant at the scales of materials, products, and construction details. Information on the specific pathways of selected materials from production to end of life provides relevant inputs for making design decisions. Similarly, product information in the form of environmental product declarations (EPDs) or other verified datasets can provide important design parameters in the selection and combination of building assemblies. The detail is one of the most critical points of a circular design, as details are where building products with different physical and chemical properties are joined and fastened together. Choices in these products and fasteners directly affect the salvageability or reusability of the component materials at the end-of-use. As a result, assessing the circularity of details is a critical step in designing for circularity. In design workflows, details are typically specified at a stage when architects and engineers have a clearer understanding of the loads, products, and fasteners that are required for a specific assembly. However, specific design intentions for developing details and constructions may well be design parameters that influence material choices and structural systems and thus play an important role in early design phases as well.

At the scale of the building, design tools for circularity can potentially have an extended impact. In addition to accounting for the circularity of materials and connections in total, such tools can also be used to track the circularity of a building over time. Materials and building passports should be updated when materials or components in a building are damaged, removed, repaired, exchanged, or renovated. Whether from manual or sensor input, circularity design tools can recompute a building's circularity score throughout a building's operation and store the information in the form of digital twins and MPs.

Surpassing the scale of the individual building, tools which enable design for circularity can have a major influence at the urban scale in closing the gap between demand and supply and in supporting green policy. As cities and regions move to decarbonise their building stocks (Root 2021), these tools can help make important decisions regarding which buildings to retrofit and in what order and how to have the greatest impact.

Exchanging fuel sources and heating equipment are often priorities in green policy. Many buildings also require envelope retrofits to compound the benefits of technical upgrades. This mandates an exchange of material, and with it a consideration of material and product impacts related to embodied carbon and the circularity of these new applications Heisel et al. (2022b).

Equally important at the urban scale is the analysis of the waste streams that result from comprehensive policy changes. The large-scale removal of building elements without a clear next-use scenario will inevitably result in an influx of construction and demolition debris sent to landfills or recycling facilities, while they could constitute a material resource. By generating this information, organisations and companies which have the physical and digital infrastructure for the salvage, reuse, and resale of building materials can be made aware of stock changes in advance. In general, computational tools for a circular economy can play a more significant role in the matching of demand and supply by aggregating available and sought-after materials and products from individual buildings within urban-scale databases.

6.3 Computational Tools for a Circular Economy

Within the narrow, slow, close, and regenerate framework, computational tools for circularity fall into two categories. These tools 'narrow' the amount of material used in buildings by providing a greater understanding of the construction specific impacts of material choices to architects and engineers. While two materials maybe seem identical in mass and volume within a design, CIs offer a wider perspective by accounting, for example, for material loss in production processes or recycling and reuse potentials at end-of-use. In creating CI metrics, computation tools for a circular economy provide the transparency in the hidden waste behind these materials to designers, enabling them to narrow their material use. Computational tools for circularity also serve to 'close' material loops. They do so in creating MPs which track the location and quantities of materials within buildings as well as their design for disassembly instructions. These are critical documents in closing material loops, as they facilitate the execution of deconstruction, reuse, and other end-of-use activities.

6.3.1 Required Flexible Inputs for CI Generation

Within a common design process, the level of detail increases with time. In early design applications, input parameters may be restricted by missing design specificity and thus need to be easily adaptable to keep pace with the iterative design process. To generate a circularity assessment, two inputs are first required: a geometry and a limited set of metadata. The geometry is user generated and can be either drawn/modelled or generated parametrically. The software does not differentiate between a new construction (e.g. massing study) and the assessment of an existing structure (e.g. a survey). Based on this initial geometry of any level of detail, volumetric and surface area information can be calculated.

A second step in generating a circularity assessment requires the pairing of this geometry with relevant metadata, which can be broken down into four subcategories:

- Materials (with associated circularity metrics)
- Constructions/ details (with associated circularity metrics)
- Shearing layers
- (Anticipated) pathways in production and end-of-use stages

While specificity is preferable during later design phases, emphasising generality within these subcategories in early design phases offers designers greater freedom to compare alternatives and make changes to increase circularity. For example, materials can be specified instead of products. Product specifications can change based on independent manufacturing aspects such as supply and demand, the geographic location, or the utilised energy mix. Materials at this stage are more general and can encompass multiple potential products.

Likewise, construction typologies are more general and representative of industry standards and regulatory frameworks than unique assemblies of building products. This allows for the use and assignment of general industry values for production and end-of-use circularity metric calculations, which supports the goal of using CIs as a design parameter in the immediate comparison of alternatives in early design additional to the specification of construction details in later stages.

Numerical user-specified parameters, such as the amount of reused materials in a design or targeted design for disassembly values, are similarly flexible and can be adjusted and refined throughout the design process, thus allowing circularity to be recalculated as the design progresses.

Another recommended input value is the assignment of geometries to shearing layers (Brand 1995). Shearing layers, or 'layers of change', provide a filter for materials, products, and components based on their anticipated performance and durability within a building and are defined as site, structure, skin, services, spaceplan, and stuff. Layers such as structure and skin have been observed to have longer use cycles and are more permanent in their arrangement within the building, whereas less determined layers such as spaceplan and stuff are subject to change more rapidly or often. Generating CIs for a breakdown of shearing layers may help understand the implications of specific systems within the overall design, may they be related to material choices, structural requirements, aesthetic preferences, or length of the use cycle.

It is also critical to understand the production and end-of-use pathways for various materials as an input. In computational tools for a circular economy, these material flows are often baked into material databases and not primarily user-inputted. These suggested values provide users with a baseline for different material types and families. If users select a product or material with a unique pathway, they should then be able to update or overwrite relevant metrics. This flexibility is important in the design of these tools, as it guides but does not constrain users in their calculations.

6.3.2 Assessing Outputs in Early-Stage Design Changes

In contrast to the adaptability of input parameters, output parameters need to be clear and give succinct guidance as to how and in what areas designers can increase circularity. This is especially true in the early design phase, when the design is still relatively conceptual and changes are easier to implement than in later stages.

Consequently, tools for circularity need to communicate a breakdown of CIs for each stage of the design, such as production and end-of-use pathways, and as described above for each shearing layer, assembly and subassembly. This can inform users on where to change parameters or redesign details in the effort to achieve higher circularity values. For instance, an early design tool could identify a subassembly with a particularly low CI as a good opportunity to incorporate design for disassembly strategies or material substitution, thereby raising the CI score of both the subassembly and the entire design.

Similarly, the outputs of such a design tool need to be visualised directly and in ways that are understood intuitively. Given the indeterminacy of these early design stages, CIs need to be displayed rapidly to provide immediate feedback to even slight changes to input parameters. This requires the ability to update relevant metrics based on changes in the formerly discussed parameters. A live feed of effects on the model's circularity score, for example, would inform the user on whether material choices or design decisions favour regenerative, reusable, or recyclable pathways, reducing the uncertainty while not burdening the designer with an overly prescriptive workflow. In doing this, the act of drawing or modelling would be elevated to an act of constant simulation, making readily apparent impacts associated with the user's decisions (May and Latour 2019).

These points can be generalised to any computational design tool but take on an increased relevance when designing for circularity. The impact of designing for circularity extends beyond the use of the building and into the entire use and life span of the building's components. Models need to be updated as changes are made to the building over its use time so that they may inform the continuous feedback loop.

6.3.3 Circularity Indicators at the End-of-Use Phase

Not only critical as feedback for the early design phase, CIs also take on important relevance at the end-of-use phase. Generating MPs and CIs for existing buildings allows for an assessment of the circularity of the present building stock and can inform salvage and deconstruction efforts for buildings which were not designed with disassembly in mind. A high CI can indicate that a material is easily reusable or salvageable and that using it might help save costs and carbon. This helps direct deconstruction and salvage efforts, as materials that are easily removable and reusable can then be prioritised for recovery, thus maximising the material value and utility of salvage and reuse from existing buildings – especially important in deconstructions with a narrow scope and tight timelines.

6.3.4 Circularity Databases

Creating a database which enables the calculation of CIs is one of the largest challenges in developing computation tools for circularity. Libraries of EPDs are expanding, but these documents are primarily concerned with carbon and production impacts and not (yet) the end-of-use of materials. Both production and end-of-use material streams vary dramatically based on location due to market factors or local legislation. For example, one municipality which requires the separation and sorting of all construction and demolition waste for recycling might have completely different circularity results than a municipality in which contractors haul all end-of-use waste to landfills because such requirements do not exist.

As a result, datasets on material circularity must be generated or adjusted at least on regional scales. Larger efforts to accomplish this are so far challenging, especially in the United States, where waste streams are often still recorded on handwritten reporting sheets (and are therefore hardly machine-readable), if such information is documented at all. Because of this, most datasets regarding the circularity of materials in the US context must make broad assumptions based on industry standards. Data availability is better in other contexts such as Europe or Asia, but it is still limited. EPEA (in collaboration with Madaster) recently published a dataset of 187 general materials and their associated circularity values, which is now accessible for users with a Madaster account (EPEA and Madaster 2023). Similarly, several other commercial providers of MPs (such as Concular) or life cycle assessment tools (such as One Click LCA) are developing their own internal databases (Campanella 2022).

As computational tools for circularity grow in popularity in the architecture, engineering, and construction industry, there will be a greater demand for certified and regionally adjusted datasets. At the same time, the use of these computational tools creates an opportunity for this data generation. As companies and organisations continue the application of CIs, they will create their own libraries of materials and assemblies and relative circularity metrics. In January 2023, the Circular Construction Lab at Cornell University published a first freely available dataset including associated sources in the hope to launch a collaborative open-source effort on the generation and collection of such circularity datasets (Heisel et al. 2023).

6.4 Examples of Computational Tools Enabling Design for Circularity

Several tools are presently available for supporting architects, engineers, and other building professionals in making decisions regarding circular design. Most of these tools generate MPs which allow for the documentation and assessment of new constructions. However, significant differences emerge when considering the requirements and considerations described above, specifically the applicability within different design phases, the required resolution of their input parameters, and the intended scale of the built environment.

6.4.1 Madaster

The Madaster platform originally developed in the Netherlands is described as an 'online registry for materials and products' (see Chap. 5 on material passports by Honic et al.). The platform is a browser-based framework for the analysis and storage of buildings and infrastructure projects utilising BIM models (.ifc files) or bill of quantities (Excel) as data input. Volume, area, or quantity information is linked to metadata providing the necessary foundation for calculations of circularity, embodied carbon, and financial material values over time. Madaster has developed into one of Europe's leading MP systems. The platform provider is involved in the creation of policy frameworks and technical tools across the European Union, has launched the platform in multiple languages and jurisdictions, and is forming partnerships with industry partners and manufacturers to create and collect a robust, up-to-date, and location-specific dataset for its calculations.

Once geometry and metadata are paired, the platform can create highly detailed building passports that allow for the tracking of materials within buildings throughout use cycles, until their eventual end-of-use scenario. Madaster's ability to track and estimate materials' (residual) monetary values is a critical feature for stakeholders interested in the deconstruction of a building at its end-of-use, as the resale value of the building's materials and components – and their ability to offset contractor costs – is a key factor in determining the profitability of a deconstruction project.

Madaster's primary use case is the documentation of buildings once their design has been finalised and a detailed BIM model can be saved and exported. Madaster is a highly evolved documentation and assessment tool and has recently set eyes on the development of design tool functionalities as described within this chapter.

6.4.2 One Click LCA

One Click LCA is a building life cycle assessment tool, allowing users to analyse the environmental impact of their buildings based on metrics such as embodied and operational carbon, to meet sustainability certifications such as LEED (Leadership in Energy and Environmental Design), BREEAM (Building Research Establishment Environmental Assessment Method), DGNB (Deutsche Gesellschaft für Nachhaltiges Bauen e.V.), or the Living Building Challenge. Its primary purpose is the assessment of carbon in buildings, but its functionalities in generating MPs are quickly expanding.

Like Madaster, One Click LCA operates as a stand-alone online platform, with its primary input being a detailed browser-based bill of quantities. It is supported by a well-organised and robust internal dataset, which uses an increasing library of EPDs as its primary resource, focusing primarily on the United States and Europe. One Click LCA also offers software plugins for CAD and BIM environments, such as Rhinoceros3D or Revit, allowing users to assess embodied carbon and circularity metrics as they design their buildings. The plugin is specifically advertised as a means to integrate the software's capabilities into early design phases and is accessible to owners of a commercial One Click LCA licence (One Click LCA 2021a; One Click LCA 2021b). These plugins are understood as extensions to the online platform to ease the linking of geometry with material and metadata. As such, the plugins display a partial amount of assessment results in the CAD environment, while the online platform computes high-resolution MPs even when starting at a low resolution of design parameters. These differences in resolution and output formats are meant to address the changing requirements and scales of the design and construction process (One Click LCA 2023).

6.4.3 RhinoCircular

RhinoCircular is an application for Rhinoceros3D and Grasshopper developed within the Circular Construction Lab at Cornell University to specifically evaluate material circularity in the early design phases, and the goals of the application match the framework outlined in Sect. 6.3 (*Circular Construction Lab* 2023). RhinoCircular's key focus is presently the assessment and visualisation of a design's environmental impact with respect to circularity: the degree to which design solutions minimise extraction and waste in favour of reusable, recyclable, and renewable material resources.

RhinoCircular allows direct and immediate feedback on design decisions regarding formal deliberations, structural considerations, material selection, and detailing based on MP and CI assessments. It can be integrated in existing and complex workflows and is compatible with industry-standard databases while providing its own starter dataset.

Figure 6.1 shows some of the potential of this tool to provide rapid and targeted feedback within the Rhinoceros3D environment. In this specific example, a detail model is assessed for circularity. CIs are generated for each element within the detail. Once these metrics are generated, they are remapped to the model geometry, demonstrating to users which elements are highly circular and which are less so.

Built as a native Grasshopper application, the tool consists of several components that can be combined or connected to suit the specific needs of a proposed project in any design phase or on any level of detail. Designed to be compatible with other applications in the Grasshopper ecosystem, RhinoCircular's circularity evaluations can be combined with structural simulation tools like Kangaroo3D or environmental systems simulation tools such as ClimateStudio. While the learning curve for the tool

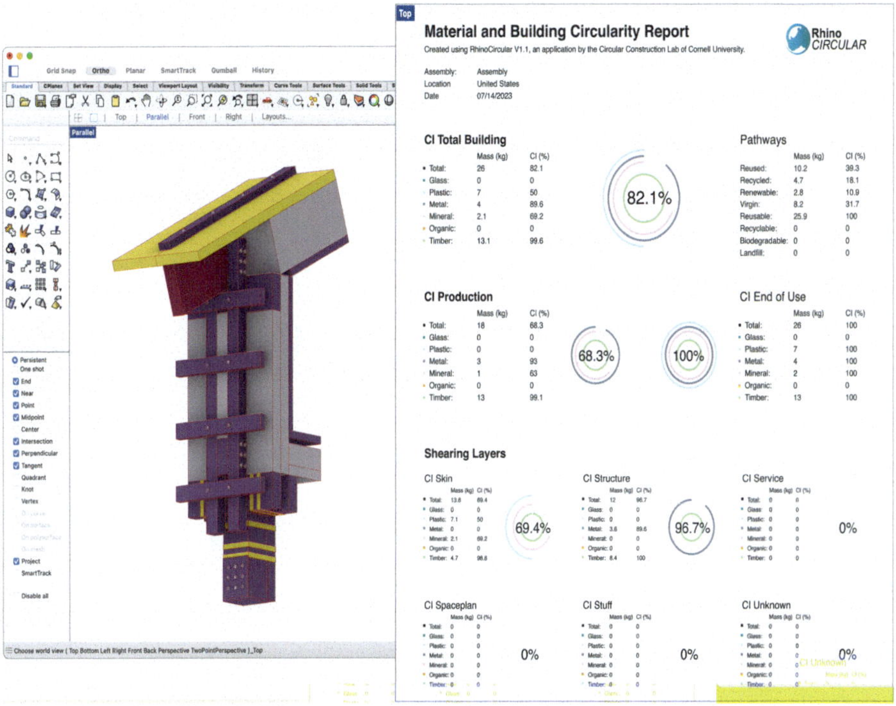

Fig. 6.1 Evaluating a detail with RhinoCircular

is relatively gentle, it assumes users have basic skills and an understanding of visual scripting in the Grasshopper ecosystem (Fig. 6.2).

Relative to the prior discussed tools, RhinoCircular generates lower-resolution outputs that represent a close approximation of a building's CI. This is because the goal of the tool is not a comprehensive MP or LCA but instead to inform designers quickly and immediately in an early-stage design where data resolution and product specification is equally lower. To support this mission, computation results are displayed directly in the modelling space and can be mapped onto the geometry, offering visual and targeted feedback to designers in the effort of informing the decision-making process.

6.5 Discussion

Assessing the relative strengths of the above MP and CI tools provides insight into where architects and other professionals can most effectively use them. When assessing the utility of each tool in various scenarios, those geared towards the earlier design phase better help to 'narrow' our present material consumption. Those which require more detail while also narrowing material consumption and emissions are more effective than earlier stage tools at closing the loop and enabling material salvage.

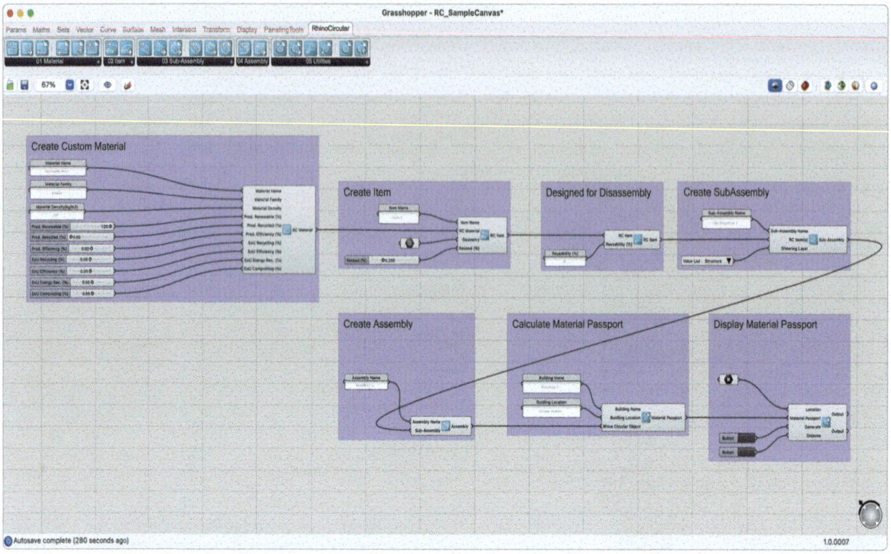

Fig. 6.2 Evaluating a structure with RhinoCircular

The software tool RhinoCircular is most relevant within early design phases when the design has yet to be finalised and rapid feedback is needed. After those phases, tools such as Madaster and One Click LCA generate detailed information on circularity and embodied carbon that can be delivered to stakeholders and compliance agencies. MPs that are produced by One Click LCA and Madaster are also relevant through the building's use and end-of-use, allowing for the recovery and reuse of building elements in the future. Madaster's platform is particularly useful when pricing materials for resale through its residual value metrics based on global material markets, which gives the user a sense of the resale value that can be realised when a building is deconstructed. Ideally, all these tools can be combined in a workflow that can leverage the benefits of CIs as a metric across scales.

Independent of the tool, circularity and CI evaluation must become a key design parameter across scales. The earlier architects, engineers, and designers can advocate and implement circularity into buildings, the greater the future impact, both in material sourcing and with respect to their end-of-use pathways. CI tools encourage circular behaviour, but it is the role of practitioners to apply and implement the feedback and optimise buildings for circularity.

As a result, greater collaboration is needed between practitioners in both sharing datasets which enable circularity assessments and their compatibility and in strengthening the accuracy and reach of computational tools for circularity. The opaqueness of supply chains and manufacturing processes inhibits professionals from having the data needed to confidently assess circularity in the built environment. These tools are only as accurate as the data which feeds them, and therefore data on materials needs to be collected and shared across all four phases (slow, close, narrow, regenerate) in order to give practitioners greater confidence in computational tools for a circular economy and their outputs.

6.6 Key Takeaways

- Tools which calculate circularity indicators have the greatest impact in the early design phase.
- Tools which calculate circularity indicators are most accurate in the construction documentation phase.
- Material passports have the greatest relevance during a building's use and end-of-use.
- More robust databases and data transparency are needed.

Acknowledgements The authors would like to thank Cornell University, the College of Architecture, Art, and Planning, and the Cornell Department of Architecture for their generous support. The development of RhinoCircular has been supported by the Special Prize for Innovation of the 2021 Zumtobel Group Award.

References

BAMB (2020) Buildings as material banks. https://www.bamb2020.eu/. Accessed 22 Feb 2023

Brand S (1995) How buildings learn: what happens after they're built, Reprint edn. Penguin Books, New York

Campanella D (2022) Concular: the digitalisation of materials in buildings. In: Heisel F, Hebel DE (eds) Building better – less – different: circular construction and circular economy. Birkhäuser, Boston, pp 114–117

Circular Construction Lab (2023) RhinoCircular. Food for Rhino. https://www.food4rhino.com/en/app/rhinocircular. Accessed 16 May 2023

Concular (2023) Concular – zirkuläres bauen. https://concular.de/. Accessed 22 Feb 2023

DGNB (2023) DGNB System https://www.dgnb-system.de/en/system/international/denmark/index.php. Accessed 22 Feb 2023

Ellen MacArthur Foundation and Granta Design (2019) Circularity indicators: an approach to measuring circularity. https://ellenmacarthurfoundation.org/material-circularity-indicator/. Accessed 16 May 2023

EPEA and Madaster (2023) Simplify measuring circularity within the built environment https://madaster.com/inspiration/epea-and-madaster-simplify-measuring-circularity-within-the-built-environment/. Accessed 22 Feb 2023

Heisel F, Hebel DE (2021) Urban Mining und kreislaufgerechtes Bauen. Fraunhofer IRB Verlag, Stuttgart

Heisel F, Nelson C (2020) RhinoCircular: development and testing of a circularity indicator tool for application in early design phases and architectural education. In: AIA/ACSA intersections res conference: CARBON, pp 154–159. https://doi.org/10.35483/ACSA.AIA.FallInterCarbon.20.24

Heisel F, Rau-Oberhuber S (2020) Calculation and evaluation of circularity indicators for the built environment using the case studies of UMAR and Madaster. J Clean Prod 243:118482. https://doi.org/10.1016/j.jclepro.2019.118482

Heisel F, Hebel DE, Webster K (2022a) Building better – less – different: circular construction and circular economy. Birkhäuser, Boston

Heisel F, McGranahan J, Ferdinando J, Dogan T (2022b) High-resolution combined building stock and building energy modeling to evaluate whole-life carbon emissions and saving potentials at the building and urban scale. Resour Conserv & Recycl 177:106000. https://doi.org/10.1016/j.resconrec.2021.106000

Heisel F, Lin A, Singh S, McGranahan J (2023) RhinoCircular material circularity database V1.0. https://doi.org/10.13140/RG.2.2.14063.92326/2

Lawson B (2005) How designers think: the design process demystified, 4th edn. Architectural Press, New York

Madaster (2023) The cadastre for materials and products. https://madaster.com/. Accessed 22 Feb 2023

May J, Latour B (2019) Signal. Image. Architecture. Columbia Books on Architecture and the City, New York

One Click LCA (2021a) COP26 Rhino & One Click LCA. https://www.oneclicklca.com/cop26/rhino/. Accessed 22 Feb 2023

One Click LCA (2021b) How to optimize embodied carbon in early design phases with One Click LCA and Rhino. https://vimeo.com/637805879/38758797fe/. Accessed 22 Feb 2022

One Click LCA (2023) The world's fastest building life cycle assessment software https://www.oneclicklca.com/. Accessed 22 Feb 2023

Root T (2021) This U.S. city just voted to decarbonize every single building. Washington Post

Rosen A (2022) The economy of urban mining: the Korbach City hall model project. In: Heisel F, Hebel DE (eds) Building better – less – different: circular construction and circular economy. Birkhäuser, Boston, pp 80–91

van Capelleveen G, Vegter D, Olthaar M, van Hillegersberg J (2023) The anatomy of a passport for the circular economy: a conceptual definition, vision and structured literature review. Resour Conserv Recycl Adv 17:200131. https://doi.org/10.1016/j.rcradv.2023.200131

WBCSD World Business Council for Sustainable Development (2023) The EU digital product passport. https://www.wbcsd.org/uxdqq. Accessed 22 Feb 2023

Felix Heisel is an assistant professor of architecture at Cornell University and the director of the Circular Construction Lab (CCL), a cross-disciplinary think-tank aiming to advance the paradigm shift from linear material consumption towards a circular industrialised construction industry. Heisel is also co-founder and partner of 2hs Architekten und Ingenieur PartGmbB in Germany, an office specialised in the development of circular prototypologies.

Joseph McGranahan is a research associate in the Circular Construction Lab at Cornell University. His research focuses on developing and utilising digital tools to assess material content, embodied carbon, and circularity within the built environment, ranging from construction assemblies to entire cities.

Chapter 7
Additive Manufacturing for the Circular Built Environment: Towards Circular Construction with Earth-Based Materials

Kunaljit Chadha, Alexandre Dubor, Edouard Cabay, Yara Tayoun, Lapo Naldoni, and Massimo Moretti

Abstract By making rapid prototyping accessible and inexpensive, additive manufacturing (AM) has transformed the fabrication industry. The adaptability of the process to various materials makes it applicable to multiple fields ranging from complex nanoscale production in the medical field to the manufacturing of large-scale structures in the construction industry. AM methods are constantly evolving, enabling the production of complex products with minimal initial investment. AM processes generate little waste and require no formwork, making them relevant to the construction industry, which conventionally produces significant amounts of waste.

This chapter provides a high-level overview of AM as an innovative technique and key developments towards its use for a circular built environment. It further delineates the viability of AM techniques using earth-based materials for implementing a circular economy in the construction sector through a series of case studies developed gradually from the scale of architectural prototypes to realised buildings. These examples address factors such as fabrication processes, techniques, and materials used and their influence on circularity through the production cycle of construction achieved using AM. Through the case studies, the chapter promotes 'closing the loop' on resources by reusing and recycling excavated construction materials. The chapter concludes with projections for AM practices and potential commercial applications of the technology. Overall, the chapter is useful for anybody interested in the built environment looking at alternative and sustainable building methods, including users, researchers, and professionals.

Keywords Excavated materials · Circular earthen construction · Additive manufacturing · On-site automation · Sustainable architecture

---

K. Chadha (✉)
ETH Zurich, Zurich, Switzerland
e-mail: chadha@arch.ethz.ch

A. Dubor · E. Cabay · Y. Tayoun
Institute for Advanced Architecture of Catalonia (IAAC), Barcelona, Spain

L. Naldoni · M. Moretti
WASP S.R.L, Massa Lombarda, RA, Italy

C. De Wolf et al. (eds.), *A Circular Built Environment in the Digital Age*, Circular Economy and Sustainability, https://doi.org/10.1007/978-3-031-39675-5_7

7.1 Introduction to Additive Manufacturing

To understand additive manufacturing (AM), it is imperative to know the context of digital fabrication within which the technology was initially developed. Digital fabrication (dfab) is a manufacturing workflow that employs computer-controlled machinery and tools to materialise objects from digital designs. Dfab is classified within the context of the third industrial revolution. The first revolution focused on mechanising manufacturing processes, while the second aimed at the mass production of parts. The third revolution centred on using digital technology, such as electronics, microprocessors, and the Internet, to change the way of working, communicating, and accessing information. It laid the groundwork for the fourth industrial revolution, which focuses on integrating physical and digital information using robotics, sensors, and artificial intelligence (Groumpos 2021).

Dfab has contributed to transforming the nature of working processes and proposed new solutions. Digital fabrication links digital technologies such as computer-aided design (CAD), computer-numerical-control (CNC) machines, and robotics, which are all part of the broader digital revolution. Dfab mainly covers three fabrication processes:

- Additive manufacturing (AM) is a computer-controlled technology for making components by depositing subsequent layers of material to form a three-dimensional object.
- Subtractive manufacturing (SM) is a process where the material is removed from a solid block or stock of material using various tools such as drills, milling machines, wire cutters, lathes, or routers to create desired shapes.
- Formative manufacturing (FM) is a range of techniques that involve the mechanical deformation, bending, forging, or shaping of a given material, with or without the use of a mould.

Due to the versatility of its fabrication process, cost-effectiveness, and accessibility, AM appears to be the preferred dfab technique for mass customisation. This process has been explored with various materials such as plastics (Wei Keat and Chow 2022), metals (Huang et al. 2023), ceramics (Chen et al. 2018), composites (Korkees et al. 2020), and biomaterials (Malik et al. 2020), to name a few.

Within the wide range of AM processes, seven subprocesses fall under the AM umbrella: binder jetting, directed energy, deposition, material extrusion, material jetting, powder bed fusion, sheet lamination, and vat photopolymerisation (Slotwinski 2014). Each AM process offers the flexibility to fabricate a wide range of complex shapes and hence was soon adopted by the industry for rapid prototyping depending on the scale and resolution of its application. Even though 3D printing (3DP) and AM are defined as the same fabrication technique (Ngo et al. 2018), a deeper understanding of the process and its parameters indicates that 3DP is a subset technique of AM processes: in contrast to 3DP, which builds three-dimensional objects by adding material in successive layers, AM creates three-dimensional objects by adding material, which may or may not be produced with consecutive layers (McCormack et al. 2020; Ming et al. 2022).

Schematically, AM utilises a computer and 3D printer to produce custom physical objects. CAD software generates 3D digital objects, and computer-aided manufacturing (CAM) or slicer produces slices of a 3D geometry, resulting in a geometric code (G-code). The G-code provides positional data, velocity, and extrusion rate values for the printer nozzle, which is moved by a motor-driven CNC system following instructions. This streamlined software and hardware infrastructure enables quick, low-cost, and highly customisable production of bespoke physical objects using AM. Several sectors, including robotics, medicine, food science, architecture, and others, have extensively used AM (Shahrubudin et al. 2019). AM has significant potential in the fields of surgery, disease modelling, organ printing, veterinary medicine, and tissue engineering (Bozkurt and Karayel 2021).

AM processes are not restricted to a particular machine configuration and can be customised to suit generic tools such as industrial robots (Pham et al. 2016). This offers advantages in terms of application scalability, operational efficiency and accuracy, versatility in executing diverse functions, and agility for multitasking. The adoption of AM process methodologies across various machine configurations has facilitated the expansion of the technology and its implementation within the built environment.

7.2 Additive Manufacturing in the Built Environment

AM processes have found their way into construction, automating dull, dirty, and dangerous site operations (Jud et al. 2021). AM methods have formed a mass-customisable production system, with 3DP as the preferred option due to its waste-free and formwork-free nature. Within the wide range of AM processes, two specific processes have been explored at the architectural building scale:

- *Material extrusion* is a fast AM method in which continuous layers of materials are deposited one on top of the other while the material is in a plastic state. The adhesive characteristics of the material and gravity determine the interlayer bonding between the layers to form a monolithic structure. The Contour Crafting technique (Khoshnevis et al. 2006) pioneered technologies to introduce material extrusion to the building industry. By linking trowelling and extrusion processes, Contour Crafting improved the surface quality faster in the built volume.
- *Binder jetting* is a high-resolution AM method in which a printer selectively deposits a liquid binder onto a bed of powder particles to build fine-resolution objects. In 2006, D-Shape (Gardiner and Burry 2010) technology presented the first demonstration of their machine to create high-resolution architectural scale structures. D-Shape introduced a new manufacturing stream of material-efficient, gravity-independent, high-resolution objects that could apply to the built environment.

While both processes have allowed for the use of a wide range of possibilities in terms of print resolution and material usage, the construction sector considers material extrusion a feasible alternative owing to the reduced number of peripherals needed for equipment installation, the faster building rate, and the scalability of the process (Puzatova et al. 2022). Another aspect influencing their choice is the ability to print directly on the construction site. In this context, there are three important criteria for adopting AM in the building industry: (a) building material, (b) machine configuration, and (c) computational design methods. The validation of these factors is made possible by the expertise of professionals and experts in the field, allowing for the effective deployment of different large-scale AM applications.

7.2.1 Materials

With the surge in the availability of large-scale construction format AM machines such as BOD II (COBOD International A/S 2017) and Crane WASP (WASP Srl 2018), the building sector was able to investigate various materials on a construction scale, including plastics, metals, and plaster. Nevertheless, material durability, size restrictions, and a slow production rate have limited these material systems to smaller building components, which has led to cementitious materials being the material of choice for structural building elements. After all, concrete is one of the most widely used building materials with superior structural properties, availability, and affordability (Crow 2008). Even though concrete processing is being improved and more automated, conventional construction activities that use concrete still generate a lot of waste and have high energy consumption.

In this context, 3D concrete printing (3DCP) has been the first AM technology to enter the construction industry, with the promise of an effective, customisable, and waste-free form of construction. Rapid growth in using 3D printers for building has highlighted the need to develop new material control systems, especially those that allow precise control over the material's hydration, rheology, and curation rate, which is critical for achieving volumetric buildup (Jones et al. 2018).

However, the consumption volume of concrete and the chemicals added to accelerate the mix to facilitate 3DP make the process less structurally capable while possibly being even more harmful to the environment per volume (Flatt and Wangler 2022), exposing the need for alternative sustainable materials for 3DP such as excavated earth and geopolymers. In particular, earth-based material offers a significant advantage in terms of transportation and sustainability, as it can be extracted and processed directly on site. It is known for allowing the construction of sustainable, healthy, and thermally efficient buildings (Minke 2013). It is also a material linked to old construction techniques requiring extensive skills and manual labour, issues that could be solved with 3D printing machines.

7.2.2 Machine Configurations for Additive Manufacturing

In addition to material control, the successful implementation of additive manufacturing (AM) operations at the building scale relies on the effective integration and accessibility of material processing machines and fabrication machine configurations. The choice of machine setups for AM in construction is contingent upon factors such as size formats and mobility. Consequently, various machine and robotic configurations have been employed in this context. Broadly, these configurations can be classified into two groups: off-site and on-site manufacturing setups.

Off-site manufacturing setups entail the construction of components within a controlled factory environment, followed by transporting prefabricated customised parts to the construction site for assembly, ultimately forming a complete building structure. Using off-site manufacturing facilities ensures regulated conditions that shield production machinery from ambient fluctuations such as temperature and humidity, thereby enabling the mass production of high-quality products. To achieve this, rigid frame Cartesian-type machines (Khoshnevis et al. 2006) are commonly employed, offering three or five degrees of freedom depending on the specific application requirements. When more intricate fabrication operations are necessary, setups incorporating a robotic arm mounted on a Cartesian gantry (Anton et al. 2020) are being implemented. This configuration allows for the gantry's robust manipulation capabilities and the robotic arm's dexterity and precision, thus accommodating large-scale construction while maintaining high-resolution detailing.

On-site manufacturing setups vary in configurations, ranging from fixed machines with predetermined footprints to autonomous setups capable of movement and localisation within the construction site. Agility and precision are crucial for these setups to respond and adapt to the dynamic site conditions. Besides large-scale Cartesian 3-axis machines, researchers are exploring using robots on mobile platforms like the In Situ Fabricator (Giftthaler et al. 2017) and digitally controlled construction machinery (Jud et al. 2021) for construction operations.

Such machines have demonstrated applicability ranging from component-based architectural structures to full-scale in situ structures. While off-site manufacturing setups require additional peripherals, it allows for the fabrication of building components in a controlled environment (Gomaa et al. 2023). Thus, it avoids delays due to dynamic site conditions and widens the potential of testing the application of novel construction materials. On the other hand, on-site machinery reduces transportation overheads and produces larger objects, often directly in situ (Dubor et al. 2018). The role of this on-site machinery includes material sourcing, material processing, and building procedures. Additional machines might be required to process materials sourced from the site and surface finishing operations.

7.2.3 Computational Methods

The paradigm shifts in architectural design, in which architects use more digital tools, parametric modelling, scripting, etc., to produce geometries providing an approach in which design-generating parameters may be changed on the go using intelligent systems such as machine learning (Guo Liang and Yeong 2022). In AM processes, such a parametric computational design approach acts as a 'middleware' in the workflow between generated digital designs and the already manufactured sequence. This allows a control to adjust relevant process parameters. The role of computational design tools is critical since most AM processes are time sensitive and need application-specific information exchange between the parameters.

Like the approach of dfab, AM processes involve integrating design and manufacturing processes within a single digital environment to reduce the gap between design and fabrication. Such control over the process in AM on a building scale is beneficial when site conditions and material qualities vary. Because of the parametric control workflow, it is now possible to model and record previously unanticipated material and site conditions change to effectively adapt to construction errors. This allows for the emergence of novel, cost-effective, and fabrication-aware individualised design solutions.

7.2.4 *Summary*

AM has established a new construction domain that can be generated digitally and has also addressed the construction sector's problems of low productivity and waste generation. Yet the materials presently used for AM have severe environmental impacts owing to the material's high embodied carbon and one-time use (Faludi et al. 2015). The following section will address the inclusion of AM processes within a circular economy framework, highlighting the essential characteristics that render AM a feasible option for transitioning towards a circular built environment.

7.3 Additive Manufacturing for a Circular Economy

7.3.1 *Advantages*

AM enables product innovation through design freedom of mass customised and cost-effective components. It further presents unique features to support circular economy initiatives, such as waste-free production and the opportunity to test novel sustainable material systems promoting product durability and reuse. Consequently, AM has been adopted by various sectors to reduce environmental impacts. The overview from the preceding section about the diverse application of AM highlights its status as a technology driving the transition to a circular production system, which results in the reconfiguration of the supply chain, laying the groundwork for attaining a circular economy. The advantages of employing AM within a circular economy include the following (Hettiarachchi et al. 2022):

- Resource efficiency and minimal waste generation due to the selective deposition of the AM processes.
- Reduction of transportation-related environmental and economic impacts through the on-demand production of customised, locally produced products.

- Diverse material applications adopting the strategy of design for disassembly with prefab components help in easy and clean disassembly for reuse and local repair.
- Flexibility to use excavated and recycled materials, which reduces the environmental impact caused due to the embodied carbon of the materials.
- Individualised production to help in the renovation and restoration of buildings.
- Flexibility to adjust and optimise each component to its individual needs and situation.
- The use of novel standard and non-standard material in the AM process.
- Adaptability for making connections that adapt to the uniqueness of various material systems.

7.3.2 Additive Manufacturing in a Circular Built Environment

Resource efficiency is a key factor in AM stepping up to a building scale in the built environment context. The article presented 'An Emerging Framework for the Circular Digital Built Environment' (Çetin et al. 2021) identifies four distinct resource strategies to achieve the CE concept in the built environment: narrowing the loop, slowing the loop, closing the loop, and regenerating the loop. Building on this framework in the context of a circular built environment, AM showcases an impactful building solution: AM process allows for optimised material use for manufacturing, thereby 'narrowing the loop' by using less material to construct and generate minimal waste. The durability of the products achieved by the computational design workflow using CAD software helps 'slow the loop' by efficiently using material through geometric optimisation and targeted component repair, instigating a longer life cycle. Additionally, the materials used contribute significantly to the efficiency of the circularity process by 'closing the loop' by allowing the use of reclaimed and recycled materials. The freedom of using naturally sourced, nonconventional materials helps in the 'regenerating loop' of the product cycle by facilitating the disposal of materials upon the completion of the cycle.

The circularity of a building in AM processes is considerably affected by activities related to both the machine configuration and the materials employed, as they are closely dependent on the environmental and economic impact linked to their sourcing, manufacturing, and transportation. These operations include transporting resources for off-site manufacturing of stock materials or equipment transportation for on-site manufacturing setup. While prefabricated parts are often preferred in the construction industry due to quicker construction times, enhanced quality control, product durability, increased safety, and decreased waste, on-site manufacturing offers several advantages, including low environmental impact and high levels of customisation in construction.

7.3.3 Summary

Consolidating science-based information on building components' greenhouse gas emissions (GHG) and related activities across their full life cycle is an important feature of implementing climate mitigation methods. In the context of the built environment, it is particularly informative to look at the emissions resulting from processes preceding the occupancy of a building early into the sourcing of construction materials: 'Quantitatively, the phases of material manufacturing, transportation, and on-site construction were responsible for 94.89%, 1.08%, and 4.03% of energy consumption, respectively, and 95.16%, 1.76%, and 3.08% of global warming potential' (Hong et al. 2014). While both off-site and on-site systems rely on transporting material or equipment from another location, the use of a kilometre-zero (km-0) approach in construction would prove to be an effective instance of AM in a circular built environment, promoting social, environmental, and economic sustainability (Farias et al. 2017). The km-0 strategy first appeared in the slow food movement to promote the consumption of local ingredients, reducing the distance between producers and consumers (Souza Eduardo 2021). In the framework of a circular built environment, this would include construction using only locally available building resources and materials benefiting from natural and low-impact processing. The following section will focus on 3D printing with earth (3DPE) as an effective solution for AM in a circular built environment using excavated sustainable and recyclable material primarily consisting of raw earth.

7.4 Case Studies

Because materials used in AM (such as cement mortars, plastics, and metals) are environmentally damaging due to their embodied carbon content and supply chain, exploring alternative materials of a more circular nature is crucial. Raw earth is a readily available material and presents a traditional precedent use in the field of construction. Traditional building techniques using unprocessed earth have evolved through centuries of local knowledge. They need minimal energy for construction, but these solutions are not competitive due to their labour-intensive nature and slow building pace (Minke 2013). Alternatively, 3DPE has upgraded the conventional earth building methods of direct shaping and extruded earth by combining them with computer-controlled machines and improved safety and control over the construction process while keeping the environment and performance benefits of traditional earth construction. The process of 3DPE demonstrates a construction system capable of minimising greenhouse gas emissions (GHG) from construction components, achieved through the use of km-0 robotic AM.

7.4.1 *Introduction to 3D Printing with Earth*

3DPE methods use the layer-based AM approach, comparable to the Contour Crafting method using earth-based materials. In contrast to other additive manufacturing (AM) processes used in the construction industry, such as 3DCP, 3DPE stands out for its ability to avoid the use of environmentally harmful chemicals to speed up the material curing process. Instead, it combines water and aggregates to achieve the necessary level of malleability for 3D printing. In 3DPE, the walls of the construction components are connected using infill. This helps create a load-bearing volume with enough structural depth. These infills provide more practical structural features, such as incorporating electrical and plumbing services, a network of air cavities for natural ventilation, adding a filler material to augment heat lag, etc. (IAAC 2022). By adopting 3DPE, there is a significant reduction in operating energy and a more efficient resource consumption loop. This is achieved by closing the cycle, which minimises the energy consumption required to operate the building. As depicted in Fig. 7.1, a crucial feature of 3DPE is the seamless integration of on-site processes, starting from excavation to the final detailed finishes in the constructed building, utilising locally sourced materials.

3DPE presents an alternative building method with a circular design-to-construction life cycle consideration of the built environment, which counteracts the tendency of excessive energy consumption in building operations. Additionally, the

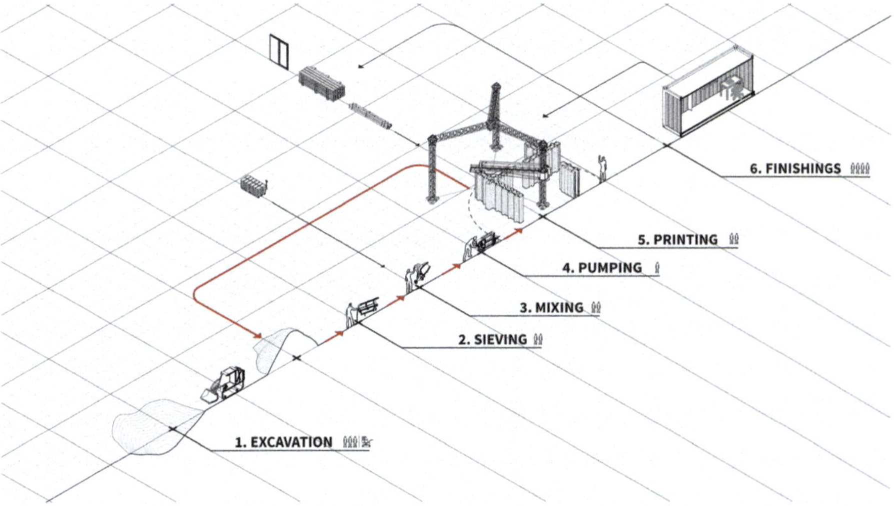

Fig. 7.1 Suggested construction scenario for on-site 3DPE excavating the material using Crane WASP. The illustration portrays the distinct phases of the supply chain, starting with material acquisition and processing, followed by the construction process, resulting in the final product of a constructed building. Highlighted in red indicates the stage at which the material can be recycled and reintroduced into production, forming a circular use of the excavated material

different aspects of a 3DPE construction process, plus the use of local labour and participation in local economies across its value chain and its accommodation of complex and innovative building models, showcase the circularity of 3DPE as a construction system that closes the loop of circularity in a building process. The following subsections present three case studies demonstrating large-scale implementation of 3DPE that indicate how the transition from an off-site to an on-site mode of a 3DPE building significantly affects the circularity of the building. In addition, the case studies also display the integration of wooden components with 3DPE in ways that add architectural functionality to the built structure.

7.4.2 Digital Adobe: Prefabricated Components Manufactured Off-Site Using Recyclable Materials

The first case study presents *Digital Adobe*, developed during the OTF 2017–2018 course at IAAC and built in the Valldaura Campus of IAAC (IAAC 2018). It is a 2-metre-wide and 5-metre-high printed clay wall with a varying thickness (0.7 m at its bottom and 0.2 m at its top) with a wooden slab resting on the wall at 2.6 m to simulate a clay/wood building unit, where the connections between two materials and the vertical load from a horizontal slab can be tested. Digital Adobe serves as the first large-scale exploration of combining 3DPE and wood elements where the 3D printed earthen component takes the compression load of the structure, and the wooden spanning element works in tension (Fig. 7.2).

The primary focus of the case study was to explore the climatic and structural performances of 3DPE. With the long-established understanding of clay's thermal properties to moderate heat transmission, the team has sought a design to enhance such properties. To limit temperature transfer from one side of the wall to the other and to improve the compressive strength of the wall, the infills were filled with unprocessed soil. A ventilated wall design reduced summer heat gain through convection between the openable top and bottom openings. It retained heat in the winter when both openings were closed.

The material mix consisted of conventional adobe mix, including clay, sand, silt, and aggregates. Vegetable enzymes are used to reach the grade of the fluidity of the material mixture needed to achieve the flow rate required to extrude the material for 3D printing. The prototype was partially built with recycled material from the preceding research of On-Site Robotics at IAAC in 2017. Around two tonnes of material from On-Site Robotics (Dubor et al. 2018) was recycled, making up almost half of the prototype's total material source. To make the recycled material usable for 3DPE again, it was crushed and rehydrated using a much-reduced number of enzymes. Finally, 'closing the loop' in resource management was proven by the recycling and reusing process.

Fig. 7.2 Digital Adobe (2018) is the outcome of research on 3DPE for a performative habitat. Design parameters in robotic construction enhance climatic and structural properties innate to the material. Thermal properties such as transmittance are regulated by robot precision through the geometry design of the global shape, surface texture, and ventilating cavities. (© Dongliang Ye)

7.4.3 TECLA: On-Site Construction Using Excavated Materials

TECLA (WASP 2022) is an innovative circular house unit built in Massa Lombarda by WASP and Mario Cucinella Architects (MCA), integrating research on vernacular building techniques with natural and regional materials. TECLA was constructed using two synchronised printer arms concurrently, utilising industrial automation protocols to optimise mobility, avoid collisions, and ensure efficient operation. Each printer unit has a printing surface of 50 square metres, allowing for the rapid construction of house modules. TECLA has a floor size of 60 square metres; it comprises a living zone with a kitchen and a night zone with services. The structure is a composition of two continuous elements that, through a sinuous and uninterrupted sine curve, culminate in two circular skylights that produce zenithal lighting (Fig. 7.3).

In addition, the composition of the earth mixture responds to local climatic conditions, and the filling of the envelope is parametrically optimised to balance thermal mass, insulation, and ventilation according to the climate needs. The materials used were local soil of 6 mm maximum aggregate size, sand, rice husk, and Mapesoil, a lime-based binder added at 5% by weight of the batch.

The proposal was centred on environmental variables, particularly solar analysis, which was the design driver behind the undulated surface and increased the total surface area of the outer facade. Using computational tools to create climate-responsive shapes to improve raw earth's physical qualities ensures increased passive energy performance of built structures.

Fig. 7.3 The sustainable, innovative TECLA model (2021) of on-site housing construction uses materials sourced from the construction site and constructed using the modular Crane WASP machine. (© WASP)

7.4.4 TOVA: On-Site Construction with Excavated and Recyclable Materials

TOVA is a building prototype (IAAC 2022) demonstrating the potential of 3DP with sustainable materials in response to increasing climate challenges and related housing emergencies. It was built in the Valldaura Labs facility in Collserola Park, on the outskirts of Barcelona. The construction spans 7 weeks and uses a Crane WASP modular printer and km-0 materials. Using local materials sourced within a 50-metre radius reduces the environmental impact of transportation and waste generation during construction. TOVA can be studied as a near-zero emissions project: the design is tested via digital and physical simulations to reduce carbon footprint, considering the life cycle assessment of the building components. The circular design approach is aimed at designing an environmentally responsive building constructed from reusable biomaterials across the construction phases as follows: a geopolymer foundation, a framework made of local earth, mixed with additives and enzymes to ensure the structural integrity and material elasticity necessary for the optimised 3DP of the house, a locally sourced timber roof structure, and wooden carpentry. To improve the material's longevity and weather resistance, a waterproof coating is added using raw extracted materials such as egg whites.

The building design of TOVA is based on a precise site condition of the Mediterranean: the volume is compact to protect from the cold in winter, yet expandable for the other three seasons. For this purpose, the wall section, composed of six earth surfaces and a network of cavities containing air or insulation, was calculated to prevent winter heat loss while protecting from summer solar radiation. The result is a climate-responsive building: the design considers digital and physical simulations to reduce construction footprints, monitor the reduction of greenhouse

Fig. 7.4 TOVA (2022) is a completed circular building prototype using locally sourced materials constructed with 3DP processes achieved by a layer-based deposition of earth material mixture and a timber wood structure and a network of cavities in the wall that participates in the climate-responsiveness of the building. (Photographs by Gregori Civera)

gas emissions, and consider the life cycle assessment of the building components. It also demonstrates the valuable knowledge of traditional material craftsmanship in informing a technology-driven association for establishing circular constructions in the built environment (Fig. 7.4).

The implementation of on-site printing techniques and the use of natural materials in 3DPE guarantee the circularity of the construction process. As no chemical modification is needed to recycle the structure at the end of its life, it effectively prevents residual waste and pollution. The printed earth layers are returned to the source of the material, completing a full loop of circularity in the construction cycle.

7.5 Discussion

Three critical developments have allowed for the widespread use of AM in the construction sector:

- *Accessibility to machines*: The advancement of lightweight and modular construction machinery, such as the Crane WASP, has led to the widespread adoption of on-site construction services utilising materials from companies like Icon3D (USA), Cobod (Denmark), Tvasta (India), and WASP (Italy). These innovations have significantly expanded their presence and usage on a global scale.
- *Material processing techniques*: In the specific case of 3DCP, rheological control of material processing machines was a key breakthrough for ensuring the 'set on demand' behaviour of the material to enable structural buildup during printing. In upcoming years, these concepts could be extended to more sustainable processes

such as 3DPE to increase build rate and construction efficiency to make the earth construction market competitive.
- *Training design professionals*: The emergence and constant evolution of new AM processes for construction require special skill sets for operators to use such technologies efficiently. From an academic standpoint, teaching and preparing the next generation of professionals is crucial for effectively managing these complex technological environments. It is crucial that the designs coming out of such a process are optimal for the technology in terms of material and structural efficiency.

The broad implementation of AM globally will be aided by the availability and accessibility of digitally controlled machines for processing material and fabricating. With the rising concerns over the sustainability of the construction industry, the focus will be on AM technologies that use sustainable materials. With a surge of technologies such as 3DPE, the applications might expand in extreme scenarios to form a sustainable, on-site, waste-free construction process. 3DPE has the potential for various uses, from emergency shelters in rural settings with plentiful local resources but limited masonry skills to commercial residences with climate-responsive designs that solve severe climatic challenges.

The complexity of AM processes stems from the interdependency of machine, material, and design characteristics. The presented projects pose limitations regarding durability and efficiency, which imply two future developments needed for research: (1) on a construction site where materials are susceptible to changes in ambient temperature, machine downtime is a significant problem that prevents the technology from being used to its total capacity and thus makes it unaffordable, and (2) in addition to dynamic building site circumstances, material variations throughout construction make it challenging to forecast the precision and effectiveness of the technology. Multi-staged diagnostics, including feedback on the integrity of the material, deformation of the structure, and maintenance areas, will help improve construction quality.

7.6 Key Takeaways

- The introduction of additive manufacturing (AM) as a unique construction method has redefined efficient processes and, when used with sustainable materials, has the potential to reduce the building sector's environmental impact.
- Machine availability and AM professional training are necessary for enabling sustainable on-site construction using sustainable AM materials.
- Using AM and integrating other building components can greatly enhance the design potential for climate-responsive building construction.
- Promoting socially and environmentally sustainable AM processes could lead to a new building system that involves closing the circularity loop via utilising local resources for construction that can be recycled and reused.

- Using local materials and resources allows for a fully inclusive construction process, supporting and boosting local economies by involving various local stakeholders in the value chain. This could lead to the rapid adoption of the AM construction system in countries most affected by climate-related housing emergencies.

Acknowledgements The authors would like to thank the architects, faculty, students, and collaborators involved in each project.

Digital Adobe is a project of the Institute for Advanced Architecture of Catalonia (IaaC), developed in the postgraduate programme Open Thesis Fabrication in 2017/2018 by A. Dubor, E. Cabay, M. Marengo, K. Chadha, S. Moreno (faculty), Y. Chang, D. Fiore, F. Sevostianov, G. Stirum, Q. Li, S. Riaz, D. Ye (students), Windmill, La Salle, Nanosystems, ArtCon, and SmartCitizen (collaborators).

Tecla House is a project developed between 2019 and 2021 by WASP (M. Moretti, A. Chiusoli, F. De Fabritiis, M. Visonà, L. Naldoni) and MCA (M. Cucinella, I. Giglio, A. Barichello, L. Rendace, L. Porcelli, L. Zillante, S. Rosso), in collaboration with SOS–School of Sustainability, Capoferri Serramenti, Cefla, Frassinago, Imola Legno, Lucifero's, Mapei, Milan Ingegneria, Tamborrino, Orange Fiber, Primat, RiceHouse, and Ter Costruzioni.

Tova is a project of IaaC developed in the postgraduate programme in 3D Printing Architecture in 2021/2022 by E. Cabay, A. Dubor, L. Tayefi, V. Huyghe, A. Foroughi, E. Chamorro Martin, E. Carnevale, G. Baraut, G. Font Basté, N. Kirova, F. Polvi, B. Ganem Coutinho, M. Papandreou, and D. Skaroupka (faculty), A. Alatassi, A. Taskin, C. Musyoki, D. El-Mahdy, E. Marais, H. Benz, J. Rodriguez Torres, L. Bin, M. A. Al-Hachami, M. Abdelrahim, M. Harrak, M. Bezik, M. A. Isoldi Campinho, M. Laalou, M. Sarnaghi, N. Saksouk, O. Pavlidis and S. Boni Dara (students), and Colette, WASP, UN Habitat, BAC Engineering, LaSalle, SmartCitizen, Scuares, and Living Prototypes Research Innovation (collaborators). The project has received funding from the German government under the ZukunftBau research innovation programme.

Declaration of Competing Interests The authors have participated in developing the case studies presented in this chapter. They are part of institutions that believe in the potential of additive manufacturing with earth for a circular built environment. The authors certify that they have no further affiliations with or involvement in any organisation or entity with any financial or non-financial interest in the subject discussed in this manuscript.

References

Anton A, Reiter L, Wangler T, Frangez V, Flatt R, Dillenburger B (2020) A 3D concrete printing prefabrication platform for bespoke columns. Autom Constr 122:103467. ISSN: 0926-5805. https://doi.org/10.1016/j.autcon.2020.103467

Bozkurt Y, Karayel E (2021) 3D printing technology: methods, biomedical applications, future opportunities and trends. J Mater Res Technol 14:1430–1450. https://doi.org/10.1016/j.jmrt.2021.07.050

Çetin S, De Wolf C, Bocken N (2021) Circular digital built environment: an emerging framework. Sustainability 13(11):6348. https://doi.org/10.3390/su13116348

Chen Z, Li Z, Li J, Liu C, Liu C, Li Y, Wang P, Yi H, Lao C, Yuelong F (2018) 3D printing of ceramics: a review. J Eur Ceram Soc 39. https://doi.org/10.1016/j.jeurceramsoc.2018.11.013

COBOD (2017) International A/S. https://cobod.com/. Accessed 19 Apr 2023

Crow JW (2008) The concrete conundrum. Chem World:62–66. https://doi.org/10.1201/b12851-5
Dubor A, Izard J-B, Cabay E, Sollazzo A, Markopoulou A, Rodriguez M (2018) On-site robotics for sustainable construction. In: Willmann J, Block P, Hutter M, Byrne K, Schork T (eds) Robotic fabrication in architecture, art and design 2018. ROBARCH. Springer, Cham, pp 390–401. https://doi.org/10.1007/978-3-319-92294-2_30
Faludi J, Hu Z, Alrashed S, Braunholz C, Kaul S, Kassaye L (2015) Does material choice drive sustainability of 3D printing? Int J Mech Aerosp Ind Mechatron Eng 9:216–223. https://doi.org/10.5281/zenodo.1098116
Farias RD, García CM, Palomino TC, Andreola F, Lancellotti I, Barbieri L (2017) Valorization of agro-industrial wastes in lightweight aggregates for agronomic use: a preliminary study. Environ Eng Manage J 16(8):1691–1699. https://doi.org/10.30638/eemj.2017.184
Flatt RJ, Wangler T (2022) On sustainability and digital fabrication with concrete. Cem Concr Res 158:106837. https://doi.org/10.1016/j.cemconres.2022.106837
Gardiner J, Burry M (2010) D_Shape – construction scale additive manufacturing reflection on current projects and the state of technology development. In: Proceedings of the 6th international conference on innovation in science and technology. ISBN: 978-609-8239-70-6 Pub
Giftthaler M, Sandy T, Dörfler K, Brooks I, Buckingham M, Rey G, Kohler M, Gramazio F, Buchli J (2017) Mobile robotic fabrication at 1:1 scale: the in situ fabricator: system, experiences and current developments. Constr Robot 2017(1):3–14. https://doi.org/10.1007/s41693-017-0003-5
Gomaa M, Schade S, Bao DW, Xie YM (2023) Automation in rammed earth construction for industry 4.0: precedent work, current progress and future prospect. J Clean Prod 398:136569. https://doi.org/10.1016/j.jclepro.2023.136569
Groumpos PP (2021) A critical historical and scientific overview of all industrial revolutions. IFAC-Pap 54(13):464–471. https://doi.org/10.1016/j.ifacol.2021.10.492
Guo Liang G, Yeong WY (2022) Applications of machine learning in 3D printing. Materials Today Proc 70:95–100. https://doi.org/10.1016/j.matpr.2022.08.551
Hettiarachchi BD, Sudusinghe JI, Seuring S, Brandenburg M (2022) Challenges and opportunities for implementing additive manufacturing supply chains in circular economy. IFAC-Pap 55(10): 1153–1158. https://doi.org/10.1016/j.ifacol.2022.09.545
Hong T, Ji C, Jang MH (2014) Assessment model for energy consumption and greenhouse gas emissions during building construction. J Manag Eng 30:226–235. https://doi.org/10.1061/(ASCE)ME.1943-5479.0000199
Huang L, Chen X, Konovalov S, Su C, Fan P, Wang Y, Xiaoming P, Panchenko I (2023) A review of challenges for wire and arc additive manufacturing (WAAM). Trans Indian Inst Metals 76: 1123–1139. https://doi.org/10.1007/s12666-022-02823-y
IAAC (2018) Digital Adobe. https://iaac.net/project/digital-adobe/. Accessed 12 May 2023
IAAC (2022) TOVA. https://iaac.net/project/3dpa-prototype-2022/. Accessed 12 May 2023
Jones S, Bentz D, Martys N, George W, Thomas A (2018) Rheological control of 3D printable cement paste and mortars. In: Wangler T, Flatt R (eds) First RILEM international conference on concrete and digital fabrication – digital concrete 2018. DC 2018, RILEM Bookseries, vol 19. Springer, Cham, pp 70–80. https://doi.org/10.1007/978-3-319-99519-9_7
Jud D, Kerscher S, Wermelinger M, Jelavic E, Egli P, Leemann P, Hottiger G, Hutter M (2021) HEAP – the autonomous walking excavator. Autom Constr 129:103783. https://doi.org/10.1016/j.autcon.2021.103783
Khoshnevis B, Hwang D, Yao K-T, Yeh Z (2006) Mega-scale fabrication by contour crafting. Int J Ind Syst Eng 1(3):301–320. https://doi.org/10.1504/IJISE.2006.009791
Korkees F, Allenby J, Dorrington P (2020) 3D printing of composites: design parameters and flexural performance. Rapid Prototype J 26(4):699–706. https://doi.org/10.1108/RPJ-07-2019-0188
Malik S, Hagopian J, Mohite S, Lintong C, Stoffels L, Giannakopoulos S, Beckett R, Leung C, Ruiz J, Cruz M, Parker B (2020) Robotic extrusion of algae-laden hydrogels for large-scale applications (Global challenges 1/2020). https://doi.org/10.1002/gch2.202070011

McCormack A, Highley C, Leslie N, Melchels F (2020) 3D printing in suspension baths: keeping the promises of bioprinting afloat. Trends Biotechnol 38:584–593. https://doi.org/10.1016/j.tibtech.2019.12.020
Ming C, Mirjan A, Medina J, Gramazio F, Kohler M (2022) Impact printing. 3D Print Addit Manuf 9(3):203–211. https://doi.org/10.1089/3dp.2021.0068
Minke G (2013) Building with Earth: design and technology of a sustainable architecture. ISBN: 9783034612623. https://doi.org/10.1515/9783034612623
Ngo T, Kashani A, Imbalzano G, Nguyen KTQ, Hui D (2018) Additive manufacturing (3D printing): a review of materials, methods, applications and challenges. Compos Part B 143:172–196. ISSN 1359-8368. https://doi.org/10.1016/j.compositesb.2018.02.012
Pham H, Hui L, Pham QC (2016) Robotic 3D-printing for building and Construction. ISSN:2424-8967, 300–305
Puzatova A, Shakor P, Laghi V, Dmitrieva M (2022) Large-scale 3D printing for construction application by means of robotic arm and gantry 3D printer: a review. Buildings 12(11):2023. https://doi.org/10.3390/buildings12112023
Shahrubudin N, Lee TC, Ramlan R (2019) An overview on 3D printing technology: technological, materials, and applications. Procedia Manuf 35:1286–1296. https://doi.org/10.1016/j.promfg.2019.06.089
Slotwinski JA (2014) Additive manufacturing: overview and NDE challenges. AIP Conf Proc 1581(1):1173–1177. https://doi.org/10.1063/1.4864953
Souza Eduardo (2021) Zero kilometer materials: preserving the environment and local cultures. https://www.archdaily.com/958893/zero-kilometer-materials-preserving-the-environment-and-local-cultures Accessed 15 May 2023
WASP Srl (2018) Crane WASP. https://www.3dwasp.com/stampante-3d-per-case-crane-wasp/. Accessed 19 Apr 2023
WASP Srl (2022) TECLA. https://www.https://www.3dwasp.com/en/3d-printed-house-tecla/. Accessed 12 May 2023
Wei Keat N, Chow W (2022) Plastics in 3D printing. In: Encyclopedia of materials: plastics and polymers, vol 4, pp 82–91. https://doi.org/10.1016/B978-0-12-820352-1.00065-1

Kunaljit Chadha is a doctoral candidate at Gramazio Kohler Research at ETH Zurich, working on 'Impact Printed Structures: Adaptive strategies for full scale on-site robotic earth construction'. His doctoral research focuses on developing computational design systems and construction strategies for upscaling the 'Impact Printing' process to form an efficient on-site, waste-free building method for earth-based materials incorporating high design and production control levels. Prior to his doctoral research Kunal worked as a lecturer and robotic fabrication expert at the Institute for Advanced Architecture of Catalonia from 2016 to 2021.

Alexandre Dubor is an architect researcher presently leading the AAG fabrication team at IAAC. Together they explore the potential of new technology, materials, and designs for a better construction industry, focusing on 3D printing and robotics for sustainable architecture.

Edouard Cabay is a program head at IAAC, where he directs the 3dPA Master's Program and the Self-Sufficient Building Research Line. He has been a senior faculty at IAAC since 2011. He has also been the director of the AA Visiting School Barcelona since 2014 and is presently an invited professor at CEDIM in Mexico. Edouard Cabay taught at the Architectural Association as a diploma unit master from 2010 to 2013, at the École Spéciale d'Architecture in Paris as an associate professor from 2012 to 2015, and at the École Polytechnique Fédérale de Lausanne as a first year studio director from 2013 to 2016.

Yara Tayoun is an architect and researcher specialising in the digital and technology-driven transition of circular processes in architecture and design and has worked on cocreating circular resource flows in cities with a focus on wood construction. She is presently the coordinator of the applied research postgraduate programme in 3D Printing Architecture at IAAC.

Lapo Naldoni (WASP) has a building engineering and architecture degree from the University of Bologna, Italy. He developed his thesis on mud constructions and robotic-informed deposition on tensioned fabrics. Presently, he is a member of the Wasp R&D team and works on 3D printing at different scales with polymers and fluid-dense materials. He is a teacher assistant of architecture planning (Prof. Alessio Erioli) at Alma Mater Studiorum, Bologna. He worked as a computational designer at Mario Cucinella Architects (IT).

Massimo Moretti (WASP) graduated in 1974 as an electronic technician, and he started developing products. Massimo Moretti collaborates with research centres, universities, R&D departments, and big national and international companies to create industrial automation projects, 3D modelling, and prototyping in the field of mechanics and interior design, cosmetics, and chemistry. In 2012, together with a group of young designers, he founded WASP (World's Advanced Saving Project), a company that designs, produces, and sells 3D printers made in Italy worldwide. Massimo Moretti, taking inspiration from nature and from the observation of the potter wasp, which builds its own nest with material recovered from the surrounding environment, moves WASP to produce large 3D printers able to build houses with natural materials and available on the territory, at a cost tending to zero.

Chapter 8
Cooperative Robotic Fabrication for a Circular Economy

Edvard Patrick Grigori Bruun, Stefana Parascho, and Sigrid Adriaenssens

Abstract In a cooperative robotic fabrication (CRF) framework, multiple industrial robots are specifically sequenced to work together, thus allowing them to execute coordinated processes with greater geometric and structural variation. In the context of the construction industry, agents in a cooperative setup can perform complementary functions such as placing or removing building components while simultaneously providing temporary support to a structure. This approach can reduce, or completely remove, the need for temporary external supports and scaffolding that would typically be required for stability during the construction of geometrically complex spanning spatial structures. For a circular economy, this means overall reductions to primary resource inputs and improvements to the disassembly, reuse, and reassembly potential of a structure at the end of its life. This chapter gives a summary of three projects that successfully demonstrate the use of cooperative robotic fabrication to promote several principles of a circular economy through different scaffold-free construction applications. The topics covered in this chapter will be of interest to researchers and professionals interested in the emergent intersection of digital fabrication, robotics, and sustainability applied to the building industry.

Keywords Robot · Cooperative · Collaborative · Construction · Assembly · Disassembly · Reuse

E. P. G. Bruun (✉)
Form Finding Lab, Civil and Environmental Engineering Department, Princeton University, Princeton, NJ, USA

Lab for Creative Computation, School of Architecture, Civil and Environmental Engineering, Swiss Federal Institute of Technology, Lausanne, Switzerland
e-mail: ebruun@princeton.edu

S. Parascho
Lab for Creative Computation, School of Architecture, Civil and Environmental Engineering, Swiss Federal Institute of Technology, Lausanne, Switzerland

S. Adriaenssens
Form Finding Lab, Civil and Environmental Engineering Department, Princeton University, Princeton, NJ, USA

C. De Wolf et al. (eds.), *A Circular Built Environment in the Digital Age*, Circular Economy and Sustainability, https://doi.org/10.1007/978-3-031-39675-5_8

8.1 Introduction

To reduce the environmental burden of the construction industry, new methods of practice must be adopted to help move away from a wasteful and resource-intensive design mentality. In this chapter, we introduce the emergent technology of cooperative robotic fabrication (CRF) and describe its potential to enable new applications that will facilitate a transition to more sustainable circular models of building design and construction. We focus on CRF as a technology strictly in the physical domain and demonstrate how such setups, when used to perform multiple tasks simultaneously in precisely choreographed sequences, can enable novel assembly, disassembly, and reuse processes.

8.1.1 What Is Cooperative Robotic Behaviour?

Robotic fabrication (RF) refers to any fabrication process that is completed with some degree of automation. CRF is a subset of RF and can be thought of as any process where the robotic agents are specifically coordinated to accomplish tasks that would not be possible if the robots were working alone. Cao et al. (1997, p. 8) state that "a multiple robot system displays cooperative behaviour if, due to [the mechanism of cooperation], there is an increase in the total utility of the system". Thus, cooperative robotic cells can fall under the category of either multi-arm individual robots, multiple single-arm robots, mechanical hands with independently controllable fingers, or a combination of these, working together in a synchronous fashion (Liu et al. 2004; Ranky 2003).

A single robotic agent, regardless of physical or digital complexity, is naturally limited in the type and number of actions it can simultaneously execute. Only in multi-robotic fabrication (MRF), where multiple agents are placed together in a work cell, does it become possible to unlock the potential of collective behaviour to achieve more complex outputs. All MRF setups exhibit some form of collective behaviour, but while cooperative behaviour is subset of collective behaviour (i.e. CRF $\subseteq$ MRF), the converse is not true (i.e. MRF CRF). A CRF process entails further utility beyond the collective behaviour that comes from a basic implementation of MRF. This hierarchy is illustrated in Fig. 8.1, where the output of an MRF setup is defined as scaling linearly with the number of agents to produce more of the same output (i.e. several robots working in parallel), as opposed to a CRF process where the output is uniquely contingent on all the agents working together.

Another important distinction is between the terms cooperative and collaborative, which are commonly used interchangeably to describe multi-agent robotic processes in the literature. To avoid ambiguity, collaborative is herein only used for a process where robot(s) work together with, or alongside, human operators. Collaborative processes exist across the entire RF hierarchy illustrated in Fig. 8.1. For example, collaborative processes are possible with a human working with a single robot

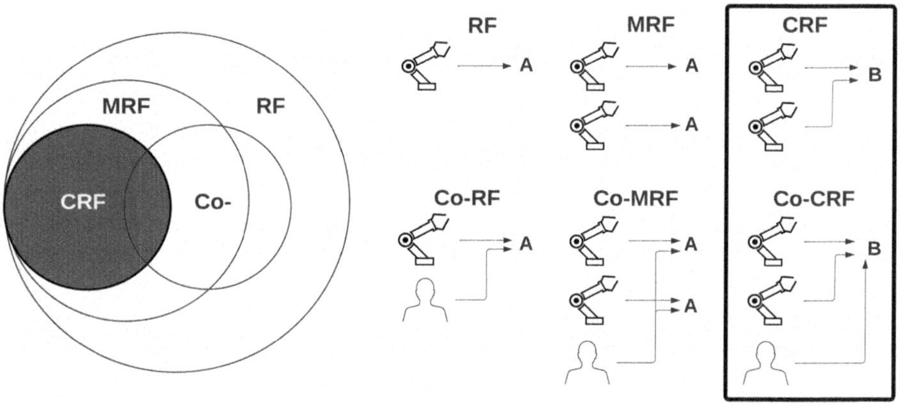

Fig. 8.1 Cooperative robotic fabrication as situated in the overall robotic setup hierarchy (*RF*-robotic fabrication, *MRF*-multi-robotic fabrication, *CRF*-cooperative robotic fabrication, *Co*-collaborative). A setup is cooperative, if by the process of cooperation, a novel output is made possible (i.e. B), as opposed to a basic MRF process which only allows more of the same output to be created in parallel (i.e. A)

(Co-RF, as in Asadi et al. (2018)), with multiple robots in series on an assembly line (Co-MRF, as in Weckenborg et al. (2020)), or to complement the cooperative function of multiple robots (Co-CRF, as in Bruun et al. (2020)).

8.1.2 *Broad Applications*

Alongside applications in the built environment, which are specifically discussed in Sect. 8.2, CRF is utilised in many industries when flexible manufacturing systems are necessary or where tasks occur in poorly structured environments (Caccavale and Uchiyama 2016). In generic manufacturing applications, CRF processes have a conceptual advantage over single robot processes with their ability to distribute the work among several potentially smaller robots and thus better control the internal forces, torques, and displacements associated with a payload (Montemayor and Wen 2005). In addition, CRF processes also allow for improved robustness against work interruptions through redundancy in the functions of the robots, improved flexibility through the ability to reconfigure a fabrication cell to fit different conditions, and improved task precision through the ability to dexterously grasp and then manipulate an object (Gudiño-Lau and Arteaga 2005; Montemayor and Wen 2005). Many generic tasks only become possible to automate when multiple robotic agents or manipulators are used cooperatively for carrying heavy loads, moving voluminous objects, avoiding obstacles through complex movements, handling flexible objects with extra degrees of freedom, and assembling multiple components without using dedicated supporting fixtures or jigs (Caccavale and Uchiyama 2016; Gan et al.

2012; Li and Zhang 2018). Different industries use CRF workflows for various industry-specific applications, for example:

- The agricultural industry has seen major adoption of automation technologies in recent years (Lytridis et al. 2021) and specifically in cooperative robotic setups for foraging and picking tasks for various fruits and vegetables (Ahlin et al. 2017; Ling et al. 2019; Sarabu et al. 2019; Sepulveda et al. 2020).
- The automotive industry has a long history of being at the forefront of automation and is a leader in developing and utilising both CRF and Co-CRF technologies (Michalos et al. 2010) for tasks such as welding (Papakostas et al. 2011; Pellegrinelli et al. 2017; Wu et al. 2000) and panel assembly (Connolly 2009).
- The fibre composite manufacturing industry has been using cooperating robots for laying and smoothing sheets of material (Malhan et al. 2018; Szcesny et al. 2017) and in filament winding (Sbanca and Mogan 2015) for fabricating high-strength, geometrically complex components.
- In heavy industry such as ship building and bridge construction, a dual-arm robot coupled with a hoist mechanism has been proposed to handle heavy workpieces (Shinohara et al. 2001).
- For generic industrial warehouse applications, cooperating mobile robots have long been used to move large and heavy objects (Hirata et al. 2000; Mataric et al. 1995).

8.2 Cooperative Robotic Fabrication in the Built Environment

The general use of robotics in the built environment is motivated by many of the same reasons as in the industries mentioned in Sect. 8.1, specifically high precision and task repeatability (Wang et al. 2021), improved productivity (Xu and Garcia de Soto 2020), improved site safety by reducing worker injuries (Chu et al. 2013), standardisation of product quality (Dritsas and Soh 2019), and the ability to conduct work remotely to facilitate any necessary social distancing (Wang et al. 2021). One of the first recorded uses of robots in the construction industry was the Motor Mason automated bricklaying machine from the 1960s (British Pathé 1967). But it was not until the 1970s, in Japan, that robots in the construction industry saw serious exploration and use, specifically for the prefabrication of modular housing components (Bock and Linner 2016). In the 1980s, more on-site robots appeared, followed by a proliferation of robots used for various specialised construction tasks over the next decades (Bock 2007). In the mid-2000s, the large-scale application of robotics in the context of architectural and building design began with the growth of the digital fabrication (DFab) movement (Bonwetsch et al. 2006; Gramazio and Kohler 2008). This movement emphasised the design and construction of geometrically complex, efficient, and bespoke structures that were often only made possible, or sufficiently productive (García de Soto et al. 2018), by combining novel digital technologies with more complex robotic setups.

A recent literature review on robots in the construction industry found that collaboration (used there to refer to both robot-robot and robot-human processes) is one of three major topics of recently published research (Xiao et al. 2022). CRF setups have been specifically demonstrated for automation, parallelisation, and scaling applied to rapid assembly and prefabrication, on-site additive manufacturing, and general task automation (Kayser et al. 2018; Petersen et al. 2019) and for future building applications in challenging environments such as space construction (Xue et al. 2021). In Sects. 8.2.1, 8.2.2, and 8.2.3, we summarise CRF applications in the construction industry organised according to the typical scale of their application (e.g. material, product, and building) and whether they originated specifically from the DFab research community or from the broader construction industry.

8.2.1 CRF at the Material Scale

CRF at the material scale is defined by small-scale processes that feature precise manipulation and subtractive/additive operations on single material units (e.g. a block of stone, a pipe, a structural member). General construction applications include the use of dual-armed table-top-sized robots, such as the IRB14000 (ABB 2015), for shaping materials and joining light building components such as small pipes (Afsari 2018). But in general such platforms suffer from limited payloads and are thus not capable of heavy lifting or manipulation of standard objects that are typical in most construction applications.

DFab applications include the use of CRF setups for cutting expanded polystyrene (EPS) foam blocks to create non-ruled and doubly curved surfaces. For example, custom concrete formwork was manufactured using a heated blade mounted on two robotic arms (Søndergaard et al. 2016). The relative displacement of the robot flanges was used to provide curvature to the blade, which shaped the cut through the workpiece as a third robot moved the foam block linearly through space. Another example used a heated wire instead, which two robots swept through a fixed foam block, using the resistance of the wire against the foam to create a non-standardised undulating surface profile for a series of wall panels (Rust et al. 2016). In the tying of knots in cables, which is a material-scale task, the creation of loops and crossings cannot be performed by a single robot (Augugliaro et al. 2015). In a project on the aerial construction of tensile rope structures, the spatial manoeuvrability of multiple flying unmanned aerial vehicles (UAVs) was utilised to tie a knot using coordinated multi-robot flight trajectories, thus establishing a structural node in three-dimensional space (Augugliaro et al. 2013; Mirjan et al. 2014).

8.2.2 CRF at the Product Scale

CRF at the product scale is common in modular construction applications, for building stand-alone components (i.e. walls, truss sections, shell panels) or transporting components as part of assembling a larger structure. In the context of prefabrication, CRF supports the goals of improving productivity, reducing labour, and maintaining a more predictable work environment (Vähä et al. 2013).

General construction applications include the assembly of a box girder structure, which was performed with a team of mobile robots that cooperated to move separate panels, align the parts, and fasten them together (Dogar et al. 2015). In another mobile robot example, NASA's Jet Propulsion Laboratory (JPL) Robot Construction Crew was used for picking and cooperatively transporting aluminium beams into an interlocking structure in the context of construction for space exploration applications (Huntsberger et al. 2005; Stroupe et al. 2005). In another space-related application, tetrahedral truss structure modules for an astronomical telescope were built on a rotating platform as a second robot placed struts into accessible regions of the structure (Doggett 2002).

DFab applications include the construction of modular components for both wood and composite fibre structures. In one project, timber modules with nonplanar geometries were constructed with two robotic arms used to place linear stud members while also supporting the corners of the structure in their unfinished state (Adel et al. 2018; Thoma et al. 2018). In another research project, prefabricated cassettes for a segmented timber shell pavilion were assembled on a rotating central turntable where one robot manipulated the unfinished module in space, while the other robot performed gluing, nailing, milling operations (Wagner et al. 2020). For composite fibre structures, a CRF process was used in the construction of a modular fibre shell pavilion consisting of 36 geometrically varying panels (Doerstelmann et al. 2015). Using the synchronised motion of two robots, a coreless filament was wound around an adaptable steel frame that defined the boundary polygon of each module (Parascho et al. 2015; Prado et al. 2014). In another filament winding project, two robots exchanged a spool of filament allowing it to reach and wind around support points in space to create varying modules for a spanning space frame structure (Duque Estrada et al. 2020).

8.2.3 CRF at the Building Scale

CRF at the building scale is common for the in situ construction of large structures or for performing work that requires complex task sequencing beyond what is possible by a single robot working alone. Processes at this scale emphasise the use of the robots to provide temporary support and guarantee stability for a structure as it is being built, and to expand the feasible work volume and reach beyond that of a standard RF setup.

General construction applications of CRF include an integrated construction robot platform featuring multiple robotic trolley hoists and mobile welding robots that are used to reach all areas of a steel structure as it is being constructed (Saidi et al. 2016). In one research project, the challenge of small payloads in aerial construction was overcome by the cooperative effort of multiple UAVs used to grasp, manipulate, and transport large structural elements into a structure on site (Mellinger et al. 2013). Several examples exist for in situ construction for space-based structures and applications. The multi-limbed Hexbot robot was designed to assemble a telescope truss structure directly in space by carrying large components that required more than one arm to grasp. The robot used its multiple limbs to simultaneously walk on the structure, stay anchored, perform the gross movement of components, and connect them to the existing structure at the point of assembly (Lee et al. 2016). In another related space construction project, the two-armed RoboSimian robot was used in a similar role as the Hexbot, for the manoeuvring and in-place assembly of a telescope truss structure (Karumanchi et al. 2018).

DFab applications of CRF at the building scale have been demonstrated for various structural typologies and typically fall under two distinct categories of material systems: continuous (e.g. filaments or cables) or discrete (e.g. rods, studs, or bricks) elements. An example of a project where a continuous material system was combined with a CRF process was in the construction of a large monocoque shell structure, where a UAV was used to pass a fibre spool between two static robotic arms placed at either end of the work volume. The filament was wound between the two robotic arms, expanding the feasible build volume by making it possible to build a structure within the interstitial space outside the reach of the two stationary robotic arms (Felbrich et al. 2017; Vasey et al. 2020). In another aerial construction project, volumetric cable structures were built in situ using two flying UAVs in a cooperative process of tying knots in space (Mirjan et al. 2013, 2016). In a final example of a continuous material system CRF process, multiple wall-climbing robots were used to pass filament between themselves, winding it around fixed anchor points to construct an in situ tensile structure (Yablonina and Menges 2019).

CRF for discrete element assembly at the building scale was first developed for the assembly of geometrically differentiated metal space frame structures (Parascho et al. 2017, 2018). This research focused on developing sequences and path-planning methods that used two robotic arms to alternate either providing temporary support or adding elements to the structure. In another project where cooperating robots were used for temporary support, a branching arch structure was built out of foam blocks without requiring scaffolding by relying on two robots as simultaneous mobile temporary supports (Wu and Kilian 2018). In the final example of a discrete element CRF process, a cooperative building-scale sequence was also demonstrated in the construction of a timber pergola roof structure, where one robot was used to support the member in space while the other performed an in situ drilling and fastening operation (Thoma et al. 2019).

8.3 Cooperative Robotic Fabrication for a Circular Economy

CRF processes can be generally used to foster a transition towards a circular economy. This discussion is situated in the context of the narrow, slow, close, and regenerate framework developed by the editors of this book (Çetin et al. 2021). To date, CRF has been applied to address objectives that are part of the narrow (Sect. 8.3.1), slow (Sect. 8.3.2), and close (Sect. 8.3.3) principles, with potential future applications discussed in Sect. 8.3.4. The regenerate principle is not yet linked to CRF but may be in the future.

8.3.1 Narrow

With respect to the narrow principle, the following objectives are specifically applicable to CRF: (1) reducing primary resource inputs, (2) designing for structural performance, and (3) improving construction efficiency. First, primary resource inputs for constructing new structures can be reduced by leveraging the potential multi-functionality of a CRF setup. For example, while one robot places structural members during construction, other robots simultaneously provide temporary support to the structure in its unfinished state. All robots can then alternate their function throughout the fabrication process. Their function at each fabrication step, as either the active robotic agent (i.e. placing material) or the passive robotic agent (i.e. supporting the structure), is determined by the operator. A structure designed based on such an alternating "support-place" cooperative robotic sequence is considered fabrication informed as the fabrication process itself explicitly shapes its design. Using such an approach allows for the reduction, or complete removal, of temporary falsework, scaffolding, and supporting structure that would normally be required to build the structure using traditional construction methods, thereby reducing the primary resource inputs associated with constructing this temporary support structure. This cooperative approach is especially relevant for spanning discrete element structures (e.g. masonry vaults and space frame structures), which often require extensive temporary supporting structures as they are only self-stable at their completion or only at specific stages during the construction process. This type of cooperative sequencing is demonstrated in each of the three projects presented in Sects. 8.4.1, 8.4.2, and 8.4.3.

The second objective of the narrow principle applicable to CRF is based on how material usage in the structure itself can also be reduced by designing its form such that it maximises structural performance. For example, form-found or topologically optimised structures are materially efficient by virtue of their shapes or connectivity being optimised for various loading conditions but often result in geometrically complex structures that are challenging to construct with traditional methods. Applied to the prefabrication of structural modules, it is possible to realise complex geometries by relying on the spatial precision of a robot to place material accurately in 3D space. This capability is augmented in a CRF setup, which allows for the simultaneous cooperative manoeuvring and repositioning of structural modules that are under construction to facilitate accessibility.

The third objective recognises efficient but geometrically complex structures can be time-consuming and require several workers to construct (García de Soto et al. 2018). A CRF process can improve construction efficiency by taking on certain material handling and movement tasks to reduce the overall time and labour resources required.

8.3.2 Slow

With respect to the slow principle, the following objectives are specifically applicable to CRF: (1) design for reversibility and (2) lifetime extension. Regarding the first objective, design for reversibility, CRF setups can be used for the disassembly of geometrically complex or spanning structures, which can thus be designed with explicit potential for reversibility from the outset. For example, the structure can be designed as an assembly of modules that can be more easily isolated and removed from the overall structure. To assist in this process, a CRF setup can be used with similar robotic task allocations as in assembly: the robots work cooperatively acting as temporary supports while simultaneously separating and removing self-rigid modules from the structure. The robots perform the physically demanding, and potentially dangerous, tasks of removing material while also indefinitely supporting and stabilising the structure in its temporary state of disassembly. The project described in Sect. 8.4.2 features a structure that is specifically designed so that it can be taken apart in a stability-preserving way when using a cooperative robotic sequence.

Regarding the second objective of the slow principle, CRF setups assisting in the task of disassembling a structure create an opportunity to start considering the use of automation for building lifetime extension. If a structure is designed with modularity in mind, damaged components can be more quickly isolated, removed, and eventually replaced without requiring large interruptions to the function of the structure (e.g. construction of temporary support or scaffolding).

8.3.3 Close

Regarding the close principle, the following activities are made possible through CRF: (1) tracking, documenting, and tracing building components and (2) reuse and reassembly. First, accurate 3D models of a structure can be created and used to build a digital twin to document geometric location and placement accuracy of structural and nonstructural components or to perform visual grading and inspection. CRF setups facilitate this process as the positional information that is inherent in a robotic platform can be used to accurately stitch together multiple 3D image captures from different cameras and perspectives. This can create a complete digital model of an existing structure, which would not always be possible with a single robot due to obstructed perspectives. In terms of the second objective, when CRF is applied to disassembly, it also facilitates the reuse and reassembly of structural components while modifying a building or recuperating material that would normally be treated as construction waste. This approach is demonstrated in the project described in Sect. 8.4.3.

8.3.4 Future Applications

CRF is typically used within laboratory environments. However, if research expands from static industrial robots towards mobile machines and large-scale construction machines, the technology could be directly applied on construction sites to enable more material-efficient construction and engender faster and more precise disassembly and reassembly processes. These developments would contribute to the slow and close principles.

In addition, integrated force-torque sensors mounted on the robot tool flange can be leveraged in a cooperative manner to carry out in situ non-destructive testing on structures to further collect data on their performance in their final state or as they are being assembled or disassembled. This wealth of data can be used to design more materially efficient structures, better evaluate overall structural performance during fabrication, and measure parameters like the stiffness or degree of damage to a member. Effectively, each robot could act as a 6-degree-of-freedom actuator capable of applying forces and moments to a structure at any location and orientation in space. If the robots are sequenced cooperatively, it would be possible to apply non-standard loading conditions, which for geometrically complex structures would be difficult to evaluate in situ using conventional load testing methods.

8.4 Examples of Cooperative Robotic Fabrication for a Circular Economy

The following section describes three recent research-based examples of how CRF is used for discrete element assembly (Sects. 8.4.1 and 8.4.2), disassembly (Sects. 8.4.2 and 8.4.3), and reassembly (Sect. 8.4.3) to target objectives related to the narrow, slow, close circular economy principles described in the previous section.

8.4.1 LightVault

The LightVault was a 3.6 × 6.5 × 2.2 m doubly curved masonry vault built with two stationary robotic arms as a demonstration of CRF applied to an assembly process (Parascho et al. 2021). In the first phase of the project, a central arch was constructed utilising the alternating cooperative robotic placement and support approach inspired by previous research on the assembly of metal space frame structures (Parascho et al. 2017, 2018). One robot continuously acted as a support to the partially completed arch, while the other was used to place additional bricks into the structure (Fig. 8.2). Thus, the arch was built from one end to the other without requiring any additional temporary supporting structure. The structural performance of the arch during construction was assessed using a discrete element modelling approach (Paris et al. 2021), and the cooperative sequencing was later theorised to setups with more than two robots to further improve the structural performance during assembly (Bruun et al. 2021). In the second phase of the project, the rest of the vault was built

Fig. 8.2 Building the central arch as the first phase in the scaffold-free cooperative robotic assembly of a masonry vault

layer by layer using the central arch as a backbone structure (Han et al. 2020; Parascho et al. 2020).

Overall, the LightVault demonstrated the potential application of CRF for scaffold-free construction of spanning structures made from heavy material. With respect to circular economy principles, the use of primary resources was reduced by eliminating temporary supporting structures and minimising the material in the structure itself by enabling the construction of a structurally efficient but geometrically complex compression-only form.

8.4.2 Remote Robotic Assemblies Workshop

In the Remote Robotic Assemblies workshop held at the 2021 Association for Computer Aided Design in Architecture (ACADIA) conference, a timber space frame arch structure was constructed using two cooperating robotic arms on linear tracks. This project was a demonstration of CRF applied to not just the assembly of the structure but extending its use for the first time to disassembly as well. Using a method based on rigidity theory, the space frame was designed explicitly to leverage cooperative robotic support sequencing to replace temporary supporting structure during both the construction and deconstruction phases (Bruun et al. 2022b). The structure was first assembled element by element, where one passive robotic agent was always required to provide support to the partially assembled structure. Following this, the structure was disassembled cell by cell, taking advantage of the fact that it was designed explicitly as an assembly of locally rigid tetrahedral cells. These cells were sequentially supported, isolated, and then removed with one robot, while the other robot supported the partially disassembled structure (Fig. 8.3). The disassembly process is an example of a collaborative-CRF (Co-CRF) process as the removal of individual elements to disconnect the rigid tetrahedral cells from the remaining structure was done in collaboration with a human.

Overall, the Remote Robotic Assemblies workshop demonstrated that CRF is a viable technology to reduce primary resource inputs in the form of scaffolding during both the assembly and disassembly of spanning space frame structures. In addition, extending the application of CRF to disassembly tasks highlighted the potential of including considerations for disassembly at the outset of a design to better facilitate the reuse and recycling of building components at the end of a structure's life.

8.4.3 ZeroWaste

ZeroWaste was a research project exploring the idea of treating existing buildings as stores of valuable reusable material in the context of a circular economy (Bruun et al. 2022a). Rather than demolishing and disposing of a building at the end of its life, the

Fig. 8.3 Isolating locally rigid cells in the scaffold-free cooperative robotic disassembly of a spanning timber space frame arch structure

goal was to leverage the use of a CRF setup to first gather data about an unknown existing structure and then use this information together with the robotic setup to disassemble and then reassemble the structure into new feasible configurations.

As the starting point, a pavilion-scale timber structure was built manually to act as a stand-in representing a generic unknown existing structure built according to standard stick framing construction practices. Next, 3D cameras were mounted on two robotic arms, which were then used to take several point cloud captures of the structure from various locations and angles. Using the accurate positional information queried from the robotic controller, the individual point cloud captures were transformed and then stitched together to create a complete spatial model of the existing structure. Creating this complete model was only possible when using multiple robots, as a single robot would not have the required reach and manoeuvrability to fully capture the structure. For an existing building, the exact geometry and spatial location of the structure is not known; thus, the as-built geometric information gathered in this imaging process was necessary when later planning the RF sequences.

Next, scaffold-free robotic cooperative disassembly and reassembly sequences were calculated algorithmically using a support hierarchy graph representation of the structure – this method is described further in Bruun et al. (2022a). These sequences were specifically planned for execution with the three robotic arms available in the fabrication cell, two on linear tracks and one stationary, without requiring external temporary formwork. The physical RF process was split into four distinct phases, targeting different objectives with respect to the cooperative robotic sequencing and the degree of disassembly and reassembly (Fig. 8.4).

Fig. 8.4 Snapshots of the four cooperative robotic fabrication phases for the ZeroWaste project. (**a**) Starting timber structure built according to traditional American stick framing construction practices; (**b**) phase 1: disassembly of a corner using a two-robot CRF process, no reassembly; (**c**) phase 2: disassembly of the front wall using a three-robot CRF process followed by reassembly of four members as a new supporting structure for the roof girder at the front of the structure; (**d**) phase 3: disassembly of a side wall using a two-robot CRF process with simultaneous parallel 1-to-1 reassembly (i.e. each member removed is reused) to create a stiff lattice configuration for the same wall; (**e**) phase 4: disassembly of all remaining walls using a three-robot MRF process with simultaneous parallel reassembly into an inclined system vertical member system

As in the project described in Sect. 8.4.2, ZeroWaste demonstrated the use of a CRF setup in providing temporary support to a structure during disassembly but further extended its use to perform scaffold-free reassembly and reuse of removed material. Improvements in construction efficiency were also demonstrated as the full fabrication process only required a single person working alongside the robots, whereas using non-robotic methods would typically require several workers to accomplish the same tasks. Overall, the successful use of CRF in the ZeroWaste project to assist in structural disassembly and reassembly tasks highlighted the potential of this technology to facilitate a more circular treatment of existing timber building stock through its reuse.

8.5 Discussion

As demonstrated in this chapter, cooperative robotic fabrication (CRF) has the potential to enable novel assembly, disassembly, and reuse processes that promote several essential principles of a circular economy. Primary resource utilisation can be reduced by minimising, or completely removing, the need for temporary scaffolding during the (de)construction of geometrically complex spanning structures. In addition, general construction efficiency can be improved by shifting certain challenging and dangerous tasks related to material handling and transport from human workers to the robotic setup. If modularity is considered and originally designed into a building, CRF can facilitate selective disassembly and removal of structural components to replace damaged elements and extend the life of a building. In the eventual decommissioning of existing buildings, CRF setups can also be used to catalogue, disassemble, and then reuse components to divert building materials away from waste streams and return them back to productive use.

Challenges with broadening the adoption of CRF technology in the construction industry relate to the complexity of implementing these setups in an on-site unstructured environment. While stationary robots, or robots with limited mobility on linear tracks, are well suited for off-site prefabrication tasks, CRF with mobile robotic setups will be required in the future to broaden the ranges of applications that are possible in larger volumes, as would be expected on a job site. Other chapters in this book describe technologies that are adjacent and relevant to CRF: scanning technologies and scan-to-BIM (Chap. 3), building information modelling (BIM) and digital twins (Chap. 1), computational design (Chap. 6), and on-site robotic fabrication (Chap. 9).

8.6 Key Takeaways

- In a cooperative robotic fabrication setup, the robotic agents are specifically coordinated to accomplish tasks that would not be possible if the robots were working alone.
- Multiple robots can be sequenced to place or remove structural components while alternating temporarily supporting the structure, performing material handling, or data acquisition operations.

- Primary resource inputs in the form of scaffolding and temporary support can be removed during construction when using a cooperative robotic fabrication setup.
- Disassembly and reuse of existing buildings is made possible when using a cooperative robotic fabrication setup.

Acknowledgements The projects described in Sect. 8.4 were generously supported by several internal Princeton University grants: the Metropolis Project, the Catalysis Initiative, and Campus as a Lab. In addition, we would like to acknowledge the support of the National Science Foundation for the "Waste-Free Robotic Construction of Spatial Discrete Element Structures" grant (CMMI-ECI 2122271) and the Natural Sciences and Engineering Research Council of Canada (funding reference number 532482).

References

ABB (2015) Product specification – IRB 14000 (No. 3HAC052982-001; p. 126). ABB. https://new.abb.com/products/robotics/collaborative-robots/yumi/irb-14000-yumi

Adel A, Thoma A, Helmreich M, Gramazio F, Kohler M (2018) Design of robotically fabricated timber frame structures. In: Proceedings of the 38th annual conference of the association for computer aided design in architecture, pp 394–403. http://papers.cumincad.org/cgi-bin/works/paper/acadia18_394

Afsari K (2018) Applications of collaborative industrial robots in building construction. In: Proceedings of the 54th ASC annual international conference, pp 472–479

Ahlin K J, Hu AP, Sadegh N (2017) Apple picking using dual robot arms operating within an unknown tree. In: 2017 ASABE annual international meeting, pp 1–11. https://doi.org/10.13031/aim.201700471

Asadi E, Li B, Chen IM (2018) Pictobot: a cooperative painting robot for interior finishing of industrial developments. IEEE Robot Autom Mag 25(2):82–94. https://doi.org/10.1109/MRA.2018.2816972

Augugliaro F, Mirjan A, Gramazio F, Kohler M, D'Andrea R (2013) Building tensile structures with flying machines. In: 2013 IEEE/RSJ international conference on intelligent robots and systems, pp 3487–3492. https://doi.org/10.1109/IROS.2013.6696853

Augugliaro F, Zarfati E, Mirjan A, D'Andrea R (2015) Knot-tying with flying machines for aerial construction. In: 2015 IEEE/RSJ international conference on intelligent robots and systems, pp 5917–5922. https://doi.org/10.1109/IROS.2015.7354218

Bock T (2007) Construction robotics. Auton Robot 22(3):201–209. https://doi.org/10.1007/s10514-006-9008-5

Bock T, Linner T (2016) Construction robots: elementary technologies and single-task construction robots, vol 4. Cambridge University Press, New York

Bonwetsch T, Kobel D, Gramazio F, Kohler M (2006) The informed wall: Applying additive digital fabrication techniques on architecture. In: Proceedings of the 25th annual conference of the association for computer-aided design in architecture, pp 489–495

British Pathé (Director) (1967) Mechanical bricklayer. https://www.youtube.com/watch?v=4MWald1Goqk

Bruun EPG, Ting I, Adriaenssens S, Parascho S (2020) Human–robot collaboration: a fabrication framework for the sequential design and construction of unplanned spatial structures. Digit Creat 31(4):320–336. https://doi.org/10.1080/14626268.2020.1845214

Bruun EPG, Pastrana R, Paris V, Beghini A, Pizzigoni A, Parascho S, Adriaenssens S (2021) Three cooperative robotic fabrication methods for the scaffold-free construction of a masonry arch. Autom Constr 129:103803. https://doi.org/10.1016/j.autcon.2021.103803

Bruun EPG, Adriaenssens S, Besler E, Parascho S (2022a) ZeroWaste: towards computing cooperative robotic sequences for the disassembly and reuse of timber frame structures. In: Proceedings of the 42nd annual conference of the association for computer aided design in architecture, p 12

Bruun EPG, Adriaenssens S, Parascho S (2022b) Structural rigidity theory applied to the scaffold-free (dis)assembly of space frames using cooperative robotics. Autom Constr 141:104405. https://doi.org/10.1016/j.autcon.2022.104405

Caccavale F, Uchiyama M (2016) Cooperative manipulation. In: Springer handbook of robotics. Springer, pp 989–1006. https://doi.org/10.1007/978-3-319-32552-1_39

Cao YU, Fukunaga AS, Kahng A (1997) Cooperative mobile robotics: antecedents and directions. Auton Robot 4(1):7–27. https://doi.org/10.1023/A:1008855018923

Çetin S, De Wolf C, Bocken N (2021) Circular digital built environment: an emerging framework. Sustainability 13(11):6348. https://doi.org/10.3390/su13116348

Chu B, Jung K, Lim MT, Hong D (2013) Robot-based construction automation: an application to steel beam assembly (Part I). Autom Constr 32:46–61. https://doi.org/10.1016/j.autcon.2012.12.016

Connolly C (2009) Motoman markets co-operative and humanoid industrial robots. Ind Robot 36(5):417–420. https://doi.org/10.1108/01439910910980132

Doerstelmann M, Knippers J, Menges A, Parascho S, Prado M, Schwinn T (2015) ICD/ITKE research pavilion 2013–14: modular coreless filament winding based on beetle elytra. Archit Des 85(5):54–59. https://doi.org/10.1002/ad.1954

Dogar M, Knepper RA, Spielberg A, Choi C, Christensen HI, Rus D (2015) Multi-scale assembly with robot teams. Int J Robot Res 34(13):1645–1659. https://doi.org/10.1177/0278364915586606

Doggett W (2002) Robotic assembly of truss structures for space systems and future research plans. In: Proceedings of the 2002 IEEE aerospace conference, vol 7, pp 1–10. https://doi.org/10.1109/AERO.2002.1035335

Dritsas S, Soh GS (2019) Building robotics design for construction: design considerations and principles for mobile systems. Constr Robot 3(1–4):1–10. https://doi.org/10.1007/s41693-018-0010-1

Duque Estrada R, Kannenberg F, Wagner HJ, Yablonina M, Menges A (2020) Spatial winding: cooperative heterogeneous multi-robot system for fibrous structures. Constr Robot 4(3):205–215. https://doi.org/10.1007/s41693-020-00036-7

Felbrich B, Frueh N, Prado M, Saffarian S, Solly J, Vasey L, Knippers J, Menges A (2017) Multi-machine fabrication: an integrative design process utilising an autonomous UAV and industrial robots for the fabrication of long span composite structures. In: Proceedings of the 37th annual conference of the association for computer aided design in architecture, pp 248–259

Gan Y, Dai X, Li J (2012) Cooperative path planning and constraints analysis for master-slave industrial robots. Int J Adv Robot Syst 9(3):88. https://doi.org/10.5772/51374

García de Soto B, Agustí-Juan I, Hunhevicz J, Joss S, Graser K, Habert G, Adey BT (2018) Productivity of digital fabrication in construction: cost and time analysis of a robotically built wall. Autom Constr 92:297–311. https://doi.org/10.1016/j.autcon.2018.04.004

Gramazio F, Kohler M (2008) Digital materiality in architecture, 1st edn. Lars Müller, Baden

Gudiño-Lau J, Arteaga MA (2005) Dynamic model and simulation of cooperative robots: a case study. Robotica 23(5):615–624. https://doi.org/10.1017/S0263574704001213

Han IX, Bruun EPG, Marsh S, Adriaenssens S, Parascho S (2020) From concept to construction: a transferable design and robotic fabrication method for a building-scale vault. In: Proceedings of the 40th annual conference of the association for computer aided design in architecture, pp 614–623. http://papers.cumincad.org/cgi-bin/works/Show?acadia20_614

Hirata Y, Kosuge K, Asama H, Kaetsu H, Kawabata K (2000) Coordinated transportation of a single object by multiple mobile robots without position information of each robot. In: Proceedings of the 2000 IEEE/RSJ international conference on intelligent robots and systems, vol 3, pp 2024–2029. https://doi.org/10.1109/IROS.2000.895268

Huntsberger T, Stroupe A, Kennedy B (2005) System of systems for space construction. In: 2005 IEEE international conference on systems, man, and cybernetics, vol 4, pp 3173–3178. https://doi.org/10.1109/ICSMC.2005.1571634

Karumanchi S, Edelberg K, Nash J, Bergh C, Smith R, Emanuel B, Carlton J, Koehler J, Kim J, Mukherjee R, Kennedy B, Backes P (2018) Payload-centric autonomy for in-space robotic assembly of modular space structures. J Field Robot 35(6):1005–1021. https://doi.org/10.1002/rob.21792

Kayser M, Cai L, Falcone S, Bader C, Inglessis N, Darweesh B, Oxman N (2018) FIBERBOTS: an autonomous swarm-based robotic system for digital fabrication of fiber-based composites. Constr Robot 2(1–4):67–79. https://doi.org/10.1007/s41693-018-0013-y

Lee N, Backes P, Burdick J, Pellegrino S, Fuller C, Hogstrom K, Kennedy B, Kim J, Mukherjee R, Seubert C, Wu YH (2016) Architecture for in-space robotic assembly of a modular space telescope. J Astron Telesc Instrum Syst 2(4):1–15. https://doi.org/10.1117/1.JATIS.2.4.041207

Li S, Zhang Y (2018) Neural networks for cooperative control of multiple robot arms. Springer, Singapore. http://link.springer.com/10.1007/978-981-10-7037-2

Ling X, Zhao Y, Gong L, Liu C, Wang T (2019) Dual-arm cooperation and implementing for robotic harvesting tomato using binocular vision. Robot Auton Syst 114:134–143. https://doi.org/10.1016/j.robot.2019.01.019

Liu G, Xu J, Wang X, Li Z (2004) On quality functions for grasp synthesis, fixture planning, and coordinated manipulation. IEEE Trans Autom Sci Eng 1(2):146–162. https://doi.org/10.1109/TASE.2004.836760

Lytridis C, Kaburlasos VG, Pachidis T, Manios M, Vrochidou E, Kalampokas T, Chatzistamatis S (2021) An overview of cooperative robotics in agriculture. Agronomy 11(9):1818. https://doi.org/10.3390/agronomy11091818

Malhan RK, Kabir AM, Shembekar AV, Shah B, Gupta SK, Centea T (2018) Hybrid cells for multi-layer prepreg composite sheet layup. In: 2018 IEEE 14th international conference on automation science and engineering, pp 1466–1472. https://doi.org/10.1109/COASE.2018.8560586

Mataric MJ, Nilsson M, Simsarin KT (1995) Cooperative multi-robot box-pushing. In: Proceedings 1995 IEEE/RSJ international conference on intelligent robots and systems. Human robot interaction and cooperative robots, vol 3, pp 556–561. https://doi.org/10.1109/IROS.1995.525940

Mellinger D, Shomin M, Michael N, Kumar V (2013) Cooperative grasping and transport using multiple quadrotors. In: Distributed autonomous robotic systems, vol 83. Springer, pp 545–558. http://link.springer.com/10.1007/978-3-642-32723-0_39

Michalos G, Makris S, Papakostas N, Mourtzis D, Chryssolouris G (2010) Automotive assembly technologies review: challenges and outlook for a flexible and adaptive approach. CIRP J Manuf Sci Technol 2(2):81–91. https://doi.org/10.1016/j.cirpj.2009.12.001

Mirjan A, Gramazio F, Kohler M, Augugliaro F, D'Andrea R (2013) Architectural fabrication of tensile structures with flying machines. In: Green design, materials and manufacturing processes, 1st edn. CRC Press, pp 513–518. https://doi.org/10.1201/b15002

Mirjan A, Gramazio F, Kohler M, Gramazio F, Kohler M, Langenberg S (2014) Building with flying robots. In: Fabricate 2014: negotiating design making, pp 266–271. https://doi.org/10.2307/j.ctt1tp3c5w.36

Mirjan A, Augugliaro F, D'Andrea R, Gramazio F, Kohler M (2016) Building a bridge with flying robots. In: Reinhardt D, Saunders R, Burry J (eds) Robotic fabrication in architecture, art and design 2016. Springer, pp 34–47. https://doi.org/10.1007/978-3-319-26378-6_3

Montemayor G, Wen JT (2005) Decentralized collaborative load transport by multiple robots. In: Proceedings of the 2005 IEEE international conference on robotics and automation, pp 372–377. https://doi.org/10.1109/ROBOT.2005.1570147

Papakostas N, Michalos G, Makris S, Zouzias D, Chryssolouris G (2011) Industrial applications with cooperating robots for the flexible assembly. Int J Comput Integr Manuf 24(7):650–660. https://doi.org/10.1080/0951192X.2011.570790

Parascho S, Knippers J, Dörstelmann M, Prado M, Menges A (2015) Modular fibrous morphologies: computational design simulation and fabrication of differentiated fibre composite building components. In: Block P, Knippers J, Mitra NJ, Wang W (eds) Advances in architectural geometry 2014. Springer, pp 29–45. https://doi.org/10.1007/978-3-319-11418-7_3

Parascho S, Gandia A, Mirjan A, Gramazio F, Kohler M (2017) Cooperative fabrication of spatial metal structures. In: Fabricate 2017: rethinking design and construction, pp 24–29. https://doi.org/10.3929/ethz-b-000219566

Parascho S, Kohlhammer T, Coros S, Gramazio F, Kohler M (2018) Computational design of robotically assembled spatial structures: a sequence based method for the generation and evaluation of structures fabricated with cooperating robots. In: Advances in architectural geometry 2018, pp 112–139. https://research.chalmers.se/en/publication/504188

Parascho S, Han IX, Walker S, Beghini A, Bruun EPG, Adriaenssens S (2020) Robotic vault: a cooperative robotic assembly method for brick vault construction. Constr Robot 4(3):117–126. https://doi.org/10.1007/s41693-020-00041-w

Parascho S, Han IX, Beghini A, Miki M, Walker S, Bruun EPG, Adriaenssens S (2021) LightVault: a design and robotic fabrication method for complex masonry structures. In: Advances in architectural geometry 2020, pp 350–375. https://thinkshell.fr/wp-content/uploads/2019/10/AAG2020_18_Parascho.pdf

Paris V, Lepore N, Bruun EPG, Ruscica G, Piccioni MD, Beghini A, Parascho S, Adriaenssens S (2021) Robotic construction of a self-balancing glass masonry vault: DEM study of stability during construction stages. Proceedings of the international conference on spatial structures 2020/21, p 12. https://doi.org/10.15126/900337

Pellegrinelli S, Pedrocchi N, Tosatti LM, Fischer A, Tolio T (2017) Multi-robot spot-welding cells for car-body assembly: design and motion planning. Robot Comput Integr Manuf 44:97–116. https://doi.org/10.1016/j.rcim.2016.08.006

Petersen KH, Napp N, Stuart-Smith R, Rus D, Kovac M (2019) A review of collective robotic construction. Sci Robot 4(28):1–10. https://doi.org/10.1126/scirobotics.aau8479

Prado M, Dörstelmann M, Schwinn T, Menges A, Knippers J (2014) Core-less filament winding. In: Robotic fabrication in architecture, art and design 2014, pp 275–289. https://doi.org/10.1007/978-3-319-04663-1_19

Ranky PG (2003) Collaborative, synchronous robots serving machines and cells. Ind Robot 30(3): 213–217. https://doi.org/10.1108/01439910310473915

Rust R, Jenny D, Gramazio F, Kohler T (2016) Spatial wire cutting: cooperative robotic cutting of non-ruled surface geometries for bespoke building components. In: Proceedings of the 21st international conference of the association for computer-aided architectural design research in Asia, pp 529–538

Saidi K, Bock T, Georgoulas C (2016) Robotics in construction. In: Springer handbook of robotics. Springer, pp 1493–1520. https://doi.org/10.1007/978-3-319-32552-1_57

Sarabu H, Ahlin K, Hu AP (2019) Graph-based cooperative robot path planning in agricultural environments. In: 2019 IEEE/ASME international conference on advanced intelligent mechatronics, pp 519–525. https://doi.org/10.1109/AIM.2019.8868747

Sbanca MP, Mogan GL (2015) Winding of carbon wire composite structures using two cooperative industrial robots. Appl Mech Mater 762:291–298. https://doi.org/10.4028/www.scientific.net/AMM.762.291

Sepulveda D, Fernandez R, Navas E, Armada M, Gonzalez-De-Santos P (2020) Robotic aubergine harvesting using dual-arm manipulation. IEEE Access 8:121889–121904. https://doi.org/10.1109/ACCESS.2020.3006919

Shinohara T, Miyawaki K, Sunayama K (2001) Heavy material handling by cooperative robot control. Adv Robot 15(3):317–321. https://doi.org/10.1163/156855301300235841

Søndergaard A, Feringa J, Nørbjerg T, Steenstrup K, Brander D, Graversen J, Markvorsen S, Bærentzen A, Petkov K, Hattel J, Clausen K, Jensen K, Knudsen L, Kortbek J (2016) Robotic hot-blade cutting. In: Robotic fabrication in architecture, art and design 2016, pp 150–164. https://doi.org/10.1007/978-3-319-26378-6_11

Stroupe A, Huntsberger T, Okon A, Aghazarian H, Robinson M (2005) Behavior-based multi-robot collaboration for autonomous construction tasks. In: 2005 IEEE/RSJ international conference on intelligent robots and systems, pp 1495–1500. https://doi.org/10.1109/IROS.2005.1545269

Szcesny M, Heieck F, Carosella S, Middendorf P, Sehrschön H, Schneiderbauer M (2017) The advanced ply placement process – an innovative direct 3D placement technology for plies and tapes. Adv Manuf Polym Compos Sci 3(1):2–9. https://doi.org/10.1080/20550340.2017.1291398

Thoma A, Adel A, Helmreich M, Wehrle T, Gramazio F, Kohler M (2018) Robotic fabrication of bespoke timber frame modules. In: Robotic fabrication in architecture, art and design 2018, pp 447–458. https://doi.org/10.1007/978-3-319-92294-2_34

Thoma A, Jenny D, Helmreich M, Gandia A, Gramazio F, Kohler M (2019) Cooperative robotic fabrication of timber dowel assemblies. In: Research culture in architecture. De Gruyter, pp 77–88. https://doi.org/10.1515/9783035620238-008

Vähä P, Heikkilä T, Kilpeläinen P, Järviluoma M, Gambao E (2013) Extending automation of building construction: survey on potential sensor technologies and robotic applications. Autom Constr 36:168–178. https://doi.org/10.1016/j.autcon.2013.08.002

Vasey L, Felbrich B, Prado M, Tahanzadeh B, Menges A (2020) Physically distributed multi-robot coordination and collaboration in construction. Constr Robot 4(1):3–18. https://doi.org/10.1007/s41693-020-00031-y

Wagner HJ, Alvarez M, Kyjanek O, Bhiri Z, Buck M, Menges A (2020) Flexible and transportable robotic timber construction platform – TIM. Autom Constr 120:103400. https://doi.org/10.1016/j.autcon.2020.103400

Wang X, Liang CJ, Menassa CC, Kamat VR (2021) Interactive and immersive process-level digital twin for collaborative human–robot construction work. J Comput Civ Eng 35(6):1–19. https://doi.org/10.1061/(ASCE)CP.1943-5487.0000988

Weckenborg C, Kieckhäfer K, Müller C, Grunewald M, Spengler TS (2020) Balancing of assembly lines with collaborative robots. Bus Res 13(1):93–132. https://doi.org/10.1007/s40685-019-0101-y

Wu K, Kilian A (2018) Robotic equilibrium: Scaffold free arch assemblies. In: Proceedings of the 38th annual conference of the association for computer aided design in architecture, pp 342–349

Wu L, Cui K, Chen SB (2000) Redundancy coordination of multiple robotic devices for welding through genetic algorithm. Robotica 18(6):669–676. https://doi.org/10.1017/S0263574799001976

Xiao B, Chen C, Yin X (2022) Recent advancements of robotics in construction. Autom Constr 144:1–16. https://doi.org/10.1016/j.autcon.2022.104591

Xu X, Garcia de Soto B (2020) On-site autonomous construction robots: a review of research areas, technologies, and suggestions for advancement. In: 37th international symposium on automation and robotics in construction, pp 385–392. https://doi.org/10.22260/ISARC2020/0055

Xue Z, Liu J, Wu C, Tong Y (2021) Review of in-space assembly technologies. Chin J Aeronaut 34(11):21–47. https://doi.org/10.1016/j.cja.2020.09.043

Yablonina M, Menges A (2019) Distributed fabrication: cooperative making with larger groups of smaller machines. Archit Des 89(2):62–69. https://doi.org/10.1002/ad.2413

Edvard Patrick Grigori Bruun is a PhD candidate in the Form Finding Lab in the Civil and Environmental Engineering Department at Princeton University. Before starting his PhD, he worked as a structural engineer at Arup. His present research interests lie in exploring how multiple industrial robotic arms can be sequenced cooperatively in the robotic fabrication of discrete element structures. He develops approaches that leverage the multi-functionality of robotic setups to design aesthetic and efficient structures that can be (dis)assembled while preserving their stability without the need for any external scaffolding.

Stefana Parascho is a researcher, architect, and educator whose work combines architecture, digital fabrication, and computational design. She is presently assistant professor at EPFL where she founded the Lab for Creative Computation (CRCL). Through her research, she has explored multi-robotic fabrication methods and their relationship to architectural design. Her goal is to strengthen the connection between design, structure, and fabrication and boost the interdisciplinary nature of architecture. Before joining EPFL, she was assistant professor at Princeton University. She completed her doctorate in 2019 at ETH Zurich and received her Diploma in Architectural Engineering from the University of Stuttgart.

Sigrid Adriaenssens directs the Form Finding Lab at Princeton University, where she also teaches courses on non-linear mechanics and design of structures and the integration of engineering and arts. Her research interests lie in the mechanics of large-span structural surfaces under extreme loading and under construction. She works on advanced analytical formulations, numerical form finding and optimisation approaches, fluid/structure interaction, and machine learning models and algorithms to open new avenues for accelerated discoveries and automated optimal designs. In 2021, she was named Fellow of the ASCE Structural Engineering Institute, elected IASS vice-president, and received the DigitalFUTURES Matthias Rippmann and the Pioneers' Award.

Chapter 9
Circular Robotic Construction

Lauren Vasey, Petrus Aejmelaeus-Lindström, David Jenny, Ryan Luke Johns, Ilmar Hurkxkens, Coralie Ming, Marco Hutter, Fabio Gramazio, and Matthias Kohler

Abstract In situ robotic construction is a type of construction where mobile robotic systems build directly on the building site. To enable on-site navigation, industrial robots can be integrated with mobile bases, while mobile, high-payload construction machines can be adapted for autonomous operation. With parallel advances in sensor processing, these robotic construction processes can become robust and capable of handling non-standard, local, as-found materials.

The potential of using autonomous, mobile robotic systems for the development of innovative circular construction processes is presented in three exemplary case studies:(i) robotically jammed structures from bulk materials, (ii) robotic earthworks with local and upcycled materials, and (iii) robotic additive manufacturing with earth-based materials. These processes exemplify key strategies for a circular industry through the utilisation of materials with low embodied greenhouse gas emissions and the implementation of fully reversible construction processes.

For each case study, we describe the robotic building process, the enabling technologies and workflows, and the major sustainability and circularity benefits compared to conventional construction methods. Moreover, we discuss the difficulty of industry transfer, considering challenges such as detailing, integration, and engineering validation. We conclude with an outlook towards future research avenues and industry adoption strategies.

L. Vasey (✉) · P. Aejmelaeus-Lindström · D. Jenny · C. Ming · F. Gramazio · M. Kohler
Gramazio Kohler Research, ETH Zurich, Zurich, Switzerland
e-mail: vasey@arch.ethz.ch

R. L. Johns
Gramazio Kohler Research, ETH Zurich, Zurich, Switzerland

Robotic Systems Lab, ETH Zurich, Zurich, Switzerland

I. Hurkxkens
Section Landscape Architecture, TU Delft, Delft, The Netherlands

M. Hutter
Robotic Systems Lab, ETH Zurich, Zurich, Switzerland

C. De Wolf et al. (eds.), *A Circular Built Environment in the Digital Age*, Circular Economy and Sustainability, https://doi.org/10.1007/978-3-031-39675-5_9

Keywords Circular robotic construction · Adaptive robotic construction · In situ robotic construction · Digital fabrication with natural materials

9.1 What Is Robotic Construction?

Robotic construction is an emerging interdisciplinary field. Robots were first introduced to the construction sector in the 1970s in Japan, in part due to the lack of skilled labour. Early initiatives, such as single-task construction robotics (Bock and Linner 2016a) and fully integrated, on-site factories for buildings, failed to be widely adopted (Bock and Linner 2016b). However, there was a clear turning point when industrial robots were appropriated for architectural application and began to be digitally programmed in 2006, enabling the direct connection of computational design processes to physical fabrication processes and the development of novel material systems (Gramazio et al. 2014). In parallel, there have been advances in the automation of existing construction machinery, where common construction machines have been adapted for digital control and autonomous operation.

In this chapter, we consider the specific case of in situ robotic construction. In contrast to robotic prefabrication, where parts are prefabricated in a factory off-site, in situ robotic construction is a type of construction where robots move directly on the construction site and produce or assemble parts directly in their final position (Helm et al. 2012, 2014). This type of robotic construction has added benefits for a circular built environment because it enables material flows and production chains that minimise transportation overhead and material processing steps. However, the implementation of in situ robotic construction faces technical, logistical, and legal challenges, described below and then presented in more detail in the case studies.

From a technical perspective, robotic systems suited for on-site operation require robust systems for mobility, navigation, and localisation. Due to the unstructured nature of construction sites, such systems also require on-board sensing, such as LiDAR sensors and global navigation satellite systems. These same sensing technologies, when coupled with robust backend computational processes, can be leveraged to enable robotic systems to simultaneously handle unstructured and natural material systems with a high degree of variability and unpredictability.

Schematically, two of the most common types of robotic systems suitable for on-site construction tasks and deployed in the presented case studies include standard industrial robots integrated with mobile bases and mobile construction equipment, such as hydraulic excavators, modified for autonomous operation. At ETH Zurich, the In situ Fabricator (IF) is a prototypical mobile robotic system consisting of an ABB IRB 4600 robotic arm with 2.55 m reach and kg payload mounted on an automated excavator base that built off precedent iterations (Giftthaler et al. 2017; Sandy et al. 2016). This unit was first developed in 2016 and has so far been deployed in mobile robotic brick stacking (Dörfler et al. 2016), custom metal formwork for non-standard concrete (Dörfler et al. 2019; Hack and Lauer 2014), and in the robotic processes discussed in Sect. 9.4.1. In contrast, HEAP is a full-scale walking excavator developed by the Robotic Systems Lab at ETH Zurich (Jud et al.

2021b), with a vertical reach up to 9 m and a maximum payload of 3 tonnes. HEAP has been utilised primarily for automating existing construction processes such as autonomous trench digging, autonomous forestry work, semi-autonomous teleoperation, and the robotic earthworks and assembly processes presented in Sect. 9.4.2.

9.2 Robotic Construction for the Built Environment

To date, robots have not been widely used in the building industry for construction. This is in part due to low profit margins for various stakeholders, minimal research funding, and a lack of vertical and horizontal integration in the construction sector (Saidi et al. 2016). Because they operate near human workers, robots on the construction site also require necessary changes in safety protocols and legislation. Moreover, for on-site conditions, gantry-based systems currently have more widespread industry use, particularly for additive manufacturing and 3D printing (Wu et al. 2016). However, mobile robotic systems have clear benefits over such rigid, fixed installations, including higher geometric freedom, lower self-weight and volume, and the possibility for operation in unstructured and variable terrain. In contrast to gantry systems, mobile robotic systems also require less drastic site modifications and therefore can minimise the need for additional foundations that contribute to construction waste, greenhouse gas (GHG) emissions, and embodied energy. Existing diesel-powered construction machines allow for construction in remote, off-grid environments, while recent developments in electrification can greatly reduce the embodied energy of these machines where infrastructure allows.

A variety of material fabrication and building systems have been robotically automated for construction (Bock 2007; Melenbrink et al. 2020; Petersen et al. 2019). Some of the most common material applications with a high level of technological readiness are described in the following paragraphs. Predominantly, workflows based on prefabrication are more technologically ready than in situ–based approaches, while on-site fabrication exists as an alternative. In this case, critical equipment for subassembly or part fabrication is transported in a mobile container and set up as a small factory on site.

For concrete and other cementitious material types, common processes include layer-based extrusion, robotic slip forming, robotic shot-creting, and robotic spraying (Burger et al. 2020; Ercan Jenny et al. 2020; Hack et al. 2021; Hack and Kloft 2020; Wangler et al. 2019). Robotic fused-deposition modelling (FDM) printing has been used for custom concrete formwork, while robotically tended rebar construction ("mesh mould") has been developed and transferred to industry through the start-up Mesh (Mirjan et al. 2022). Robotic wire cutting of foam formwork for concrete casting has been developed in academia and then also transferred to industry through the Danish company Odico (Feringa 2014; Søndergaard 2014; Søndergaard et al. 2016).

Robotic fibre composite manufacturing techniques, including coreless filament winding and tape laying, are processes that originated in the aerospace industries but have been modified for architecture applications (Prado et al. 2014; Vasey et al. 2015, 2020; Bodea et al. 2021, 2022). These fabrication methods have transferred into industry through the start-up Fibr (Dörstelmann n.d.).

A variety of approaches and building systems have been achieved in timber and wood construction, including multi-robotic assembly processes for timber framing systems (Willmann et al. 2016; Apolinarska 2018; Thoma et al. 2018; Leung et al. 2021), long-spanning robotically fabricated plate systems (Li and Knippers 2015; Schwinn and Menges 2015), and multi-layer cassette-based systems (Alvarez et al. 2019; Wagner et al. 2020a). A fully equipped multi-robot mobile factory for on-site prefabrication of timber modules has also been developed (Wagner et al. 2020b). Companies such as Intelligent City and Design-to-Production leverage robotic technologies for customised timber structures and housing (Scheurer et al. 2005; intelligent city 2023).

Robotic masonry construction has been explored extensively in both academic and industry contexts for facades and load-bearing walls (Bonswetch et al. 2006; Helm et al. 2012; Gramazio and Kohler 2014; Piškorec et al. 2019). Custom brickwork has been successfully transferred to the industry by companies such as ROB technologies and Keller Ziegeleien (Keller Systeme 2023; ROB Technologies 2023). Custom robotic systems such as the Hadrian X® mobile robotic block laying machine have achieved a high level of technical readiness for the on-site assembly of concrete masonry units (CMUs) (FBR 2023).

Several other efforts focus on automating existing manual tasks, such as dry wall installation, curtain wall installation, drilling, and welding, among others (Brosque et al. 2020, 2021; Iturralde et al. 2022).

9.3 Robotic Construction for a Circular Economy

In situ robotic construction can address the needs of a circular building industry primarily by slowing the consumption of resources through the following strategies:

- Enabling the use of natural and as-found materials with low embodied energy and GHG emissions.
- Minimising extra transportation steps of material, components, or assemblies to and from external processing or production sites.
- Minimising peripheral supporting elements such as formwork or falsework.
- Minimising material use through structural optimisation and realisation of complex geometries.
- Enabling reversible construction processes with minimal material downgrading.

In situ robotic construction can slow the consumption of resources using locally available, natural, and upcycled materials. Natural materials and local materials both exhibit high geometric and mechanical variability. Scanning, on-site robotic processing, and assembly can enable the use of completely natural materials with lower embodied energy and greenhouse gases. With design systems that adapt to these variable geometries of existing material or upcycled material stock, extra resource- and energy-intensive processing steps can be avoided. Energy and embodied GHG emissions due to transportation can also be minimised by utilising materials that are available near the construction site.

In situ robotic construction can also minimise peripheral equipment and extra site work. For example, formwork, which is a major component of construction and demolition waste (Shen et al. 2004), can be avoided, as demonstrated in the following case studies. Scaffolding, falsework, framing systems, and other stabilisation elements that enable the lifting of subassemblies or components are also made unnecessary. Furthermore, subassemblies do not have to be designed for the unique load cases incurred during lifting and transportation, leading to over-dimensioning. As mentioned in the previous section, mobile on-site robots potentially require less custom foundation work in contrast to more extensive on-site gantry-based systems.

Another important criterion is the reversibility of construction processes. On-site robotic assembly processes, such as robotic dry-stone masonry, which can achieve load-bearing behaviour without mortar or other adhesives, can also be largely reversible with minimal material downgrading and are therefore more circular. Robotic additive manufacturing with earth-based material mixtures, composed of materials like clay, gravel, sand, and silt, but without chemical stabilisers, can also be reversible with some additional, but minimal, processing steps.

The following exemplary case studies demonstrate the potential of in situ robotic construction towards enabling a circular building industry. These academic projects emerged out of a half-decade of interdisciplinary research at ETH Zurich. These projects are situated in their local economic context: sourcing materials from both local suppliers and the construction waste stream and engaging with industry partners offering material processing and construction services. Moreover, these full-scale and sometimes permanent demonstrators required collaboration with geotechnical engineers, structural engineers, and general contractors, thus engaging questions relating to implementation and long-term industry adoption.

9.4 Examples of in Situ Circular Robotic Construction

9.4.1 Robotic Construction of Jammed Architectural Structures (JAS) from Bulk Material

The combination of robotic fabrication and structural health monitoring enables the construction of jammed architectural structures (JAS) composed of gravel, a common bulk material, and twine. Jamming is a physical phenomenon where loose granular materials are compacted into self-stable configurations through externally applied pressure, self-weight, and/or confinement. Jammed materials behave fundamentally differently than conventional construction materials as they can change back and forth between a jammed, solid state and a loose, malleable state. In robotic fabrication of JAS, crushed porphyry is held in place by robotically placed twine (Aejmelaeus-Lindström et al. 2016). In 2018, a full-scale architectural structure, Rock Print Pavilion, was built to demonstrate the potential of JAS (Aejmelaeus-Lindström et al. 2017, 2020). It was opened to the public in the historic city centre of Winterthur, Switzerland, and then fully deconstructed (Fig. 9.1).

Similar to dry masonry, JAS requires in situ fabrication and cannot be prefabricated, as the structural properties of the material change due to small changes of the confinement. Thus, the pavilion was built by the IF introduced in Sect. 9.1. The IF's tracked base enabled it to move on the construction site: an unpaved square covered with gravel at a slight (approximately 2-degree) angle and with significant surface irregularities. The robotic positioning system is based on a Hilti POS 150 robotic total station, a reflector prism mounted on the end-effector and custom software (Sandy et al. 2016). The robot arm was moved to a series of positions, which were automatically registered by the total station and used to calculate the transformation from the tool coordinate frame to the world coordinate frame. The IF is equipped with a custom, multi-purpose end-effector consisting of a gravel dispensing tool, a compacting tool, and a reinforcement-laying tool. First, it lays the twine in layers of aligned, interlocking circular loops, after which gravel is measured and placed inside the string loop and compacted. The compacting of the crushed rock and twine displaces the particles concentrically, which in turn tensions the reinforcement loops, providing the confinement necessary for jammed vertical structures.

The pavilion is designed as five tapered elements that are wall shaped at the base. Towards the top, they branch into 11 columns that carry an 8.7-tonne cantilevering steel roof. Each element is designed to fit within the work envelope of the robotic arm. The steel roof is temporarily mounted on pillars during the construction to protect the construction site from rain. The structure was fabricated from 36 tonnes of porphyry gravel and 85 km of string. After being exhibited for 6 weeks, the steel roof was dismantled and the string was pulled out, returning the raw material to its original state. A structural health monitoring approach was developed where the movement of the steel roof was monitored daily to ensure minimal movement, required by the supervising engineer. Additionally, deformation inside the structure was measured with a fibre optic strain measuring device (LUNA Sensor) to identify any internal changes to the structure. No major movement of the roof was recorded during the six-week lifespan. Custom detailing between the steel roof and top of the columns allowed for height adjustments and load redistribution in the case of asymmetric creep of the structure.

To conclude, JAS is a highly experimental robotic building process but with advantageous sustainability and circularity metrics, as it uses simple, widely available raw materials, and the resulting structures can be fully reversed without downgrading. For the demonstrator, the aggregates were sourced from a quarry located within 30 km of its construction site and returned after the life span of the pavilion. However, the material system is significantly different from conventional

Fig. 9.1 (i) The Rock Print Pavilion is a full-scale robotically jammed structure composed of gravel aggregates and twine. The enabling technologies facilitating the on-site adaptive construction process include a custom robotic end-effector (ii) for extruding twine, depositing aggregates, and compacting layers, a structural health monitoring approach for the movement of the structure and the roof over time (iii), and mobility and localisation of the IF enabling the production of a larger structure on uneven ground (iv). The structure can be easily deconstructed with no material downgrading by removing the twine (v). (© Georg Aerni)

and standardised construction material: it is significantly anisotropic and sensitive to surface erosion. Future work is required both to understand and monitor the structural behaviour and to increase the surface strength. In terms of possible applications, JAS might be suitable for infrastructure construction, with particular utility in increasing the stiffness and longevity of road substrates. This research area has been explored in collaboration with the Swiss Federal Laboratories for Materials Science and Technology (Empa 2017).

9.4.2 Robotic Earthworks with Local and Upcycled Materials

The application of roboticised heavy hydraulic machines has enabled recent advances in on-site excavation and assembly. Methods for robotic landscaping and the robotic assembly of dry-stone masonry walls have been integrated towards the construction of digitally designed earthworks and soil-retaining structures – executed in the form of a full-scale, publicly accessible Circularity Park that features a permanent stone retaining wall, terraced landscapes, and a public circulation trail (Fig. 9.2).

Robotic landscaping is a process for forming natural granular materials like sand, soil, and gravel utilising HEAP, the autonomous excavator. The process can realise geometrically complex landscape formations with high precision, with an estimated average error of 3–5 cm (Jud et al. 2021a). Digital terrain modelling tools based on signed distance functions enable the balancing of cut and fill volumes for material-neutral, on-site construction, while incremental LiDAR scanning enables digital reconstruction of the site and current ground condition (Hurkxkens et al. 2019; Jud et al. 2017).

Large-scale dry-stone masonry structures are constructed by the Mobile Robotic Aggregation of Found Objects, a robotic construction method that enables robotic construction from highly irregular local boulders and waste concrete. The process can realise mortar-free masonry walls as both free-standing and soil-retaining structures. One of the significant technical challenges of the process is that the geometry of the material stock is not known ahead of construction, and thus the walls cannot be designed ahead of time. A scanning routine was developed to locate and digitise individual stones utilising HEAPs cabin-mounted LiDAR: accumulating points that are meshed using Poisson reconstruction to provide a full 3D model of each stone with a resolution suitable for manipulation and construction. A custom geometric planner was developed within the scope of the project, and it algorithmically determines where stones can be placed within a designer-specified volume, given an inventory of available boulders and concrete debris (Johns et al. 2020). A robotic grasp-planning workflow uses 3D mapping and collision constraints to reliably grasp and reorient irregularly shaped stones, using the excavator's 2-jaw gripper (Mascaro et al. 2021), allowing for solutions from the geometric planner to be placed on the wall. The locations of these stones are incrementally updated using the LiDAR

Fig. 9.2 (i) The Circularity Park is a full-scale and publicly accessible landscape park built with robotic landscaping and autonomous robotic dry-stone masonry, utilising HEAP. The material includes locally sourced boulders and waste concrete (ii). The main enabling technologies facilitating the on-site adaptive construction process included (iii) an adaptive planning computational design and tool and (iv) a scanning process for digitising the individual stones. Robotic landscaping enabled precise landscaping of the surrounding terraces (v). (© Gramazio Kohler Research. Drone Videography: Girts Apskalns. Photography: Mark Schneider)

scanner, ensuring that shifting and settling is accounted for in subsequent construction steps.

The developed construction method has several sustainability and circularity benefits when benchmarked against conventional methods of construction, particularly when compared to reinforced concrete retaining walls. For the case of retaining walls, previous research has suggested the sustainability advantages of masonry when benchmarked against concrete in terms of GHG consumption and energy footprint (Farcas et al. 2015). Significantly, dry-stone masonry surpasses these performances, considering that the construction process takes advantage of locally sourced materials, and the structures are produced without mortar, rendering them fully reversible with little downgrading. This robotic assembly process also includes no secondary processing, such as cutting the stones into shapes that more easily fit together. Additionally, the developed method of construction incorporates recycled concrete debris and thus could be used to upcycle a portion of the estimated 2.6 million tonnes of concrete recycled each year from demolished houses (Guerra and Kast 2015). In Zurich, for example, this has particular significance, as the approved landfill volume for recycled concrete will only be sufficient for the next 10 years (Guerra and Kast 2015). The design tool for the landscape design further enhances the sustainability of the developed methods, as the designer can balance cut and fill volumes, proactively avoiding transporting extra material to or from the site.

The two robotic construction processes were integrated into a workflow for the production of the full-scale demonstrator in collaboration with Eberhard AG, a Swiss construction and material processing company that operates the recycling facility where the park was built. To expedite construction, a rough cut of the landscape was first executed with a large, manually operated excavator within approximately 1 m of the target digital landscape. A minimal foundation for the wall was provided by compacting the local soil and further reinforcing it with a low-cement stabiliser. The robotic construction process was then staged accordingly to maintain the accessibility of HEAP to the area of construction. First, the upper terraces were autonomously and precisely excavated in accordance with the 3D digital blueprint. The construction of the retaining wall was then executed incrementally in stable layers. An inventory of approximately 25 stones was scanned and stored on-site and within reach of the excavator until it was replenished by truck-based material delivery. The wall was constructed from boulders from a local quarry, erratics unearthed during construction in nearby Eberhard building sites, and concrete debris from demolished structures around Zurich. Robotic landscaping final passes were then alternated with placing stones until the structure was complete. Finally, the whole site was scanned and additional details, such as stairs, railings, finishing layers, and benches, were put in place through conventional manual methods. Some details, such as the stair, had to be especially designed on site to fit with the stone wall dimensions. Here, the adjacent rocks were scanned, and an old concrete stair from the west side of the site was scanned, cut-to-size, reassembled, and fixed with mortar.

The Circularity Park occupied the private land of Eberhard but was intended to ultimately be permanently accessible to the public, so it was critical that safety measures were put in place. In addition to the academic research team, the project was supported by external contractors for permitting and a team of geotechnical engineers who oversaw and guided construction. Ultimately, because no existing building codes can certify robotically constructed walls, the construction elicited a high degree of risk. To mitigate risks, the structural engineer over-dimensioned the thickness of the wall, specifying an additional layer of backfilling stones that were collectively digitised such that the robotic process could also adapt. Additional manual-stability testing methods were ordered and executed at the end of construction to assess the stability of individual stones. One additional research trajectory investigated was to utilise HEAPs force-torque sensing to apply targeted point load cases on the wall. Currently, this method can realise similar conclusions as manual testing methods by identifying unstable and non-load-bearing stones that slip at low threshold forces. Loose stones discovered manually or robotically had to be mechanically fastened to neighbouring stones.

As the client, Eberhard assumed all liability for any issues with the function, serviceability, and safety of the retaining wall. Long-term industry adoption and implementation would necessitate new building codes and codified methods of validation and in situ testing. Being able to validate the structural ability to withstand typical retaining wall load cases would be a key hurdle to proving the technological soundness of the given construction process. Only then would the developed method be able to serve as a viable alternative for infrastructure such as concrete gravity retaining walls.

In summary, the developed method of robotic construction updates a vernacular building process and enhances it through a digital toolset. The main circular attributes include the use of locally sourced natural stones and waste material and the reversible nature of the construction process. However, detailing and engineering validation remain significant challenges for long-term industry adoption.

9.4.3 Robotic Additive Manufacturing with Earth-Based Materials

Earth-based materials, such as soil, gravel, sand, silt, and clay have great relevance for circular and sustainable construction. Yet conventional earth-based construction methods such as rammed earth construction have high costs, low levels of digitalisation, and high dependency on manual labour. Rapid Clay Formations is an additive robotic fabrication process that reinterprets the traditional construction method for cob walls, where discrete parts of malleable earth blocks are manually aggregated to form a solid mass. The robotic process was developed to produce a full-scale and permanent demonstrator, the Clay Rotunda, a cylindrical structure constituting the outer soundproof shell of the electroacoustic auditorium SE MusicLab (Fig. 9.3).

For the robotic process, malleable cylindrical "soft bricks" were pre-produced off-site through an extrusion-based process within the standard brick production facilities of the industry partner, Brauchli Ziegelei, a local brick manufacturer. In the

Fig. 9.3 (i) The Clay Rotunda is a cylindrical structure constituting the outer, soundproof shell of the electroacoustic auditorium SE MusicLab. (ii) The soft bricks were produced externally with an industry partner. (iii) The robotic pressing process was realised with the IF, which could be relocated on a temporary scaffold to realise a two-story structure. (iv) Detail of structure showing the bonding and interlocking between adjacent elements. (© Gramazio Kohler Research)

additive robotic process, the soft bricks were grabbed by the robotic arm with a pneumatic end-effector from a picking station, precisely positioned and oriented, and sequentially pressed into their final position, thus bonding with the previous layers through material cohesion and geometric interlocking.

The hardware setup consists of the custom robotic platform – the IF – which allows relocation and navigation on a temporary scaffold after every built segment and therefore enables the construction of larger structures on site. The overall precision of the structure is achieved by monitoring the sequential buildup using both LiDAR scanning and point measurements, digitised with a robotic total station. Deformations due to shrinkage are partially compensated through a predictive computational workflow that estimates the expected deformations of a given subassembly of parts. A lean design-to-construction pipeline allows subsequent control code to be regenerated based on these tolerances and re-output to the robot control setup.

This first full-scale robotic clay pressing process addresses sustainability and circularity through several aspects. The material used for the soft bricks is a mix of 40% clay, 45% sand, 15% stones, and 16% water. The clay is sourced from a clay pit located in eastern Switzerland, right next to the brick production facility, which provided the sand for the mix. Stones were provided by Eberhard AG. The material thus has low embodied energy compared to concrete or bricks as it is locally sourced, minimally processed, and unfired.

The Clay Rotunda was designed for permanent long-term use. However, these structures can hypothetically be completely recycled, and the material can be completely reused. Once the structure is demolished, the material can be crushed, sieved to extract desired granulometry, rehydrated, and re-processed into soft bricks. In other additive manufacturing processes for cementitious materials such as concrete, chemical additives have been shown to be detrimental to both embodied GHG and recyclability (Flatt and Wangler 2022). A critical distinction to other earth-based additive manufacturing processes is that no chemical stabilisers such as lime or cement were used.

The digital design and additive manufacturing process enables the construction of highly efficient, thin, and complex structures without custom formwork, which allows the structures to be built with minimal waste produced. Reusable scaffolding and tension elements were used in some cases to stabilise the structure during construction. Besides the plastic sheets reused to maintain the malleability of the soft bricks during storage and transport, the presented project did not produce any significant waste.

The Clay Rotunda measures almost 11 m in diameter and reaches a height of 5 m with a (median) width of only 15 cm of earth. Rammed earth walls have a typical minimum thickness of 20 cm, so this is a material saving of approximately 25%. The single-layer, load-bearing, and free-standing wall is unique in its complex and structurally stiffened, undulating, and doubly curved geometry. This structure demonstrates how the soft-brick robotic pressing process can build highly efficient structures at the architectural scale that are fully recyclable. It shows that by combining digital design and fabrication methods with traditional earthen building methods, new and radically sustainable construction methods can be developed. In addition, it shows that highly efficient structures can be built from natural, nearly unprocessed, and circular materials systems.

Despite its success, several adaptations should be considered for future constructions of this type. Material shrinkage was a significant issue that resulted in high tolerances in addition to cracks that had to be filled in manually. This issue can for instance be improved by a further reduction of the water percentage or by introducing natural (mineral) additives or fibres to the mix. Further steps could be taken to source the material even more locally. In a different setup, excavation material, typically unused during construction processes, could be sieved, mixed, extruded, and used on site. By processing the material directly, the redundant transportation steps to and from processing facilities could also be minimised or excluded to lower the embodied energy and GHG emissions. For a viable integration of this additive manufacturing process in the building industry, the construction speed and level of automation should be dramatically increased; the Clay Rotunda had an average cycle time of 25 s per 1.5-kg brick, approximately 0.1 m^3 per hour, not accounting for initial material processing, other manual tasks causing machine downtime, or the robotic platform relocalisation time of 1 h.

Currently in development as a next research step is an alternative additive manufacturing process based on high-velocity discrete deposition, or "impact printing," which was first explored on a prototypical scale (Ming et al. 2022). The process is being developed for implementation on HEAP to realise full-scale earth-based structures in situ. The project explores the added values of integrated material processing and rheological control, and it has the goal to streamline the integration of scan data for automatic adjustment to the as-built conditions.

9.5 Discussion

The presented projects demonstrate that in situ robotic processes have reached technological maturity and that they can offer significant benefits for a circular building industry, but several hurdles must still be solved before these building methods are embraced in the construction sector. Regarding engineering validation: materials that are as-found or natural are highly heterogeneous and thus pose problems to verification or calculation methods that rely on standardised or isotropic properties. Moreover, adaptive design workflows based on available materials result in structures that cannot fully be pre-designed and pre-calculated. These construction techniques require new methods of analysis and design workflows which compensate for uncertainty and tolerances and consider a high number of unknowns. Here, data-driven analysis methods and in situ non-destructive testing suggest high potential and relevance for verifying the structural performance of both components and structures. Non-standard materials with emergent geometric boundary conditions from adaptive robotic processes also pose challenges for detailing and interfacing with other standardised building systems. Downstream and subsequent construction tasks would need to be adjusted to the resulting geometry only emerging at the end of construction. Thus, truly adaptive robotic building methods are not compatible with fragmented and compartmentalised production chains where there is a lack of transfer of digital information between multiple actors and stakeholders. All three example projects exhibited long-term issues with durability, requiring both monitoring overtime for quality assurance and structural performance, while yearly maintenance was also required. Thus, industry adoption is inextricably tied to other developments such as structural health monitoring.

In summary, on-site robotic construction can be deployed towards novel methods of circular construction. The key circular strategies employed include the utilisation of highly natural and local material; minimisation of site work, peripheral equipment, and formwork; and robotic assembly for reversibility. These strategies primarily align with slowing the consumption of resources. In addition to mobile robotic platforms, the main enabling technologies include sensor-based methods for geometry acquisition of material stock and as-built global conditions, suggesting that there could be strong overlaps with other technological developments, including scan-to-BIM workflows and material passports. Lean and adaptive computational design-to-fabrication workflows are also essential to enable just-in-time adjustments and adaptive planning due to material, construction, and on-site variability.

9.6 Key Takeaways

- In situ robotic construction is a type of construction where robots move directly on the construction site and build structures in their final position.
- The key circular strategies implemented in the presented robotic construction methods include (i) utilising locally sourced or natural material; (ii) minimising site work, peripheral equipment, and formwork; and (iii) implementing reversible processes.
- LiDAR scanning and other sensor-based methods can be used for geometry acquisition of material stock and as-built global conditions.
- Lean and adaptive design-to-fabrication workflows can also enable just-in-time adjustments and adaptive planning due to material, construction, and on-site variability.
- Several barriers prevent robotic methods from being embraced in the building sector, including engineering validation, integration, detailing, and safety.

Acknowledgements The authors would like to thank clients, collaborators, experts, and sponsors involved in each project.

Rock Print Pavilion was designed for Gewerbemuseum Winterthur (client) in the framework of an ETH Zurich Foundation research program, in collaboration with Petrus Aejmelaeus-Lindström, Gergana Rusenova, Hannes Mayer, Ammar Mirjan, Esther Lombardini, Jesús Medina Ibáñez, Selen Ercan, Sandro Meier, Michael Lyrenmann, Philippe Fleischmann as well as selected experts Josef Meyer Stahl, Metall AG, Dr. Lüchinger + Meyer Bauingenieure AG, Daniel Meyer, Reto Furrer. The project was funded by Dr. Lüchinger + Meyer Bauingenieure AG, Migros Kulturprozent, Keller Systeme AG, Toggenburger AG, Förderverein Gewerbemuseum Winterthur.

Circularity Park was designed for Eberhard Unternehmungen AG (client) in the framework of a Swiss National Science Foundation research program through the National Centre of Competence in Digital Fabrication (NCCR dFAB), in collaboration with Gramazio Kohler Research, ETH Zurich (Matthias Kohler, Fabio Gramazio, Lauren Vasey, Ryan Luke Johns); Robotic Systems Lab, ETH Zurich (Marco Hutter, Martin Wermelinger, Dominic Jud, Varin Buff, Vuk Pakovic, Mads Albers); Chair of Landscape Architecture, ETH Zurich (Christophe Girot, Ilmar Hurkxkens).

Clay Rotunda Park was designed for SE MUSICLAB AG (Jürgen Strauss, Jost Kutter, Manuel Frick, Lorenzo Zanetta, Filippo Melena, Anna Imfeld-Aebischer, Markus Imfeld) (client), in collaboration with Coralie Ming, David Jenny, Hannes Mayer, Edurne Morales, Anton Johansson, Indra Santosa, Jomana Baddad, Nicolas Feihl, Selen Ercan Jenny, Jesus Medina, Karol Wojtas, with support from Mike Lyrenmann and Philippe Fleischmann (Robotic Fabrication Laboratory, ETH Zurich), Andi Reusser (Institute for Building Materials, ETH Zurich) as well as selected experts Seforb Sàrl and Joerg Habenberger, Gotham design studio; selected contractors Felix Hilgert, LEHMAG AG and industry partners Brauchli Ziegelei AG, Wirz AG Bauunternehmung. The project was funded by Wirz AG Bauunternehmung, Welti Furrer, Eberhard, Siemens, Geberit, ETH Zürich Foundation.

Declaration of Competing Interests The authors have participated in developing the case studies presented in this chapter. The authors certify that they have no further affiliations with or involvement in any organisation or entity with any financial or non-financial interest in the subject discussed in this chapter.

References

Aejmelaeus-Lindström P, Willmann J, Tibbits S et al (2016) Jammed architectural structures: towards large-scale reversible construction. Granul Matter 18:1–12

Aejmelaeus-Lindström P, Mirjan A, Gramazio F et al (2017) Granular jamming of loadbearing and reversible structures: rock print and rock wall. Archit Des 87:82–87

Aejmelaeus-Lindström P, Rusenova G, Mirjan A et al (2020) Rock print pavilion: robotically fabricating architecture from rock and string. Construction Robotics 4:97–113

Alvarez M, Wagner H, Groenewolt A, et al (2019) The BUGA Wood Pavilion: integrative interdisciplinary advancements of digital timber architecture. In: 2019 39th ACADIA conference, ubiquity and autonomy. Austin, Texas, pp 1–14

Apolinarska AA (2018) Complex timber structures from simple elements: computational design of novel bar structures for robotic fabrication and assembly. PhD Thesis, ETH Zurich

Bock T (2007) Construction robotics. Auton Robot 22:201–209

Bock T, Linner T (2016a) Site automation. Cambridge University Press

Bock T, Linner T (2016b) Construction robots: elementary technologies and single-task construction robots, 1st edn. Cambridge University Press

Bodea S, Zechmeister C, Dambrosio N et al (2021) Robotic coreless filament winding for hyperboloid tubular composite components in construction. Autom Constr 126:103649

Bodea S, Mindermann P, Gresser GT, Menges A (2022) Additive manufacturing of large coreless filament wound composite elements for building construction. 3D Print Addit Manuf 9:145–160

Bonswetch T, Kobel D, Gramazio F, Kohler M (2006) The informed wall: applying additive digital fabrication techniques on architecture. In: Synthetic landscapes proceedings of the 25th annual conference of the association for computer-aided design in architecture, pp 489–495

Brosque C, Skeie G, Orn J, et al (2020) Comparison of construction robots and traditional methods for drilling, drywall, and layout tasks. In: 2020 international congress on Human-Computer Interaction, Optimization and Robotic Applications (HORA). IEEE, Ankara, Turkey, pp 1–14

Brosque C, Skeie G, Fischer M (2021) Comparative analysis of manual and robotic concrete drilling for installation hangers. Constr Eng Manag 147:05021001

Burger J, Lloret-Fritschi E, Scotto F et al (2020) Eggshell: ultra-thin three-dimensional printed formwork for concrete structures. 3D Printing and Additive Manufacturing 7:48–59. https://doi.org/10.1089/3dp.2019.0197

Dörfler K, Sandy T, Giftthaler M et al (2016) Mobile robotic brickwork. In Reinhardt D, Saunders R, Burry J (eds) Robotic fabrication in architecture, art and design 2016. Springer International Publishing, Cham, pp 204–217

Dörfler K, Hack N, Sandy T et al (2019) Mobile robotic fabrication beyond factory conditions: case study Mesh Mould wall of the DFAB HOUSE. Constr Robot 3:53–67

Dörstelmann M (n.d.) Unique Filament Structures I Germany I FibR GmbH. In: fibr-current. https://www.fibr.tech. Accessed 20 Apr 2023

Empa (2017) Empa – Concrete & Asphalt – Completed Project: ASTRA 2017/005 Robotic reinforcement concrete. https://www.empa.ch/web/s308/astra-2017/005-robotic-reinforcement-concrete. Accessed 3 Feb 2023

Ercan Jenny S, Lloret-Fritschi E, Gramazio F, Kohler M (2020) Crafting plaster through continuous mobile robotic fabrication onsite. Constr Robot 4:261–271. https://doi.org/10.1007/s41693-020-00043-8

Farcas V, Ilies N, Cot R (2015) Co2 and energy footprint of different retaining walls solutions. Masonry Retaining Wall VS. cantilever Retaining Wall. Environmental Engineering & Management Journal (EEMJ) 14

FBR. Hadrian X® (2023). https://www.fbr.com.au/view/hadrian-x. Accessed 21 Apr 2023

Feringa J (2014) Entrepreneurship in architectural robotics: the simultaneity of craft, economics and design. Arch Des 84:60–65

FibR (n.d.). https://www.fibr.tech. Accessed 20 Apr 2023

Flatt RJ, Wangler T (2022) On sustainability and digital fabrication with concrete. Cem Concr Res 158:106837

Giftthaler M, Sandy T, Dörfler K et al (2017) Mobile robotic fabrication at 1: 1 scale: the in situ fabricator: system, experiences and current developments. Construction Robotics 1:3–14

Gramazio F, Kohler M (2014) Made by robots: challenging architecture at a larger scale. Wiley

Gramazio F, Kohler M, Willmann J (2014) The robotic touch: how robots change architecture: Gramazio & Kohler Research ETH Zurich 2005–2013. Park Books

Guerra F, Kast B (2015) Bauabfälle in der Schweiz-Hochbau Studie 2015. Bundesam t für Umwelt BAFU, Zürich

Hack N, Lauer WV (2014) Mesh-mould: robotically fabricated spatial meshes as reinforced concrete formwork. Archit Design 84:44–53. https://doi.org/10.1002/ad.1753

Hack N, Kloft H (2020) Shotcrete 3D printing technology for the fabrication of slender fully reinforced freeform concrete elements with high surface quality: a real-scale demonstrator. In: Second RILEM international conference on concrete and digital fabrication: digital concrete 2020. Springer, pp 1128–1137

Hack N, Bahar M, Hühne C et al (2021) Development of a robot-based multi-directional dynamic fiber winding process for additive manufacturing using shotcrete 3D printing. Fibers 9:39

Helm V, Ercan S, Gramazio F, Kohler M (2012) Mobile robotic fabrication on construction sites: DimRob. In: 2012 IEEE/RSJ international conference on intelligent robots and systems. IEEE, pp 4335–4341

Helm V, Willmann J, Gramazio F, Kohler M (2014) In-situ robotic fabrication: advanced digital manufacturing beyond the laboratory. In: Röhrbein F, Veiga G, Natale C (eds) Gearing up and accelerating cross-fertilization between academic and industrial robotics research in Europe. Springer International Publishing, Cham, pp 63–83

Hurkxkens I, Mirjan A, Gramazio F et al (2019) Robotic landscapes: designing formation processes for large scale autonomous earth moving. In: Design Modelling Symposium. Springer, Berlin, pp 69–81

Intelligent City (2023). https://intelligent-city.com/. Accessed 20 Apr 2023

Iturralde K, Feucht M, Illner D et al (2022) Cable-driven parallel robot for curtain wall module installation. Autom Constr 138:104235. https://doi.org/10.1016/j.autcon.2022.104235

Johns RL, Wermelinger M, Mascaro R et al (2020) Autonomous dry stone: on-site planning and assembly of stone walls with a robotic excavator. Constr Robot 4:127–140. https://doi.org/10.1007/s41693-020-00037-6

Jud D, Hottiger G, Leemann P, Hutter M (2017) Planning and control for autonomous excavation. IEEE Robot Autom Lett 2:2151–2158. https://doi.org/10.1109/LRA.2017.2721551

Jud D, Hurkxkens I, Girot C, Hutter M (2021a) Robotic embankment. Construction. Robotics 5: 101–113

Jud D, Kerscher S, Wermelinger M et al (2021b) HEAP - the autonomous walking excavator. Autom Constr 129:103783. https://doi.org/10.1016/j.autcon.2021.103783

Keller Unternehmungen (2023). https://robmade.com. Accessed 20 Apr 2023

Leung PY, Apolinarska AA, Tanadini D, et al (2021) Automatic assembly of jointed timber structure using distributed robotic clamps. In: 'PROJECTIONS'–Proceedings of the 26th international conference of the association for Computer-Aided Architectural Design Research in Asia (CAADRIA 2021), pp 583–592

Li J-M, Knippers J (2015) Segmental timber plate shell for the Landesgartenschau exhibition hall in Schwäbisch Gmünd – the application of finger joints in plate structures. Int J Space Struct 30: 123–139

Mascaro R, Wermelinger M, Hutter M, Chli M (2021) Towards automating construction tasks: large-scale object mapping, segmentation, and manipulation. Journal of Field Robotics 38: 684–699

Melenbrink N, Werfel J, Menges A (2020) On-site autonomous construction robots: towards unsupervised building. Autom Constr 119:103312. https://doi.org/10.1016/j.autcon.2020.103312

Ming C, Mirjan A, Medina Ibanez J et al (2022) Impact printing. 3D Printing and Additive Manufacturing 9:203–211

Mirjan A, Mata-Falcón J, Rieger C et al (2022) Mesh mould prefabrication. In: Third RILEM international conference on concrete and digital fabrication: digital concrete 2022. Springer, pp 31–36

Petersen KH, Napp N, Stuart-Smith R et al (2019) A review of collective robotic construction. Sci Robot 4:eaau8479. https://doi.org/10.1126/scirobotics.aau8479

Piškorec L, Jenny D, Parascho S et al (2019) The brick labyrinth. In: Willman J, Block P, Hutter M et al (eds) Robotic fabrication in architecture, art and design. Springer, Cham, pp 489–500. https://doi.org/10.1007/978-3-319-92294-2_37

Prado M, Dörstelmann M, Schwinn T, et al (2014) Core-less filament winding: robotically fabricated fiber composite building components. In: McGee W, Ponce de Leon M (eds) Robotic fabrication in architecture, art and design, pp 275–289. https://doi.org/10.1007/978-3-319-04663-1_19

ROB Technologies (2023). https://rob-technologies.com. Accessed 20 Apr 2023

Sandy T, Giftthaler M, Dörfler K et al (2016) Autonomous repositioning and localization of an in situ fabricator. In: 2016 IEEE international conference on robotics and automation (ICRA). IEEE, pp 2852–2858

Scheurer F, Schindler C, Braach M (2005) From design to production: three complex structures materialised in wood

Schwinn T, Menges A (2015) Fabrication agency: Landesgartenschau exhibition hall. Arch Des 85: 92–99. https://doi.org/10.1002/ad.1960

Shen LY, Tam VWY, Tam CM, Drew D (2004) Mapping approach for examining waste management on construction sites. J Constr Eng Manag 130:472–481. https://doi.org/10.1061/(ASCE)0733-9364(2004)130:4(472)

Søndergaard A (2014) Odico formwork robotics. Arch Des 84(3):66–67. https://doi.org/10.1002/ad.1756

Søndergaard A, Feringa J, Nørbjerg T et al (2016) Robotic hot-blade cutting: an industrial approach to cost-effective production of double curved concrete structures. In: Reinhardt D, Saunders R, Burry J (eds) Robotic fabrication in architecture, art and design. Springer, Cham, pp 150–164. https://doi.org/10.1007/978-3-319-26378-6

Thoma A, Adel A, Helmreich M et al (2018) Robotic fabrication of bespoke timber frame modules. In: Willmann J, Block P, Hutter M et al (eds) Robotic fabrication in architecture, art and design: foreword by Sigrid Brell-Çokcan and Johannes Braumann, Association for Robots in architecture. Springer, Cham, pp 447–458

Vasey L, Baharlou E, Dörstelmann M et al (2015) Behavioral design and adaptive robotic fabrication of a fiber composite compression shell with pneumatic formwork. In: Computational ecologies: design in the anthropocene, Proceedings of the 35th annual conference of the Association for Computer Aided Design in Architecture (ACADIA), University of Cincinnati, Cincinnati, pp 297–309

Vasey L, Felbrich B, Prado M et al (2020) Physically distributed multi-robot coordination and collaboration in construction: a case study in long span coreless filament winding for fiber composites. Constr Robot 4:3–18

Wangler T, Roussel N, Bos FP et al (2019) Digital concrete: a review. Cem Concr Res 123:105780. https://doi.org/10.1016/j.cemconres.2019.105780

Wagner HJ, Alvarez M, Groenewolt A, Menges A (2020a) Towards digital automation flexibility in large-scale timber construction: integrative robotic prefabrication and co-design of the BUGA wood pavilion. Constr Robot 4:187–204

Wagner HJ, Alvarez M, Kyjanek O et al (2020b) Flexible and transportable robotic timber construction platform: TIM. Autom Constr 120:103400. https://doi.org/10.1016/j.autcon.2020.103400

Willmann J, Knauss M, Bonwetsch T et al (2016) Robotic timber construction: expanding additive fabrication to new dimensions. Autom Constr 61:16–23

Wu P, Wang J, Wang X (2016) A critical review of the use of 3-D printing in the construction industry. Autom Constr 68:21–31. https://doi.org/10.1016/j.autcon.2016.04.005

Lauren Vasey is a post-doctoral researcher at ETH Zurich, within Gramazio Kohler Research and National Centre of Competence in Research (NCCR) digital fabrication, focusing on adaptive robotic construction. Previously she was a research associate at Institut für computerbasiertes Entwerfen (ICD) Stuttgart, where she taught within the Integrative Technologies & Architectural Design Research M.Sc. Program (ITECH) master's program and collaborated with the European Space Agency, Autodesk, and Kuka.

Petrus Aejmelaeus-Lindström is an architect interested in sustainable construction enabled by computational design, robotic fabrication, and material innovation. Since 2011 he has been part of Gramazio Kohler Research, chair of architecture and digital fabrication at ETH Zurich, where he worked on interdisciplinary research projects and conducted his doctoral research.

David Jenny is a practising architect and researcher focusing on digital fabrication and computational design methods for sustainable constructive systems. He is a senior research associate at the Centre for Building Technologies and Processes, Zurich University of Applied Sciences. At ETH Zurich, he was leading the teaching projects of the postgraduate program MAS ETH in Digital Fabrication.

Ryan Luke Johns conducted his doctoral research with Gramazio Kohler Research, together with the Robotic Systems Lab at ETH Zurich. He is the co-founder of the design research studio GREYSHED and has taught as a lecturer at Princeton University, Rensselaer Polytechnic Institute, and Vassar College, and as an adjunct assistant professor at Columbia GSAPP.

Ilmar Hurkxkens is head of the Geographic Design cluster at Boskalis and lecturer at the Section Landscape Architecture of TU Delft. His work focuses on the form and processes of natural granular materials using advanced surveying, modelling, and fabrication techniques. He is the co-founder of design research labs LANDSKIP and UngenauRobotics.

Coralie Ming holds an M.Arch from the University of Melbourne and an MAS in Digital Fabrication from ETH Zürich. She is currently a consultant with Boston Consulting Group (BCG) in the area of Digital, Technology, and Data Consulting. Coralie Ming previously worked in the industry as an architect and held positions in teaching and research at ETH Zürich with Gramazio Kohler Research where she led the Clay Rotunda, SE MusicLab project.

Marco Hutter is Associate Professor and Director of the Robotic Systems Lab at ETH Zurich and founder and advisor of anybotics AG. The Robotic Systems Lab builds up control technologies for autonomous operation in challenging environments, with a strong focus on applications for construction robotics with mobile manipulation systems.

Fabio Gramazio is an architect and Professor of Architecture and Digital Fabrication at ETH Zurich. As a co-founder of the group Gramazio Kohler Research and initiator of the NCCR Digital Fabrication, his work focuses on the integration of digital fabrication and robotic technologies into architectural design and construction. He has lectured and exhibited his work worldwide and has received numerous awards for his contributions to the field.

Matthias Kohler is an architect and Professor of Architecture and Digital Fabrication at ETH Zurich. As a co-founder of the group Gramazio Kohler Research and initiator of the NCCR Digital Fabrication, his work focuses on the integration of digital fabrication and robotic technologies into architectural design and construction. He has lectured and exhibited his work worldwide and has received numerous awards for his contributions to the field.

Chapter 10
Extended Reality as a Catalyst for Circular Economy Transition in the Built Environment

Ranjith K. Soman, Dragana Nikolić, and Benjamin Sanchez

Abstract Extended reality (XR) technologies refer to mixed reality and virtual reality configurations that augment real or represent fully virtual information in an intuitive and immersive manner, transforming the way we plan, design, construct, and operate built environment assets. XR offers great potential to support and accelerate the transition of built environment practices to a circular economy by supporting decisions based on narrow, slow, close, and regenerate strategies. Narrow strategies use XR to simulate the building process to identify potential issues, reduce material waste, and avoid costly mistakes. Slow strategies use XR to enable construction with durable materials and designing for adaptability to extend the lifespan of buildings. Close strategies use XR to facilitate material recovery and support repurposing and reuse, thus reducing waste. Regenerate strategies use XR as a motivational tool to engage citizens, communities, and professionals in design and management decisions. However, applying XR is not without challenges, including technical and process-related limitations, potential misuse, and a lack of rich digital twins. Future research opportunities include the development of rich and accurate digital twins, ethical and sustainable use of XR technologies, and overcoming technical and logistical challenges through interdisciplinary collaboration and user-friendly and accessible XR hardware and software.

Keywords Extended reality (XR) · Mixed reality (MR) · Virtual reality (VR) · Immersive experiences · Built environment · Circular economy · Digital twins

R. K. Soman (✉)
Delft University of Technology, Delft, the Netherlands
e-mail: r.soman@tudelft.nl

D. Nikolić
University of Reading, Reading, UK

B. Sanchez
Appalachian State University, Boone, NC, USA

C. De Wolf et al. (eds.), *A Circular Built Environment in the Digital Age*, Circular Economy and Sustainability, https://doi.org/10.1007/978-3-031-39675-5_10

10.1 Introduction

Extended reality (XR) is an umbrella term for the kinds of technologies that mediate user perception of digital information. The overarching aim of XR is to augment human perception by giving users compelling, intuitively interactive, and often immersive experiences with little to no awareness on the part of the user of the interference (LaValle 2016). XR technologies can be conceptualised and classified using the virtuality continuum (Milgram and Kishino 1994), a continuous scale spanning from entirely virtual to real worlds (see Fig. 10.1), encompassing varying extents to which real and virtual objects overlap in a mediated environment. We broadly refer to this middle as mixed reality (MR) approaches, although these approaches are also referred to as augmented reality (AR) (the virtual augments the real), augmented virtuality (the real augments the virtual), and diminished reality (removing content from a user's visual environment). XR thus includes various configurations of MR and virtual reality (VR) situated on the virtuality continuum.

10.1.1 Need for XR

We are witnessing unprecedented ways of how we generate, visualise, and share information through more intuitive, wearable, and ever more powerful devices. Within the built environment practices, technologies have long held a promise of offering ways to improve the design and delivery of assets at a greater quality and improved performance. With an urgent call to respond to the challenges of climate change, reduce carbon emissions, and eliminate waste, basing decisions on how to design for future uses increasingly depends on understanding the implications of the status and planned interventions. A network of sensors and real-time data that users generate with their mobile devices begin to give us clues for detecting patterns and simulating and predicting future needs and plan design interventions accordingly (Whyte and Nikolić 2018). While technologies have already supported these kinds of simulations, the trajectory is towards automating these processes by making a more direct link to the readily available sources of data. These novel digital capabilities can transform design and delivery, increase off-site manufacturing, and alter design practices where the delivery is not only for the physical but also for digital assets or digital twins.

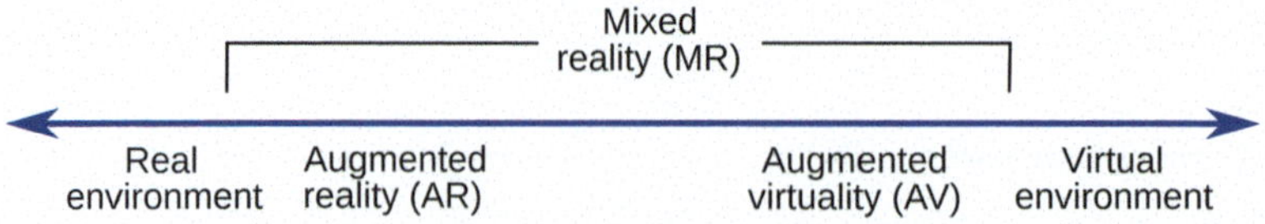

Fig. 10.1 Reality–virtuality continuum. (Adapted from Milgram and Kishino 1994)

When human experiences and behaviours do not lend themselves easily to automation, technologies can furthermore be used to engage people in conversations to share ideas, knowledge, and experiences that may otherwise remain elusive to the designers. In such instances, using visualisation and communication technologies, such as VR or AR, can present a design as something tangible and shareable with clients and prospective users. We can broadly refer to these types of technologies that replicate, present, extend, or augment the real environment and information as XR technologies.

10.1.2 Components of XR System

XR can be viewed as computer-generated environments that are displayed using an array of special hardware to give users compelling, intuitively interactive, and – when relevant – immersive experiences. The XR medium is a combination of output and input devices and plays a large part in shaping user experience. For example, input devices enable users to interact with the virtual world and may also track the user's movement, while output devices display information and respond to the user's input. These include (LaValle 2016; Whyte and Nikolić 2018):

1. Displays (output): screens for visuals, or auditory, olfactory, or haptic displays for other sensory experiences.
2. Sensors (input): devices that track the user data (especially movement), such as cameras, accelerometers, gyroscopes, temperature sensors, etc.
3. Control devices (input-interaction): devices that take inputs from user interaction with the virtual world, such as keyboards, mouse, joysticks, game controllers, haptic devices, etc.
4. Computing devices: devices that generate and continuously align or change the virtual world and project the virtual world to the displays, maintaining the correspondence between users' actions in the real world and the virtual world using sensors and control devices.

The choice of XR configuration will largely depend on the context of its application and the nature of the tasks at hand. Discerning the content and perceptual characteristics of the system will further inform the appropriate hardware configuration as the most optimal for achieving the specific outcomes (Nikolić et al. 2019; Whyte and Nikolić 2018).

10.1.3 Working Principle of an XR System

XR or VR has always conveyed an aim of the "realness" of the virtual experience, although the synthetic nature of this "realness" has been the subject of debates. In all XR applications, choices are made about the salient features that offer compelling user experiences such as sense of presence or immersion. For example, as users navigate and experience the real world through the complex interplay of perceptual and sensory inputs where depth perception, movement, smell, or sound all play an important part, XR approximates and replicates this experience by replacing these sensory inputs with the artificial ones provided by visual, olfactory, sound, and motion tracking devices. However, most of the time, multisensory experience of the real world is replaced by a predominantly visual XR. The prevalence of a visual sense in XR has led to a broad classification of XR configurations based on the extent to which the user's field of view is enveloped in virtual information.

10.1.4 Types of XR Systems

Depending on the extent to which the user's field of view is covered by the digital display or the extent to which the user is visually immersed, XR systems can be broadly characterised as fully immersive, semi-immersive, or non-immersive configurations (Whyte and Nikolić 2018). Fully immersive and semi-immersive systems provide virtual experiences by enclosing the user's field of view, either entirely or partially. Specialised hardware, such as head-mounted or surround-screen displays, stereo views, and position-tracking capabilities, are used to create compelling simulations, but this requires a lot of computing power. Systems that are not immersive use common hardware, such as standard monitors and 3D glasses. They are known as "fishtank" or "window-on-a-world" systems and employ the same software as immersive systems, but without covering the user's field of view. These systems may include some immersive features like stereoscopic viewing, but they do not provide the same level of experience as fully or semi-immersive systems.

XR systems can be for single or multiple users. Head-mounted displays are an example of a single-user XR, which supports the navigation of virtual spaces using tracked controllers and body movements. These systems range from high-end fully immersive displays, like HTC Vive, to lower-end hardware, like Google Cardboard. Conversely, projection-based VR systems that employ large-screen display configurations allow multiple users to experience an XR environment simultaneously. They can also offer fully immersive experiences like enclosed CAVE systems or have more open-footprint semi-immersive applications. Recent software developments in MR have enabled multi-user experiences in head-mounted displays, where users can share the same virtual environment and interact with each other in real time. Tracking user movements and gaze enhances these experiences. Choosing the level of immersion and user participation will depend on the intended users, goals, and tasks for the XR system. Understanding these factors will inform the development of effective XR experiences.

10.2 Existing XR Applications in the Built Environment Life Cycle

The use of immersive and augmented visualisation is changing how we interact with the built environment, presenting new challenges for professionals in planning, architecture, engineering, construction, operations, and deconstruction. Collaboration across diverse roles, goals, and expertise can be facilitated using XR technologies, which enable intuitive and clear visualisation of project information, leading to easier consensus on complex projects (Nikolić and Whyte 2021). The following subsections describe how XR technologies have been used in different life cycle stages of a construction project.

10.2.1 Design Phase

XR technologies have been mostly used to support teams in making decisions during the design phase. For example, 3D interactive visualisation using XR has been more effective in increasing the users' accuracy, perception, and memory in understanding the designs (Calderon-Hernandez et al. 2019; Roupé et al. 2016). From a cognitive viewpoint, XR offers more effective visualisation compared to viewing 3D information on 2D monitors, deemed to be cognitively more demanding (Hermund et al. 2018). This principle has been used in collaborative visualisation for participatory design, crowd simulations, and interactive visualisation of simulation data such as structural simulation, lighting simulations, and fluid dynamics simulations (Safikhani et al. 2022). Studies have revealed that the use of XR is effective in design for presenting different design configurations.

XR provides a realistic, safe, and fully manipulable testing environment through interactions with the virtual environment. BIM workflows can be completed with real-time interactive visualisation enabling communication and collaboration with stakeholders who may not possess BIM skills (Prouzeau et al. 2020). For example, lighting design done based on the designer's previous experience or simulation results (Natephra et al. 2017) may not account for the intended users' experience. Instead, XR can leverage user feedback from occupants during the lighting design process (Natephra et al. 2017; Wong et al. 2019). Similarly, quantitative metrics, such as movement speed and directions, can also be incorporated in VR-based human-building interaction studies to improve the design. Biometric sensors such as electroencephalograms, galvanic skin responses, and facial- or vision-based electromyography can be used as measurements. Further, risk situations and the safety of workers can be effectively assessed with VR. Specific aspects that imperil virtual labour can be easily identified and adjusted (Casini 2022). How people experience these spaces and interact with each other can directly influence people's health and comfort. Data observed during XR interactive simulations can, in turn, be collected and analysed (Casini 2022; Wong et al. 2020).

During the design phase of construction, VR technology is used more often than other XR configurations. VR offers immersive, 1:1 scale representations of the final structure, which can be experienced by a single user wearing a head-mounted display, or by multiple users viewing the environment on large screens. The level of immersion in these VR environments can vary, with some using stereoscopic displays for a highly realistic experience and others using monoscopic displays for a less intense but still informative experience. Ultimately, the choice of VR technology used during the design phase of construction will depend on the specific application and the users' needs.

10.2.2 Construction Phase

XR-supported communication can be used to extend or replace the traditional concept of face-to-face communication in projects. For example, in a study by Du et al. (2016), a cloud-based multi-user XR communication platform enhanced interpersonal interactions and supported users in performing better on assigned construction tasks than users in the traditional desktop application. The use of collaborative VR configurations offers communication experiences comparable to traditional face-to-face communication, particularly in terms of discussion quality (level of effectiveness and satisfaction experienced), communication richness (detailed responses and compelling messages), and openness (enjoyableness and open-mindedness). However, in-person communication still tends to outperform remote alternatives with higher accuracy due to a strong reliance on social cues and the weak human–human interaction in the current generation of XR (Abbas et al. 2019). Prior studies have identified that XR could lead to better problem-finding performance (Wu et al. 2019), improving communication efficiency among stakeholders and motivating them to share a common vision for the project as a joint walk-through (Du et al. 2016; Shi et al. 2016).

In addition to collaborative tasks, XR has been used in construction planning, site planning, and execution visualisation. VR, for example, can play a significant role in the design of construction workplace scenarios (Yu et al. 2019) and can be used both as a learning environment for workers and as a planning tool for construction managers. To create a virtual construction site for information sharing between disciplines, construction teams can visualise the execution methods on a construction site to better understand the procedures (Tran 2019). The 4D simulation workflow in VR can provide a supportive environment for constructability analysis meetings (Boton 2018). However, a proper workflow is required to update the virtual environment according to the current design state (Vincke et al. 2019).

AR is widely used during the construction phase to visualise the design of a building or infrastructure before it is built on the site, allowing stakeholders to identify any issues or discrepancies early on and make necessary changes. This can help reduce the risk of costly delays or rework during construction. AR can also be used to create virtual mock-ups of the construction site, which can be helpful for planning logistics and identifying potential issues (Chalhoub and Ayer 2019). Additionally, there are AR applications that provide stepped instructions and guidance to workers on site, helping to reduce the risk of errors or accidents (Kwiatek et al. 2019; Lin et al. 2020). Furthermore, AR can be used to check that construction work is being completed to the correct specifications and standards, enabling quality control and ensuring that the final product meets the desired standards (Zhou et al. 2017).

Single-user and multi-user VR systems are commonly used in the office to visualise the design of a building or infrastructure, identify potential issues, and create virtual mock-ups of the construction. This allows stakeholders to assess the design and plan for any necessary changes before the construction phase begins. In contrast, single-user AR is typically used on the construction site, overlaying digital information onto the real world and assisting with various tasks. Head-mounted displays and window-on-the-world displays can deliver AR content on the construction site. While head-mounted displays offer a more immersive experience, window-on-the-world displays are often preferred due to cost and safety considerations on the construction site. These devices allow workers to see digital information while maintaining a clear view of their surroundings.

10.2.3 Operations and End-of-Life

The use of XR in the operations and end-of-life stage lags behind its use in design and construction stages but offers great potential for facility managers to enhance information retrieval and visualisation of maintenance-related issues. In operations and maintenance use cases, AR has been explored more than VR to support tasks that require an overlay of virtual information over existing assets. For example, during maintenance interventions, technicians can use AR to augment their view of the physical world with overlaid digital content. AR presents information in a context-aware and more comprehensible manner, allowing more effective operations and flexibility in workers' deployment. Manuri et al. (2019) proposed an AR-based system to help the user detect and avoid errors during the maintenance process.

One of the most common applications of XR is certainly that of remote maintenance. Colleagues and experts can see the direct view captured by the operator's AR device on site and send back augmented support information back to the operator along with voice instructions. XR allows facility managers to enhance data visualisation by displaying information right on the field. XR allows the device to estimate the location and orientation of the user. Localisation is performed via global navigation satellite system (GNSS) positioning and/or by comparing the user's perspective to BIM based on deep learning computation (Casini 2022). Wearable devices can enable a collaborative workspace between different professionals. On-site and remote team members can consult with each other. XR allows the creation of collaborative environments where several people, who may even be in different places, can walk around and simultaneously interact with a virtual 3D model (Lee and Yoo 2021).

In addition, AR is being used to preview renovations and retrofit interventions (Casini 2022). With AR, the user can use the screen of a smartphone or tablet to project a "digital window" that overlays the BIM model of an object. Mobile AR applications such as AirMeasure and MeasureKit enable direct measurement of objects directly on the screen of the device. AR solutions can also support the scheduling and planning of building renovations. AR-enhanced visualisation of non-visual data can be a useful cognitive aid for identifying the information needed for decision-making (Meža et al. 2014). Chung et al. (2018) presented a study in which AR-based smart facility management systems demonstrated faster and easier access to information. Alonso-Rosa et al. (2020) presented a monitoring energy system based on mobile AR. AR systems can also overlay the results of building thermal or fluid dynamics simulations on the virtual model or project those in the real environment.

XR technologies can aid in the efficient disassembly and material recovery process during end of life. They are used to support reversible BIM and BIM-based selective disassembly planning for buildings (SDPB). Reversible BIM is a virtual platform that estimates and visualises the degree of reversibility at a component level (Durmisevic et al. 2021). SDPB evaluates BIM disassembly models to optimise disassembly sequence plans and program deconstruction works (Sanchez et al. 2021). In these approaches, XR can provide step-by-step instructions and guidance to workers, reducing the risk of errors and accidents (Kwiatek et al. 2019; Lin et al. 2020). XR can also be used for disassembly sequencing and communication, helping workers identify recovered materials and how they can be reused or recycled (Frizziero et al. 2019). Additionally, XR can provide visual cues and feedback during the disassembly process, assisting workers in identifying the correct tools and techniques for specific building components (Eswaran et al. 2023).

10.3 Leveraging XR for Circular Strategies

This section discusses how XR technologies could foster a transition to a circular economy, especially when applied in tandem with the circular strategies of regenerate, narrow, slow, and close (see Table 10.1).

10.3.1 Regenerate

The regenerate principle focuses on creating sustainable systems that actively restore and enhance their environments, and XR technologies can play a crucial role in enabling such strategies. By facilitating collaboration among professionals, users, and citizens, XR can be utilised for various purposes, including design, participation,

Table 10.1 Summary of existing XR research categorised by circular strategies

		Design	Construct	Operate	Deconstruct
Regenerate	Stimulate human nature and biodiversity	Ball et al. (2008) and Chandler et al. (2022)			
	Use healthy and renewable resources	Kamel Boulos et al. (2017)			
Narrow	Reduce primary input	Parry and Guy (2021) and Wibranek and Tessmann (2023)	Farghaly et al. (2021)	Wibranek and Tessmann (2023)	Farghaly et al. (2021)
	Design for performance	Fukuda et al. (2019), Rezvani et al. (2023) and Banfi et al. (2022)			
	Improve efficiency	Natephra et al. (2017)	Chen and Huang (2013)	Banfi et al. (2022) and Scorpio et al. (2020)	
Slow	Design for long life	Dembski et al. (2019)			
	Design for reversibility				Kunic and Naboni (2022)
	Lifetime extension	Li et al. (2022)		Alavi et al. (2021) and Corneli et al. (2019)	Carbonari et al. (2022), Li et al. (2022) and Gheisari et al. (2016)
	Smart use of space			Kunic and Naboni (2022)	
	Deliver access and performance			Issa and Olbina (2015)	
	Reuse				Parry and Guy (2021)
Close	Recycle			Do et al. (2020) and Mohamad et al. (2021)	
	Urban mining	O'Grady et al. (2021)	Calderon-Hernandez (2018) and Lin et al. (2019)		Frizziero et al. (2019) and Eswaran et al. (2023)
	Track and trace resources			Munaro and Tavares (2021)	

and promoting pro-environmental behaviours. In urban planning and design, head-mounted displays such as OculusRift paired with the powerful Esri CityEngine have been used to engage citizens and communities in evaluating neighbourhood walkability and street noise levels (Kamel Boulos et al. 2017), as well as urban resource allocation, disaster planning, and environmental protection (Chen et al. 2013; Vanegas et al. 2009). In environmental planning applications, for example, fully immersive single-user VR configurations may be used to create a stronger sense of presence to evoke emotional responses when the goal is for participants to act or make behaviour-related decisions (Ball et al. 2008). In the context of land use and biodiversity, Chandler et al. (2022) have explored dynamic audio-temporal virtual landscapes simulating seasonal changes in VR to offer users a visceral experience of these complex dynamics and build stakeholder empathy. In informing and aligning often diverse perceptions on environment and landscape values, XR and simulation technologies can support a constructive debate about alternative options for design and management decisions (Griffon et al. 2011).

10.3.2 Narrow

Narrow strategies focus on optimising specific aspects of a system or process, often leading to incremental improvements. In the context of sustainable architecture and construction, this can involve enhancing resource efficiency, building performance, and user engagement, among others (Çetin et al. 2021). XR technologies can contribute to these improvements, including supporting decision-making with data-intensive simulations, creating high-performing buildings, increasing user engagement, and facilitating renovations for better resource use.

To improve resource efficiency, multiple scenarios can be developed to improve resource efficiency using state-of-the-art modelling and machine learning methods. However, the results of these simulations and models are data-intensive and multi-dimensional, adding complexity for the array of stakeholders in the decision-making process. XR technology can represent these data-intensive results more meaningfully to support decision-making (Dembski et al. 2019). For example, complex computational fluid dynamics models can be simplified using AR to enable users to create high-performing buildings without compromising design requirements (Fukuda et al. 2019).

XR technologies have been used to create high-performing buildings. They have been used to convey data on energy performance, thermal comfort, and lighting from real environments and the BIM models and present them to users more intuitively to improve building performance and user comfort (Banfi et al. 2022). XR technologies combined with BIM and game engines can offer interactive visualisation to support the design of highly efficient lighting by comparing multiple lighting configurations. This can then be combined with user engagement studies to create the best lighting scenarios both at a building and a city scale (Scorpio et al. 2020), where users can change, move, and rotate fixtures (Natephra et al. 2017). Furthermore, the efficiency of new building construction has also been influenced by XR technologies through logistics and construction simulations (Chen and Huang 2013).

XR also contributes to high-performance designs through increased user engagement by offering novel ways to perceive BIM information. VR can present the BIM model through the lens of human experience and perception, allowing for design assessments of issues such as traffic sign sensitivity, road marking, highway landscaping, traffic safety, lane glare, and more (Rezvani et al. 2023). As AR combines the real world with virtual information, mobile applications of AR can scan the real world and measure and identify the materials of existing stock. This information is then connected to design algorithms to promote the integration of pre-used components in a new building, thereby reducing the primary resource needs (Wibranek and Tessmann 2023). Finally, combining XR technologies and image processing techniques creates as-built representations of existing assets, generating material databases for future buildings (Sato et al. 2016).

XR technology has also increased existing building reuse through better renovation. As it can display virtual models of alternative design scenarios superimposed over the existing physical facilitation, combining BIM with MR can speed up and improve the quality of renovation design processes (Carbonari et al. 2022). Old buildings are often renovated with complicated site constraints, multiple interests, and limited capital costs. Therefore, the transformation process has always encouraged stakeholders to participate in improving the design's effectiveness. Tangible user interfaces made up of physical models further simplify the operation. However, most designs are projected, which does not provide a realistic interactive experience. Interactivity and clear visualisation are two advantages of XR technology. Studies have established that using the participatory design approach of XR technology with tangible models will provide a powerful platform for engaging stakeholders in renovating old buildings (Li et al. 2022). In this regard, Gheisari et al. (2016) developed a methodology of a semi-augmented-reality tool, using BIM and panorama, for a building renovation project. They concluded that superimposing the building information models using an augmented panoramic environment provides construction personnel with a simple way to access their required information in a natural, interactive, and location-independent virtual environment.

10.3.3 Slow

As with the narrow strategy, XR use encompasses improving assets' life cycle, design for reversibility, adaptability, and reconfigurability to support the slow strategy.

XR technologies can act as an interface for building digital twins and be used in facility management to extend the asset's service life. Maintenance activities are improved by providing the location of malfunctioning equipment and appropriate and reliable information, and downtime is reduced. Such integration will help the facility managers in optimising building maintenance strategies and decision-making (Alavi et al. 2021). The main challenge in intensifying asset use and extending their valuable service life is the retrieval of specific data during the life cycle of buildings. However, generating and updating information required for operating buildings is costly and the inventory requires thousands of person-hours. To address these concerns, XR combined with deep learning techniques has been used to retrieve the asset data for a real-time check on the status. The proposed system aims to achieve some degree of automation in the data collection process, particularly compared to current inventory procedures that still require lengthy post-processing (Corneli et al. 2019). The applications of AR in facility management include intelligent fault diagnosis, visualised operation guidance, situational awareness, and building performance monitoring (Issa and Olbina 2015).

Another example is the use of XR for design for reversibility, adaptability, and reconfigurability. Kunic and Naboni (2022) developed a methodology for collaborative design and construction of reconfigurable wood structures in an AR environment. They concluded that using AR can drastically increase the efficiency in the process of assembly and reassembly of reconfigurable systems.

10.3.4 Close

The close strategy in the built environment revolves around efficient resource use, recycling, and reusing materials, ultimately minimising waste and promoting sustainability. The following paragraphs delve into how XR technologies can facilitate recycling by training people to identify waste types, track resources throughout a built asset's life cycle, and support urban mining by visualising the bill of quantities and material stock. Moreover, XR can aid in disassembly sequencing and communication, providing visual cues and feedback to ensure effective recovery and implementation of circular economy design.

For recycling, XR has been effective in training people to identify different wastes and dispose of them accordingly. This application uses interaction cues and visual and auditory feedback to help the user learn proper recycling behaviour (Do et al. 2020). The users are guided by these visual interaction cues as to which elements they can interact with, where to go in navigation, and what information is available about the content in the proposed AR application called Recycl-AR. The visual interaction cues framework comprises four components: task, markedness, trigger, and characteristic (Mohamad et al. 2021). In addition, there is work on using AR to use construction waste in new projects effectively. This has been tested to create a wooden structure using scrap timber beams of different cross sections. Furthermore, the same method has been tested to work with different lengths of waste material too. Instead of the more traditional method of designing and documenting, the designer had a more flexible relationship with the design and the digitised inventory of parts. This technique reflects a fundamental shift in the design paradigm, where designers work with a blank slate of materials where cost or structural competence is the only constraining factor (Parry and Guy 2021).

In addition to recycling, XR can help track resources over the life cycle of built assets. This helps to create a material database with high provenance. The materials passports can be used to obtain a comprehensive set of information and tracking in order to reuse and recycle building materials (Munaro and Tavares 2021). However, it is difficult to maintain up-to-date material passports. As stated earlier, AR has already been used for facility management tasks (Alavi et al. 2021; Chung et al. 2018). The materials passports can thus be integrated with hybrid reality-based facility management systems to improve data maintenance. IT will also be easy to retrieve the details of an asset as it is localised and contextualised in an XR environment.

Furthermore, XR technologies can effectively close and support urban mining. For example, studies have begun to explore how VR tools can allow building designers to see and implement their plans for improving CE design. The XR tools can support users to visualise the bill of quantities and material stock embedded within the studied building, furthering our knowledge of concepts such as buildings as material banks. Furthermore, they allow building designers to see and implement their plans for improving CE design (O'Grady et al. 2021). In addition, XR can be used for disassembly sequencing and communication (Frizziero et al. 2019). The construction sector has already used AR-based construction sequencing for assembling buildings (Calderon-Hernandez 2018; Lin et al. 2019). Lessons from the manufacturing sector show that creating disassembly sequences for construction and embedding the (dis)assembly sequence in the materials passports for effective recovery hold great potential. XR can provide value as it can provide visual cues and feedback during the disassembly (Eswaran et al. 2023).

10.4 Circular Economy Examples of XR in Construction Practice

The upcoming section presents three examples of how XR is used in the construction industry to enable circular transition strategies.

10.4.1 Collaborative Visualisation of Design

Collaborative design visualisation in the construction industry is a process that brings together different experts, such as architects, engineers, contractors, and clients, to work together in the design and construction of buildings and infrastructure projects. Collaborative design visualisation in the construction industry is achieved using XR semi-immersive technologies enabling real-time collaboration among experts. By working together, these experts can identify areas of the design that can be optimised for resource efficiency and the use of fewer inputs in products.

For example, through collaborative visualisation, they can detect and eliminate redundancies, reduce waste, and identify areas where fewer resources could be used while maintaining the desired functionality and performance. This contributes to the narrow strategy of circular transition by reducing the overall environmental impact of the built environment and enabling the sustainable use of resources.

An example of collaborative design visualisation is 3D MOVE (Mobile Visualisation Environment), developed at the University of Reading. It is a collaborative tool for multiple users to interact with and explore full-scale 3D models and built environments (see Fig. 10.2). A study of this technology showed that using collaborative VR environments like 3D MOVE can give project teams more ability to question, evaluate, and justify design decisions (Nikolić et al. 2019), resulting in improved design performance and efficiency and reducing resources needed to build the asset. There are commercial offerings that provide collaborative design visualisation capabilities for the construction industry. For example, Mission Room and Fulcro Fullmax are two companies that offer semi-immersive hardware solutions for construction projects. They use software solutions such as Revitzo, Unity Reflect, BIM 360, and Fuzor for their software workflows.

Fig. 10.2 Collaborative visualisation in 3D Move. (Nikolić et al. 2019)

10.4.2 Construction Production Control Rooms

Construction production control rooms are collaborative digital interfaces that offer real-time project information and efficient construction management. These non-immersive XR-based rooms serve as interfaces to the construction stage digital twins. They use fishtank displays, which present information in a two-dimensional manner like a regular computer screen. Construction production control rooms play an essential role in implementing narrow and close strategies in construction projects. They provide real-time project information, including Gantt charts, schedules, and resource allocation, leading to efficient use of resources, time, and budget while minimising unnecessary delays or rework. By reducing rework, they help to minimise waste and maximise the value of materials, closing resource loops. They optimise resource use by identifying underutilised or overused areas and making necessary adjustments. Overall, production control rooms are valuable for promoting narrow and close strategies in construction projects by enabling real-time data tracking, interpretation, and collaboration for efficient construction management. The AEC Production Control Room (see Fig. 10.3) is an example of a collaborative visualisation platform that aims to make the UK construction industry more efficient and proactive by providing a scalable and repeatable platform for construction management and reporting (Farghaly et al. 2021).

10.4.3 Construction AR

Construction AR enables construction professionals to interact with digital models and information overlaid onto the physical environment, which can play a significant role in helping the industry achieve circular transition strategies that include narrow, slow, and close concepts. In the narrow approach, AR can optimise resource use by

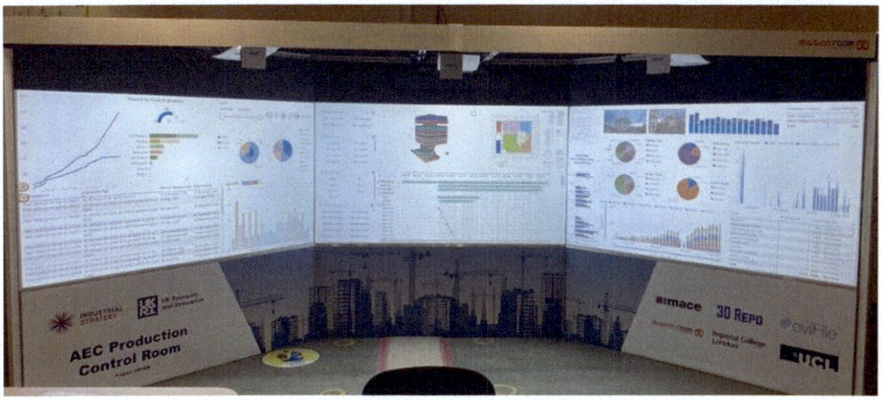

Fig. 10.3 Collaborative data visualisation in AEC Production Control Room

providing real-time data on resource availability, allocation, and utilisation, improving resource efficiency, narrowing resource flows, and reducing waste. In the slow approach, AR can help construction teams improve project planning, prevent delays, rework, and material waste, and provide progress tracking and documentation features that ensure all team members capture project progress at the exact same location over time. In the close approach, AR can help with the recovery and reuse of materials from construction sites by providing real-time data on the location and condition of building materials, which can support the identification and tracking of materials for potential reuse or recycling, closing the loop of material use and contributing to a circular transition.

There are several companies offering AR solutions for the construction industry. Arvizio provides features such as 3D model and LiDAR scan import, processing, optimisation, and hybrid rendering to build digital twins and facilitate use cases such as design reviews, spatial data management, marketing demos, and quality assurance inspections. Innovative construction technology (ICT) offers ICT Tracker, an AR software for contractors to streamline project installation tracking and reporting. This app provides comprehensive data in easy-to-read reports and allows contractors to compare BIM or 3D models against current installations, preventing margin slip. VisualLive offers AR solutions on HoloLens 1 & 2, Android, and iOS, enabling AEC professionals to push design models onto their AR devices and bring their computer-aided design (CAD) or BIM onto the construction site. Lastly, the XYZ Reality Atom is a powerful engineering tool that combines a construction safety headset, AR displays, and in-built computing power (see Fig. 10.4). It uses laser-based tracking technology to position 3D design models with millimetre accuracy, allowing users to view holograms of models positioned within construction tolerances.

Fig. 10.4 Augmented visualisation and construction site using XYZ Reality Atom

10.5 Discussion

XR technologies have significant, yet untapped potential to support and accelerate the transition to a circular economy. XR provides intuitive, immersive, and interactive experiences of imagined futures that can support more informed decision-making and optimise the use of resources. XR can help the construction sector transition to a circular economy by enabling more effective and informed design and construction processes while reducing waste and extending the lifespan of buildings. With narrow strategies, VR and AR can be used to simulate the building process and identify potential issues, helping construction companies avoid costly mistakes and reduce material waste. Slow strategies, such as building with durable materials and designing for adaptability, allow buildings to have a longer lifespan, reducing the need for new construction. Close strategies, such as using XR to facilitate deconstruction and material recovery at the end of a building's life, help ensure that materials can be repurposed and reused, further reducing waste and supporting a circular economy. Finally, for regeneration strategies, XR can be used as a powerful participatory and motivational tool to engage citizens, communities, and professionals in the design and management decisions of the built environment.

However, there are several limitations to applying XR that warrant further and careful consideration. One major limitation is the slow development of rich digital twins of buildings, infrastructure, and other physical assets. Digital twins are crucial for the success of XR in the built environment because they provide detailed and accurate representations of physical assets, which are essential for creating realistic and useful XR experiences. Rich digital twins enable XR technologies to provide dynamic and responsive representations of the built environment, which can enable more efficient and effective design, construction, and operation of the built environment. Without access to rich digital twins, XR technologies may not be able to provide the necessary level of detail and accuracy for effective use in the circular economy. Another limitation is the potential for misuse or abuse of XR technologies. It is essential to ensure that the virtual models of proposed circular systems created by XR technologies reduce domain-specific biases but are based on realistic and sustainable principles, which reflect the needs and preferences of stakeholders, such as workers, consumers, and the environment. Misleading or ineffective designs can have negative impacts on the success of the circular economy and may even be harmful. Finally, there are technical and logistical challenges associated with implementing XR technologies in the circular economy. Small and medium-sized enterprises may face barriers to adoption and use due to the need for specialised hardware, software, and training, which require significant resources and expertise. There may also be challenges in integrating XR technologies with existing systems and processes in the circular economy, which need to be addressed to ensure that the technologies can be implemented effectively and generate the desired benefits. However, with careful planning and investment, these challenges can be overcome, and XR technologies can play a key role in accelerating the transition to a circular economy.

Future research opportunities for XR in the circular economy include the development of rich and accurate digital twins, ensuring ethical and sustainable use of XR technologies, and overcoming technical and logistical challenges. To address these challenges, interdisciplinary collaboration is necessary, involving experts in architecture, engineering, computer science, sustainability, and social sciences. Additionally, the development of user-friendly and accessible XR hardware and software is crucial, with the potential for Metaverse-like technologies to provide inclusive and intuitive interfaces for a wider range of stakeholders. However, the development of XR software must be guided by the principles of circularity and sustainability and be evaluated in terms of their social, economic, and environmental impacts to ensure that they contribute to the broader goals of a more circular and resilient built environment. It is essential to remain sceptical of the potential for XR technologies to address the challenges of the circular economy and to design and evaluate them through a participatory and inclusive process that reflects the diversity of perspectives and experiences of those who will be using the technologies.

10.6 Key Takeaways

- Extended reality (XR) can help implement the regenerate principle by fostering collaboration, promoting pro-environmental behaviours, and supporting constructive debates on design and management decisions.
- XR can enable data-driven decision-making in the narrow strategy for a circular economy, contributing to high-performing buildings, increased user engagement, and better resource utilisation.
- XR can support the slow strategy for a circular economy by enhancing asset life cycle management and promoting design for reversibility, adaptability, and reconfigurability.
- XR can facilitate recycling efforts by improving waste identification, resource tracking, and urban mining visualisation, while also aiding in disassembly sequencing and communication.
- The lack of rich digital twins of physical assets, inadequate use of XR technologies, and technical and logistical challenges still tend to limit the effective implementation of XR in the circular economy and need to be addressed.

Acknowledgements Research work of the first and third authors contributing to this chapter was conducted at the Future Cities Lab Global at Singapore-ETH Centre and ETH Zurich. Future Cities Lab Global is supported and funded by the National Research Foundation, Prime Minister's Office, Singapore, under its Campus for Research Excellence and Technological Enterprise (CREATE) programme and ETH Zurich, with additional contributions from the National University of Singapore, Nanyang Technological University, Singapore, and the Singapore University of Technology and Design.

References

Abbas A, Choi M, Seo J, Cha SH, Li H (2019) Effectiveness of immersive virtual reality-based communication for construction projects. KSCE J Civ Eng 23:4972. https://doi.org/10.1007/s12205-019-0898-0

AirMeasure. http://airmeasure.com/. Accessed 15 May 2023

Alavi H, Forcada N, Fan SL, San W (2021) BIM-based augmented reality for facility maintenance management. In: Proceedings of the 2021 European conference on Computing in Construction, pp 431–438. https://doi.org/10.35490/EC3.2021.180

Alonso-Rosa M, Gil-de-Castro A, Moreno-Munoz A, Garrido-Zafra J, Gutierrez-Ballesteros E, Cañete-Carmona E (2020) An IoT based mobile augmented reality application for energy visualization in buildings environments. Appl Sci 10(2). https://doi.org/10.3390/app10020600

Ball J, Capanni N, Watt S (2008) Virtual reality for mutual understanding in landscape planning. Int J Soc Sci 2(2):78–88

Banfi F, Brumana R, Salvalai G, Previtali M (2022) Digital twin and cloud BIM-XR platform development: from scan-to-BIM-to-DT process to a 4D multi-user live app to improve building comfort, efficiency and costs. Energies 15(12):4497. https://doi.org/10.3390/en15124497

Boton C (2018) Supporting constructability analysis meetings with immersive virtual reality-based collaborative BIM 4D simulation. Autom Constr 96:1–15. https://doi.org/10.1016/j.autcon.2018.08.020

Calderon-Hernandez C (2018) Lean, BIM and augmented reality applied in the design and construction phase: a literature review. Int J Innov Manag Technol 60–63. https://doi.org/10.18178/ijimt.2018.9.1.788

Calderon-Hernandez C, Paes D, Irizarry J, Brioso X (2019) Comparing virtual reality and 2-dimensional drawings for the visualization of a construction project. In: Computing in civil engineering 2019, pp 17–24. https://doi.org/10.1061/9780784482421.003

Carbonari A, Franco C, Naticchia B, Spegni F, Vaccarini M (2022) A mixed reality application for the on-site assessment of building renovation: development and testing. Sustainability 14(20). https://doi.org/10.3390/su142013239

Casini M (2022) Extended reality for smart building operation and maintenance: a review. Energies 15(10). https://doi.org/10.3390/en15103785

Çetin S, De Wolf C, Bocken N (2021) Circular digital built environment: an emerging framework. Sustainability 13(11):6348. https://doi.org/10.3390/su13116348

Chalhoub J, Ayer SK (2019) Exploring the performance of an augmented reality application for construction layout tasks. Multimed Tools Appl 78(24):35075–35098. https://doi.org/10.1007/s11042-019-08063-5

Chandler T, Richards AE, Jenny B, Dickson F, Huang J, Klippel A, Neylan M, Wang F, Prober SM (2022) Immersive landscapes: modeling ecosystem reference conditions in virtual reality. Landsc Ecol 37(5):1293–1309. https://doi.org/10.1007/s10980-021-01313-8

Chen HM, Huang PH (2013) 3D AR-based modeling for discrete-event simulation of transport operations in construction. Autom Constr 33:123–136. https://doi.org/10.1016/j.autcon.2012.09.015

Chen M, Lin H, Hu M, He L, Zhang C (2013) Real-geographic-scenario-based virtual social environments: integrating geography with social research. Environ Plan B 40(6):1103–1121. https://doi.org/10.1068/b38160

Chung S, Kwon S, Moon D, Ko T (2018) Smart facility management systems utilizing open BIM and augmented/virtual reality. In: 34th International Symposium on Automation and Robotics in Construction, Taipei, Taiwan. https://doi.org/10.22260/ISARC2018/0118

Corneli A, Naticchia B, Cabonari A, Bosché F (2019) Augmented reality and deep learning towards the management of secondary building assets. In: 36th International Symposium on Automation and Robotics in Construction, Banff, AB, Canada. https://doi.org/10.22260/ISARC2019/0045

Dembski F, Wössner U, Letzgus M (2019) The digital twin tackling urban challenges with models, spatial analysis and numerical simulations in immersive virtual environments. Blucher Des Proc:795–804. https://doi.org/10.5151/proceedings-ecaadesigradi2019_334

Do TD, Yu DS, Katz A, McMahan RP (2020) Virtual reality training for proper recycling behaviors. In: ICAT-EGVE 2020 – international conference on Artificial Reality and Telexistence and Eurographics Symposium on Virtual Environments. https://doi.org/10.2312/EGVE.20201284

Du J, Shi Y, Mei C, Quarles J, Yan W (2016) Communication by interaction: a multiplayer VR environment for building walkthroughs. In: Construction research congress 2016, pp 2281–2290. https://doi.org/10.1061/9780784479827.227

Durmisevic E, Guerriero A, Boje C, Domange B, Bosch G (2021) Development of a conceptual digital deconstruction platform with integrated reversible BIM to aid decision making and facilitate a circular economy. In: Proceedings of the Joint Conference CIB W78-LDAC, Luxembourg, pp 11–15

Eswaran M, Gulivindala AK, Inkulu AK, Raju Bahubalendruni MVA (2023) Augmented reality-based guidance in product assembly and maintenance/repair perspective: a state-of-the-art review on challenges and opportunities. Expert Syst Appl 213:118983

Farghaly K, Soman RK, Whyte J (2021) Visualizing real-time information through a construction production control room. In: Proceedings of the 2021 European conference on Computing in Construction, vol 2, pp 415–422. https://doi.org/10.35490/EC3.2021.169

Frizziero L, Liverani A, Caligiana G, Donnici G, Chinaglia L (2019) Design for disassembly (DfD) and augmented reality (AR): case study applied to a gearbox. Mach Des 7(2):29. https://doi.org/10.3390/machines7020029

Fukuda T, Yokoi K, Yabuki N, Motamedi A (2019) An indoor thermal environment design system for renovation using augmented reality. J Comput Des Eng 6(2):179–188. https://doi.org/10.1016/j.jcde.2018.05.007

Gheisari M, Foroughi Sabzevar M, Chen P, Irizzary J (2016) Integrating BIM and panorama to create a semi-augmented-reality experience of a construction site. Int J Constr Educ Res 12(4):303–316

Griffon S, Nespoulous A, Cheylan JP, Marty P, Auclair D (2011) Virtual reality for cultural landscape visualization. Virtual Reality 15(4):279–294. https://doi.org/10.1007/s10055-010-0160-z

Hermund A, Klint L, Bundgaard T (2018) BIM with VR for architectural simulations building information models in virtual reality as an architectural and urban design tool. In: 6th annual international conference on Architecture and Civil Engineering (ACE 2018). https://doi.org/10.5176/2301-394X_ACE18.152

Issa RRA, Olbina S (eds) (2015) Building information modeling: applications and practices. Am Soc Civil Eng. https://doi.org/10.1061/9780784413982

Kamel Boulos MN, Lu Z, Guerrero P, Jennett C, Steed A (2017) From urban planning and emergency training to Pokémon Go: applications of virtual reality GIS (VRGIS) and augmented reality GIS (ARGIS) in personal, public and environmental health. Int J Health Geogr 16(1):7. https://doi.org/10.1186/s12942-017-0081-0

Kunic A, Naboni R (2022) Collaborative design and construction of reconfigurable wood structures in a mixed reality environment. In: Proceedings of the 25th SIGraDi conference: critical appropriations

Kwiatek C, Sharif M, Li S, Haas C, Walbridge S (2019) Impact of augmented reality and spatial cognition on assembly in construction. Autom Constr 108:102935. https://doi.org/10.1016/j.autcon.2019.102935

LaValle S (2016) Virtual reality. Cambridge University Press

Lee Y, Yoo B (2021) XR collaboration beyond virtual reality: work in the real world. J Comput Des Eng 8(2):756–772. https://doi.org/10.1093/jcde/qwab012

Li J, Lo TT, Guo X, Chen F (2022) Integrating XR and tangible model to enhance old building renovation design. In: Proceedings of ECAADe 2022: co-creating the future – inclusion in and through design, pp 207–216. https://doi.org/10.52842/conf.ecaade.2022.2.207

Lin Z, Petzold F, Hsieh SH (2020) 4D-BIM based real time augmented reality navigation system for tower crane operation. In: Construction research congress 2020, November, Reston, pp. 828–836. https://doi.org/10.1061/9780784482865.088

Lin Z, Petzold F, Ma Z (2019) A real-time 4D augmented reality system for modular construction progress monitoring. In: 36th international symposium on Automation and Robotics in Construction, Banff, AB, Canada. https://doi.org/10.22260/ISARC2019/0100

Manuri F, Pizzigalli A, Sanna A (2019) A state validation system for augmented reality based maintenance procedures. Appl Sci 9(10). https://doi.org/10.3390/app9102115

MeasureKit. https://measurekit.com/. Accessed 15 May 2023

Meža S, Turk Ž, Dolenc M (2014) Component based engineering of a mobile BIM-based augmented reality system. Autom Constr 42:1–12. https://doi.org/10.1016/j.autcon.2014.02.011

Milgram P, Kishino F (1994) A taxonomy of mixed reality visual displays. IEICE Trans Inf Syst 77(12):1321–1329

Mohamad NHR, Ijab MT, Hamid S (2021) Visual interaction cues framework for recycling education using augmented reality. In: 2021 International Conference on Computer Science and Engineering (IC2SE), pp 1–8. https://doi.org/10.1109/IC2SE52832.2021.9791879

Munaro MR, Tavares SF (2021) Materials passport's review: challenges and opportunities toward a circular economy building sector. Built Environ Proj Asset Manag 11(4):767–782. https://doi.org/10.1108/BEPAM-02-2020-0027

Natephra W, Motamedi A, Fukuda T, Yabuki N (2017) Integrating building information modeling and virtual reality development engines for building indoor lighting design. Vis Eng 5(1):19. https://doi.org/10.1186/s40327-017-0058-x

Nikolić D, Maftei L, Whyte J (2019) Becoming familiar: How infrastructure engineers begin to use collaborative virtual reality in their interdisciplinary practice. J Inf Technol Constr 24:489–508. https://doi.org/10.36680/j.itcon.2019.026

Nikolić D, Whyte J (2021) Visualizing a new sustainable world: Toward the next generation of virtual reality in the built environment. Buildings 11(11):Article 11

O'Grady TM, Brajkovich N, Minunno R, Chong HY, Morrison GM (2021) Circular economy and virtual reality in advanced BIM-based prefabricated construction. Energies 14(13):4065. https://doi.org/10.3390/en14134065

Parry C, Guy S (2021) Recycling construction waste material with the use of AR. In: Yuan PF, Yao J, Yan C, Wang X, Leach N (eds) Proceedings of the 2020 DigitalFUTURES. Springer, Singapore, pp 57–67. https://doi.org/10.1007/978-981-33-4400-6_6

Prouzeau A, Wang Y, Ens B, Willett W, Dwyer T (2020) Corsican twin: authoring in situ augmented reality visualisations in virtual reality. In: Proceedings of the international conference on advanced visual interfaces, pp. 1–9. https://doi.org/10.1145/3399715.3399743

Rezvani M, Taghaddos H, Sobhkhiz S, Noghabaei M, Ghorab K (2023) User engagement for sustainable development: how can virtual reality help? In: Walbridge S, Nik-Bakht M, Ng KTW, Shome M, Alam MS, el Damatty A, Lovegrove G (eds) Proceedings of the Canadian Society of Civil Engineering Annual Conference 2021, vol 251, pp 239–249. Springer, Singapore. https://doi.org/10.1007/978-981-19-1029-6_18

Roupé M, Johansson M, Tallgren MV, Jörnebrant F, Tomsa PA (2016) Immersive visualization of building information models. In: Living systems and micro-utopias: towards continuous designing, Proceedings of the 21st International Conference of the Association for Computer-Aided Architectural Design Research in Asia (CAADRIA 2016), pp 673–682

Safikhani S, Keller S, Schweiger G, Pirker J (2022) Immersive virtual reality for extending the potential of building information modeling in architecture, engineering, and construction sector: systematic review. Int J Digit Earth 15(1):503–526. https://doi.org/10.1080/17538947.2022.2038291

Sanchez B, Rausch C, Haas C, Hartmann T (2021) A framework for BIM-based disassembly models to support reuse of building components. Resour Conserv Recycl 175:105825
Sato Y, Fukuda T, Yabuki N, Michikawa T, Motamedi A (2016) A marker-less augmented reality system using image processing techniques for architecture and urban environment. In: 21st International Conference of the Association for Computer-Aided Architectural Design Research in Asia (CAADRIA): Living Systems and Micro-Utopias: Towards Continuous Designing, 11
Scorpio M, Laffi R, Masullo M, Ciampi G, Rosato A, Maffei L, Sibilio S (2020) Virtual reality for smart urban lighting design: review, applications and opportunities. Energies 13(15):3809. https://doi.org/10.3390/en13153809
Shi Y, Du J, Lavy S, Zhao D (2016) A multiuser shared virtual environment for facility management. Procedia Eng 145:120–127. https://doi.org/10.1016/j.proeng.2016.04.029
Tran N (2019) Exchange information enhancement in execution method by using virtual reality. ISPRS Int Arch Photogramm Remote Sens Spatial Inf Sci XLII-5(W2):71–75. https://doi.org/10.5194/isprs-archives-XLII-5-W2-71-2019
Vanegas CA, Aliaga DG, Benes B, Waddell P (2009) Visualization of simulated urban spaces: inferring parameterized generation of streets, parcels, and aerial imagery. IEEE Trans Vis Comput Graph 15(3):424–435. https://doi.org/10.1109/TVCG.2008.193
Vincke S, de Lima HR, Bassier M, Vergauwen M (2019) Immersive visualisation of construction site point cloud data, meshes and BIM models in a VR environment using a gaming engine. Int Arch Photogramm Remote Sens Spatial. Inf Sci XLII-5-W2:77–83. https://doi.org/10.5194/isprs-archives-XLII-5-W2-77-2019
Whyte J, Nikolić D (2018) Virtual reality and the built environment, 2nd edn. Routledge. https://doi.org/10.1201/9781315618500
Wibranek B, Tessmann O (2023) Augmented reuse: a mobile app to acquire and provide information about reusable building components for the early design phase. In: Gengnagel C, Baverel O, Betti G, Popescu M, Thomsen MR, Wurm J (eds) Towards radical regeneration. Springer, pp 411–423. https://doi.org/10.1007/978-3-031-13249-0_33
Wong JY, Chun Chieh Y, Yong ST, Chan A, Kok ST, Lau T, Ali M, Gouda E (2020) BIM-VR framework for building information modelling in engineering education. Int J Interact Mob Technol (IJIM) 14:15. https://doi.org/10.3991/ijim.v14i06.13397
Wong MO, Du J, Zhang ZQ, Liu YQ, Chen SM, Lee SH (2019) An experience-based interactive lighting design approach using BIM and VR: a case study. IOP Conf Ser Earth Environ Sci 238(1):012006. https://doi.org/10.1088/1755-1315/238/1/012006
Wu TH, Wu F, Kang SC, Chi HL (2019) Comparison of virtual communication environment for remote BIM model review collaboration. In: ISARC Proceedings, pp 1149–1154
Yu Z, Peng H, Zeng X, Sofi M, Xing H, Zhou Z (2019) Smarter construction site management using the latest information technology. Proc Inst Civil Eng – Civil Eng 172(2):89–95. https://doi.org/10.1680/jcien.18.00030
Zhou Y, Luo H, Yang Y (2017) Implementation of augmented reality for segment displacement inspection during tunneling construction. Autom Constr 82:112–121. https://doi.org/10.1016/j.autcon.2017.02.007

Ranjith K. Soman is an Assistant Professor at TU Delft specialising in digital construction and project management, and an Honorary Research Associate at Imperial College London. His research combines information technology with civil engineering, with a particular focus on using extended reality, knowledge graphs, and artificial intelligence to improve infrastructure sustainability.

Dragana Nikolić is an associate professor in Digital Built Environment at the University of Reading. Her work focuses on advanced digital and visualisation technologies to support and improve design and construction.

Benjamin Sanchez is an Assistant Professor at Appalachian State University, North Carolina, USA. Benjamin's research is focused on the development and implementation of digital technologies (BIM, IoT, LCA, 3D scan-to-BIM) for increasing the performance of construction building projects in terms of sustainability and circular economy.

Part III
Business and Governance

Chapter 11
Digital Technology Use Cases for Deconstruction and Reverse Logistics

Marc van den Berg

Abstract The transition towards a circular built environment challenges dismantling firms to revisit their practices. These firms traditionally demolish buildings with crushing force, essentially creating poorly recyclable waste. This practice leads to a loss of economic value and has several negative social and environmental consequences. Deconstruction, defined as construction in reverse, represents an alternative practice in which as many materials are recovered as possible. Deconstruction is particularly challenging because responsible firms need to process more information to organise various reverse logistics options efficiently. This chapter, therefore, reviews reverse supply chain practices in construction and illustrates how digital technologies could support dismantling firms and their partners during essential deconstruction activities. Through evidence-based insights and examples from practice, the chapter presents a state-of-the-art overview of digital deconstruction technology use cases for identifying, harvesting, and distributing reusable building elements. It shows that digital technologies have been developed for separate deconstruction activities but are rarely used in an integrated manner. Further integration through aligning the digital technologies with practitioners' information needs will, accordingly, unlock new opportunities for closed-loop material flows.

Keywords Circular Economy · Deconstruction · Digital Technologies · Information Needs · Reverse Logistics · Reuse

11.1 Introduction

The transition towards a circular built environment challenges the construction industry to rethink and reorganise building end-of-life practices. The fate of almost every obsolete building is conventional demolition, during which dismantling firms essentially convert it into waste (Thomsen et al. 2011). Dismantling firms typically

M. van den Berg (✉)
Department of Civil Engineering & Management, Faculty of Engineering Technology, University of Twente, Enschede, The Netherlands
e-mail: m.c.vandenberg@utwente.nl

C. De Wolf et al. (eds.), *A Circular Built Environment in the Digital Age*, Circular Economy and Sustainability, https://doi.org/10.1007/978-3-031-39675-5_11

use heavy equipment and crushing force to efficiently tear down buildings at the end of their service life. These demolition practices generate huge amounts of waste. It is estimated that demolition waste, together with waste generated during construction and renovation, accounts for approximately 30–40% of all solid waste (Cheshire 2016; Li et al. 2020). The sheer volume of construction and demolition waste has high environmental impacts, particularly associated with its logistics and land occupation (Gálvez-Martos et al. 2018). Traditional landfilling of the waste can also cause space problems in densely populated areas and may lead to contamination of nearby water bodies (Cooper and Gutowski 2015). These problems thus call for novel end-of-life approaches.

New approaches require rethinking materials hidden in the built environment as attractive alternatives for raw ones. Andersson and Buser (2022, p.488), for instance, illustrated how dismantling firms started renaming the materials generated during their dismantling practices as "products" or "resources" and how they viewed waste only as "a state in a never-ending transformation." Buildings can, likewise, be seen as material banks where materials are only temporarily stored (Debacker and Manshoven 2016). A building can, in this view, be used to mine resources for new constructions (Koutamanis et al. 2018). Gorgolewski (2018, p.1), likewise, envisioned how "new urban vernacular may emerge if we focus on previously used materials and components that come from the local area." This "urban mining" is an important circularity strategy for the construction industry as it can offer significant economic savings and reuse benefits (Arora et al. 2021).

The circularity strategy also calls for reorganising end-of-life practices. Conventional demolition typically marks the end-of-life phase of a building and its parts. Yet to enable reuse, dismantling firms must embrace an alternative dismantling method, called deconstruction, that is oriented towards retaining the value of building materials. Deconstruction has been described as "construction in reverse in which the building and its components are dismantled for the purpose of reusing them or enhancing recycling" (Kibert 2016, p.480). It is the first stage in reverse logistics, which is concerned with the movement of materials from the building dismantling point to the point of new construction (Hosseini et al. 2015b). Reverse logistics is nonetheless complicated in construction due to particular uncertainties, information deficiencies, and uncoordinated material flows (Tennakoon et al. 2022). Digital technologies seem particularly promising to that end as these enable data collection, integration, and analysis (Çetin et al. 2022).

This chapter describes how digital technologies could support deconstruction and reverse logistics. It first discusses challenges and information needs in reverse construction supply chains. The next section then presents an overview of how digital technologies can be used to support three essential deconstruction activities, namely: identifying, harvesting, and distributing reusable building elements. The chapter ends with an in-depth discussion of remaining technology adoption challenges and an outlook on future developments.

11.2 Reverse Supply Chains in Construction

Reverse logistics can become an effective sustainable practice with many benefits to the construction industry. This potential is not yet fully exploited though. Reverse logistics deals with products at the end of their life cycle. A general definition – originally formulated for manufacturers, wholesalers, retailers, and service firms – is that reverse logistics concerns "the process of planning, implementing, and controlling the efficient, cost effective flow of raw materials, in-process inventory, finished goods, and related information from the point of consumption to the point of origin for the purpose of recapturing value or proper disposal" (Rogers and Tibben-Lembke 1999, p. 2). These reverse flows differ in maturity per industry, but are generally quite well-developed in the manufacturing industries while often overlooked in construction (Hosseini et al. 2014). Instead, researchers and practitioners have devoted much of their attention to the classical, forward supply chain approach that "does not feel any responsibility for end-of-life" products (Govindan and Soleimani 2017, p.371). Connecting both forward and reverse supply chains into closed-loop material flows has, consequently, become at the forefront of much strategy- and policymaking (Ghaffar et al. 2020).

Closed-loop construction industries direct materials from deconstruction sites towards new construction sites, either directly or indirectly. A building owner or principal contractor usually selects a firm specialised in dismantling once a decision is made to deconstruct or renovate. That firm then initiates a range of collection, separation, sorting, treatment, reuse, and recycling activities aimed at removing a building or parts of it (Brandão et al. 2021). The end-of-life strategy can differ per individual building element, such as doors, floors, or installations. Reuse entails that an element is transferred to another location where a principal contractor assembles it again in a new construction without structurally changing it (Allwood et al. 2011). This may be achieved directly, without an intermediate party, but often reuse is indirect. It is then temporarily stored at a storage facility that serves as a buffer between reverse and forward material flows. Sometimes small repairs are also conducted at such storage facilities. Recycling entails the structural reduction of an element to its constituent materials. This is typically done by specialised waste processors, although some materials can also be recycled on site (like crushing of concrete). Suppliers can, subsequently, replace virgin materials with these recycled materials to produce new building elements. Reverse logistics practices thus represent different ways in which materials are brought back into the loop.

Possible benefits of reverse logistics practices include economic, social, and environmental aspects. Economic benefits could be achieved by cost savings offered by reusing salvaged elements instead of virgin materials (Hosseini et al. 2015b). While revenue can be made with recovered materials, the practice also saves landfill disposal costs (Diyamandoglu and Fortuna 2015). Since deconstruction is generally more labour-intensive than demolition, this can furthermore generate many new jobs. Other social benefits are the mitigation of noise, dust and compaction (Iacovidou and Purnell 2016) and an improved "green" image and reputation of the companies involved (Chileshe et al. 2018). Environmental benefits mainly include a reduction in both the use of virgin materials and in the waste generated (Del Río Merino et al. 2010). Since this reduces associated emissions and environmental impacts, reverse logistics finally represents a major climate mitigation strategy (Arora et al. 2021).

Yet compared with forward supply chains, reverse supply chains are more complex and affected by a wide range of uncertainties (Tennakoon et al. 2022). Buildings consist of heterogeneous materials and are typically immobile and not designed for deconstruction (Schultmann and Sunke 2007). The quality and size of reclaimed building elements varies widely (Iacovidou and Purnell 2016). Specifications may also be unclear, which can prompt the need to recertify them. Deconstruction thereby appears, on average, financially less attractive than conventional demolition, particularly because of the higher associated costs (Dantata et al. 2005; Coelho and De Brito 2011). Recovery facilities, infrastructure, and second-hand material markets are simultaneously underdeveloped, particularly in comparison with the manufacturing industries (Hosseini et al. 2014). Sourcing of reusable building elements therefore often requires individual searching and negotiation (Allwood 2014). Moreover, updates to rules and standards can limit the reuse potential of existing elements. A principal problem with reuse thus concerns matching supply with demand, as reclaimed materials may not show up at the right time, in the right amount, or with the right dimensions (Gorgolewski 2008).

The lack of information is a root cause of these challenges (Hosseini et al. 2015a; Chileshe et al. 2019; Wu et al. 2022). Uncertainty has been defined as a lack of information required to make a project decision (Winch 2015). Reverse supply chain operations need to deal with uncertainties related to the building, workflow, and environment (Van den Berg et al. 2020a). A typical building-related uncertainty that dismantling firms often face concerns the lack of information about the current conditions of any elements that potentially can be reused. It may, for example, be unclear which manufacturer produced a certain element, under what conditions, and according to which quality standards. Yet also any later damage or wear and tear of the elements may not have been documented well. Information, or the lack thereof, thus plays a crucial role in determining the actual conditions of building elements. But information must also be processed during a wide range of other organisational activities, such as coordinating site work or maintaining interorganisational relationships. Reverse material flows must thus be supported with sound information flows (Jayasinghe et al. 2019).

These information flows nonetheless appear to be hampered due to the complex, fragmented, cross-functional, and multi-disciplinary nature of reverse supply chains (Wijewickrama et al. 2021a). There are many actors involved in reuse and recycling processes. Their activities are typically dispersed and disordered. Information could strengthen the coordination among these activities, but is often poorly shared between different actors (Chileshe et al. 2019). Systemic information-sharing gaps were identified at links between the forward and reverse supply chains (Wijewickrama et al. 2021b). This was explained because of limited collaboration and connections between key actors involved in building operation and end-of-life stages. According to Wu et al. (2022), the most important barriers for sharing information are a lack of certainty in market environments, limited trust among actors, and a lack of government support. Poorly connected information flows consequently hinder the successful implementation of circularity in the built environment.

Digitalisation efforts across the sector are nevertheless opening new possibilities to support reverse supply chain practices. Innovative solutions are needed to address the various wicked barriers. To that end, information and communication technologies (ICT) are increasingly recognised and prioritised as critical circularity enablers (Demestichas and Daskalakis 2020). Yu et al. (2022), as such, mapped the readiness and effectiveness of ICT-based decision support tools throughout the building life cycle. They related end-of-life research with ICT solutions based on building information modelling (BIM), geographic information systems (GIS), radio-frequency identification (RFID), modelling and simulation (MS), and big data analytics (BDA). Such solutions are still rarely implemented during deconstruction and reverse logistics though. The evidence base of potential digital technology usages has thus far remained limited.

11.3 Digital Deconstruction Technology Use Cases

This section illustrates circularity-oriented use cases of digital technologies that support deconstruction and reverse logistics. These usages are structured along activities that dismantling firms and their partners follow: identifying, harvesting, and distributing reusable building elements.

11.3.1 Identify Reusable Building Elements

One of the first activities in any deconstruction project is identifying reusable materials. Wassenberg (2011) listed several reasons for dismantling a building, such as physical decay, a surplus of similar buildings, changed needs or expectations, quality-of-life problems, or social engineering processes. These different reasons suggest that at least some of the building elements may still be reusable. When there is a demand for such elements, it can be attractive to recover and resell those (Van den Berg et al. 2020b). A dismantling firm will therefore analyse existing building conditions to identify any such reusable building elements. Building owners may also stimulate this by mandating that the selected dismantling firm ensures the reuse and/or recycling of a certain number of elements.

Dismantling firms need information to make sense of existing building conditions. Basic project information about the building type, floor area, and primary materials used provides input for quick waste estimations based on waste rates per unit, like kg/m^2 or m^3/m^2 (Mah et al. 2016). Waste audits, site visits, dismantling contracts, and as-built or construction drawings fulfil most of these information needs (Tennakoon et al. 2022). However, more accurate building information is warranted to determine the reuse potentials of distinct elements – and that is most often incomplete, obsolete, or fragmented for many existing buildings (Volk et al. 2014). Dismantling firms will want to know about the material composition and the aesthetic and structural performance of distinct elements. Information about the number, type, and accessibility of the way those elements are connected to other elements is needed to assess whether any such elements can be reclaimed without damaging them or not. Furthermore, relevant market information from waste processors and material suppliers is needed to become aware which elements are demanded in secondary markets.

Several types of building capture and auditing technologies have emerged in response to as-is information needs. Such technologies aim to provide accurate insights into the geometric dimensions and other material properties of existing building elements (Han et al. 2021). BIM-based representations can, for example, be used to review how constructions were built (Van den Berg et al. 2020a). Inventory methods that combine photography with digital forms to record relevant characteristics of building elements and to assess their reuse potentials are also used more and more often by dismantling firms (see Wahlström et al. 2019) – or by partnering firms to which such activities may be outsourced (like Rotor or Sloopcheck). Honic et al. (2021, p. 1) demonstrate that material passports also provide "an outstanding advantage" to that end. Material passports essentially give elements an identity by digitally describing their characteristics, location, history, and ownership status (Luscuere 2017; Çetin et al. 2021). Such passports could inform dismantling firms about reuse potentials and enable them to extract exact quantities, but they are mainly being developed for new buildings rather than existing ones (Chap. 5 on Material Passports by Honic et al.).

More automated digital modelling methods have also emerged, though these still demand significant effort and cause high costs (Rašković et al. 2020). Laser scanners can capture dense 3D measurements of any building's as-is conditions, and the resulting point cloud can be processed to create a BIM that reflects the current situation (Tang et al. 2010). Using geometry as a foundation, modellers then attempt to augment the building representation with object metadata (semantics) related to any facet of the built environment. Since this can be a time-consuming and error-prone process, much research has been devoted to automating parts of it (Fathi et al. 2015; Che et al. 2019). As such, object recognition algorithms have been developed for walls (see Ochmann et al. 2016) and some other common building elements. Such algorithms are still infrequently combined into scalable and contextualised methods (Czerniawski and Leite 2020). Further advances in scan-to-BIM techniques that rely on low-cost, accessible hardware can nevertheless promise "a logistical base for complex reuse analyses" (Gordon et al. 2023, p.14).

Digital technologies are also used to support waste management decision-making. Having acquired insight into the as-is conditions of a to-be-deconstructed building, dismantling firms need to estimate how much waste will be generated. Lifetime analyses, which are based on a mass balance principle, assume that waste can be quantified based on the initial mass of constructed buildings and reasonable projections of material life cycles (Wu et al. 2014). Alternatively, more recent approaches attempt to quantify waste and its associated impacts based on BIM (Cheng and Ma 2013; Ge et al. 2017). For example, Kang et al. (2022) developed a conceptual framework that integrated BIM with advanced technologies, such as Internet of Things (IoT), to assist in planning alternative reuse and recycling scenarios; Su et al. (2021) combined BIM, GIS, and life cycle assessment (LCA) to develop a waste estimation and evaluation system. Works like these attempt to promote more informed waste management decisions and help to identify which building elements could be reused through closed material loops.

11.3.2 Harvest Reusable Building Elements

The next deconstruction activity concerns harvesting those building elements that were identified as reusable. The Dutch architectural firm Superuse Studios coined the term "harvesting" in reference to the practice of reclaiming valuable elements from the existing built environment – with the aim to reuse those in new buildings (Jongert et al. 2011). Dismantling firms typically do not reuse building elements themselves: they enable reuse through this harvesting. The intention to reuse implies that damage to selected elements must be minimised. Harvesting (or reclaiming) those elements hence usually requires non-destructive techniques and more skilled labour over a longer duration (Coelho and De Brito 2013). This implicates that the site work must be reorganised accordingly.

Information needs for harvesting building elements originate mainly from the workflow on site. The sequence and time allocated for deconstruction tasks are essential variables that dismantling firms need to control (Chileshe et al. 2019). Site work starts with disconnecting services and removing any present hazardous materials, like asbestos. Reusable elements can then be disassembled and (temporarily) stored somewhere on- or off-site. Dismantling firms process planning information (e.g., Gantt charts or timetables) and other project management documentation to coordinate these interdependent tasks (Van den Berg et al. 2020a). To ensure compliance with regulatory frameworks, the firms thereby need information regarding government planning requirements, health and safety guidelines, and waste handling procedures (Tennakoon et al. 2022). Information is furthermore needed to sort and prepare transportation of any harvested building elements to the next destination.

Digital technologies can support coordinating deconstruction workflows. BIM is particularly suited to facilitate the planning and organisation of site work. It can, at the outset, provide input for handling instructions and procedures to minimise possible damage during disassembly. Information may be retrieved regarding, for instance, the thickness of the cover concrete of an embedded steel connection to be removed (Akbarnezhad et al. 2014). BIM could also be used to analyse and visualise deconstruction sequencing. It is crucial to understand interdependencies and physical relationships between different elements. To that end, Marzouk and Elmaraghy (2021) used a BIM plugin to illustrate how mechanical, electrical, and plumbing

(MEP) elements intersect with walls (embedded, ending, or passing). Such insights can be used to determine in which order the elements need to be disassembled. Other existing BIM functionalities related to spatiotemporal site analyses could support managing where and when specific tasks, such as crane operations (Tak et al. 2021) or storage of deconstructed elements, need to be done. Evidence of dismantling firms using BIM for purposes like the above is nevertheless still scarce though.

Robotic technologies are likewise only occasionally used on deconstruction sites. These technologies are being developed with the intention to perform deconstruction more efficiently and precisely (Bademosi and Issa 2022). For example, Lee et al. (2015) presented a prototyping process for automated and robotised disassembly of high-rise buildings. As another example, Chen et al. (2022) described a compact robot prototype for automatic waste recycling. Robotic technologies like these are much more common in industrialised construction settings though. In end-of-life contexts, they may prove particularly suitable for repetitive deconstruction tasks. However, a general downside from a sustainability perspective is that they require the additional consumption of a significant amount of energy for operating tasks.

Other digital technologies prepare for future use. Dismantling firms will need to generate or update reusability information about the selected elements, for example, through a material passports platform. That information can then be made available to other actors in the reverse supply chain (see Wijewickrama et al. 2021b). Information systems may furthermore be needed to label harvested building elements so that those can then be tracked to new construction sites or intermediate storage facilities. The technologies can, accordingly, lead to more informative harvesting practices.

11.3.3 Distribute Reusable Building Elements

Deconstruction ends with activity regarding distributing the harvested building elements. Dismantling firms organise the diverging movements of materials away from a site. They can do this on their own or together with a transportation partner. Destinations for the different building elements typically differ. Depending on the planned end-of-life strategies, elements are transported to a new construction site (for direct reuse), an intermediate warehouse/hub (for indirect reuse), a reprocessing facility (for recycling), or a landfill site (for disposal/incineration). Dismantling projects, accordingly, lead to a large number of transport movements and associated environmental impacts, which is a primary reason that construction and demolition waste is a priority for most environmental programmes around the world (Gálvez-Martos et al. 2018).

Distribution activities depend on information to facilitate matchmaking between the supply and demand for reusable building elements (Van den Berg et al. 2020a). When a dismantling firm is contracted, that firm usually obtains ownership of the focal building and will attempt to resale reusable elements to contractors or other potential buyers. This triggers information needs. Dismantling firms need information about the current market conditions, such as prevailing prices and price volatility (Wijewickrama et al. 2021a). Buyers also need information about reusable elements, such as where and when certain elements are (or will become) available. Information is furthermore used to organise logistics or, in other words, to make sure that harvested elements arrive at the right destination at the right time (Chap. 2 by Tsui et al. on GIS). For organising closed-loop material flows, it is thus essential that material flows are accompanied with supportive information flows (Jayasinghe et al. 2019).

Various e-commerce initiatives have emerged that aim to connect supply and demand for harvested building elements. Online marketplaces for local or global trade in salvaged construction materials are growing rapidly (Caldera et al. 2020). Most dismantling firms in the Netherlands, for example, maintain their own online stores on which they showcase reclaimed building elements for sale. Common elements that can be found on such online stores include doors, timber beams, windows, insulation materials, furniture, and heating systems. Elements are typically accompanied with a picture and some information about relevant characteristics (e.g., type and dimensions), including indications of any wear or damage. Some of that information could also be retrieved from an accurate BIM. As such, Jayasinghe and Waldmann (2020) demonstrated a web-based tool that links elements to their digital counterparts in BIM. An additional benefit from linking an online store to BIM would be that the deconstruction sequencing could be automatically updated based on the demand for elements (Marzouk and Elmaraghy 2021).

Other e-commerce initiatives attempt to move beyond the project level to benefit from the advantages of scale. A particular type of online marketplace was pioneered by Jongert et al. (2011, p.56). They created (and later sold) a "harvest map," which highlights the geographic locations of reclaimed elements by plotting those on a map. This map supports resource-based design practices. It aims to serve as a regional material catalogue that a design firm can use to locate the available supply of materials in the vicinity of a new building project. Another example is the initiative "Insert," which was founded by several collaborating dismantling firms in the Netherlands. Their online platform bundles elements that were harvested by its partnering firms. The initiative also offers hubs where elements can be stored for indirect reuse and small repairs are conducted.

Digital technologies can furthermore support distribution with tracking methods. Several technologies were identified for tracking elements from an obsolete building to a new one. Van den Berg et al. (2021) experimented with a BIM-based method where site personnel simply wrote down numbers on pieces of tape attached to reusable facade elements. More advanced methods also make use of technologies to identify and index information of physical elements, such as RFID. This is a technology that uses tags and readers to make wireless communication possible (Yu et al. 2022). The technology has been coupled with BIM in efforts to develop various digital tracking systems, such as for steel components specifically (Ness et al. 2015). Xing et al. (2020) integrated RFID with a cloud-based BIM platform to allow bidirectional data exchange between physical building elements and their virtual counterparts. The general idea of such systems is that they allow the exact status (e.g., ownership or location) of individual elements to be checked and updated over time. Blockchain technologies, which save and link data records using cryptography, thereby appear promising as they offer transparency in tracing back status changes over time (Shojaei et al. 2021) (Chap. 12 by Shojaei and Naderi on blockchain technology). Actual implementations of tracking systems in reverse logistics still remain fairly limited though.

11.4 Discussion

This chapter presented a state-of-the-art overview of digital technology use cases for supporting deconstruction and reverse logistics. To realise circularity targets in the built environment, it is essential to rethink demolition waste as resources and to reorganise traditional end-of-life practices. Digitalisation advancements provide dismantling firms and their partners new possibilities to that end. With evidence-based insights and examples from practice, the present chapter illustrated how digital technologies can be used in identifying, harvesting, and distributing reusable building elements. The implications are profound, but several challenges and future perspectives remain.

The illustrated digital deconstruction technology use cases imply that reverse logistics could benefit from more informed practices. Deconstruction is an exciting life cycle stage: it can be seen as a restart rather than an end in closed-loop material flows. The practice thereby reduces the demand for raw materials. Circularity measures can be taken during design, construction, or operation stages, but these are often intended to pay off only during deconstruction. Exemplary measures listed by Benachio et al. (2020, p. 7) include "design for disassembly of building structures" (during design), "off-site construction" (during construction), and "minimise recuperative maintenance with preventive maintenance" (during operation). Measures like these merely promote reuse; actual reuse prerequires that end-of-life activities are organised accordingly. Those practices appear to be information intensive. That is, dismantling firms need information to organise reverse logistics and to realise reuse. Digital technologies have emerged with the potential to inform those practices.

Various technologies are so becoming available to dismantling firms and their partners. BIM technologies seem most prevalent: evidence of (pioneering) uses was found for identifying, harvesting, and distributing reusable elements. This may nevertheless be surprising given that dismantling firms are not acknowledged as potential BIM users in established handbooks (Eastman et al. 2011) and taxonomies (Kreider and Messner 2013). BIM models are also not available for most existing buildings (Volk et al. 2014), although that is likely to change as the methodology becomes increasingly widespread in the industry. Advances in scanning and automated digital modelling methods can thereby speed up the process of recreating accurate as-is models for existing buildings. More possibilities are also likely to emerge through ongoing efforts to develop low-cost scan-to-BIM solutions (Gordon et al. 2023) and to establish real-time connections between BIM and IoT applications into digital twins (see Deng et al. 2021). These BIM developments seem well aligned with circularity trends to replace demolition with a deconstruction alternative.

Other digital technologies can support specific tasks in deconstruction. Robotic solutions are most suitable to replace heavy and repetitive manual labour. Material passports seem particularly useful in understanding both present conditions and past history of potentially reusable building elements (Debacker and Manshoven 2016; Honic et al. 2021). These passports can inform dismantling firms during activity to identify reusable elements. They could also be linked to GIS systems and blockchain technologies, which would enable tracing building elements across space and over time (Xing et al. 2020). A particular challenge for developing any such tracing systems concerns the relatively long lifespan of building elements, which implies that robustness and future-proofness need to be taken into account. Simpler labelling solutions, such as those described by Van den Berg et al. (2021), can therefore be a pragmatic choice for distribution activities in the near future. Online stores and other e-commerce initiatives are essential to inform designers and general contractors about the (direct) supply of harvested materials, although their misalignment with demand remains a challenge (Çetin et al. 2022). Indirect reuse, where building elements are brought to and from a storage point, could improve supply predictability and create advantages of scale.

Several challenges persist that limit the uptake of digital technologies in circular end-of-life contexts though. Information is poorly shared between actors due to the fragmented, unorganised, cross-functional, and multi-disciplinary nature of reverse supply chains (Wijewickrama et al. 2021a). The industry is furthermore characterised by limited trust and governmental support (Wu et al. 2022). Dismantling firms typically face significant building uncertainty. Moreover, it is often still too costly or time-consuming to recreate (BIM) models that accurately represent as-is conditions (Czerniawski and Leite 2020). Actors may also lack the knowledge or skills to adopt certain technologies, like BIM. Other technologies, such as material passports, are only started to get standardised in the industry (see Platform CB'23 2022) and require changes in the way certain work activities are organised. Challenges in adopting digital technologies are closely related to general barriers in adopting digital technology and specific barriers that emerge from organising circular material flows (Jayasinghe et al. 2019; Çetin et al. 2022).

11.5 Outlook

Digital technologies can support dismantling firms and their partners with deconstruction and reverse logistics practices. Potential use cases for various technologies have been pioneered during information-intensive tasks in identifying, harvesting, and distributing reusable elements. Material passports and building capture and auditing technologies, most of which use an existing or recreated BIM model, can be used to identify reusable building elements. The planning and organisation of site work focuses on harvesting such elements, which can be supported with BIM, robotic technologies, and labelling methods. Distribution activities can make use of various types of e-commerce initiatives, BIM, and tracking technologies. Most of these technologies are not yet widely adopted in circular end-of-life contexts due to persistent industry and reverse supply chain challenges. Implementation of any digital technology hence requires adaptation of the technology to local project routines and vice versa. More research and development efforts are necessary to meet both practitioners' information needs and the potentials of illustrated digital technologies for promoting circular closed-loop material flows.

11.6 Key Takeaways

- Reverse logistics intends to close material loops, starting from the point of deconstruction.
- Deconstruction challenges dismantling firms to process more information for organising reverse logistics.
- Dismantling firms can use digital technologies in identifying, harvesting, and distributing reusable building elements.
- Reverse material flows remain poorly supported with information flows, as digital technologies tend to focus on separate activities only.
- Aligning digital technology use cases with practitioners' information needs could unlock new circularity opportunities.

References

Akbarnezhad A, Ong KCG, Chandra LR (2014) Economic and environmental assessment of deconstruction strategies using Building Information Modeling. Autom Constr 37:131–144

Allwood JM (2014) Squaring the circular economy: the role of recycling within a hierarchy of material management strategies. In: Worrell E, Reuter MA (eds) Handbook of recycling. Elsevier, Boston, pp 445–477

Allwood JM, Ashby MF, Gutowski TG, Worrell E (2011) Material efficiency: a white paper. Resour Conserv Recycl 55:362–381. https://doi.org/10.1016/j.resconrec.2010.11.002

Andersson R, Buser M (2022) From waste to resource management? Construction and demolition waste management through the lens of institutional work. Constr Manag Econ 40:477–496. https://doi.org/10.1080/01446193.2022.2081989

Arora M, Raspall F, Fearnley L, Silva A (2021) Urban mining in buildings for a circular economy: planning, process and feasibility prospects. Resour Conserv Recycl 174:105754. https://doi.org/10.1016/j.resconrec.2021.105754

Bademosi FM, Issa RRA (2022) Automation and robotics technologies deployment trends in construction. In: Jebelli H, Habibnezhad M, Shayesteh S, Asadi S, Lee S (eds) Automation and robotics in the architecture, engineering, and construction industry. Springer, pp 1–30

Benachio GLF, do Carmo Duarte Freitas M, Tavares SF (2020) Circular economy in the construction industry: a systematic literature review. J Clean Prod 260:121046. https://doi.org/10.1016/j.jclepro.2020.121046

Brandão R, Edwards DJ, Hosseini MR, Silva Melo AC, Macêdo AN (2021) Reverse supply chain conceptual model for construction and demolition waste. Waste Manag Res 39:1341–1355. https://doi.org/10.1177/0734242x21998730

Caldera S, Ryley T, Zatyko N (2020) Enablers and barriers for creating a marketplace for construction and demolition waste: a systematic literature review. Sustainability 12:9931. https://doi.org/10.3390/su12239931

Çetin S, De Wolf C, Bocken N (2021) Circular digital built environment: an emerging framework. Sustainability 13:6348. https://doi.org/10.3390/su13116348

Çetin S, Gruis V, Straub A (2022) Digitalization for a circular economy in the building industry: multiple-case study of Dutch social housing organizations. Resourc Cons Recyc Adv 15: 200110. https://doi.org/10.1016/j.rcradv.2022.200110

Che E, Jung J, Olsen MJ (2019) Object recognition, segmentation, and classification of mobile laser scanning point clouds: a state of the art review. Sensors 19:810. https://doi.org/10.3390/s19040810

Chen X, Huang H, Liu Y, Li J, Liu M (2022) Robot for automatic waste sorting on construction sites. Autom Constr 141:104387. https://doi.org/10.1016/j.autcon.2022.104387

Cheng JCP, Ma LYH (2013) A BIM-based system for demolition and renovation waste estimation and planning. Waste Manag 33:1539–1551. https://doi.org/10.1016/j.wasman.2013.01.001

Cheshire D (2016) Building revolutions: applying the circular economy to the built environment. RIBA Publishing, Newcastle upon Tyne

Chileshe N, Rameezdeen R, Hosseini MR, Martek I, Li HX, Panjehbashi-Aghdam P (2018) Factors driving the implementation of reverse logistics: a quantified model for the construction industry. Waste Manag 79:48–57. https://doi.org/10.1016/j.wasman.2018.07.013

Chileshe N, Jayasinghe RS, Rameezdeen R (2019) Information flow-centric approach for reverse logistics supply chains. Autom Constr 106:102858. https://doi.org/10.1016/j.autcon.2019.102858

Coelho A, De Brito J (2011) Economic analysis of conventional versus selective demolition—a case study. Resour Conserv Recycl 55:382–392. https://doi.org/10.1016/j.resconrec.2010.11.003

Coelho A, De Brito J (2013) Conventional demolition versus deconstruction techniques in managing construction and demolition waste (CDW), Handbook of recycled concrete and demolition waste. Woodhead Publishing, pp 141–185

Cooper DR, Gutowski TG (2015) The environmental impacts of reuse: a review. J Ind Ecol 21:38–56. https://doi.org/10.1111/jiec.12388

Czerniawski T, Leite F (2020) Automated digital modeling of existing buildings: a review of visual object recognition methods. Autom Constr 113:103131. https://doi.org/10.1016/j.autcon.2020.103131

Dantata N, Touran A, Wang J (2005) An analysis of cost and duration for deconstruction and demolition of residential buildings in Massachusetts. Resour Conserv Recycl 44:1–15. https://doi.org/10.1016/j.resconrec.2004.09.001

Debacker W, Manshoven S (2016) Synthesis of the state-of-the-art: key barriers and opportunities for materials passports and reversible building design in the current system. Buildings As Material Banks

Del Río Merino M, Izquierdo Gracia P, Weis Azevedo IS (2010) Sustainable construction: construction and demolition waste reconsidered. Waste Manag Res 28:118–129. https://doi.org/10.1177/0734242x09103841

Demestichas K, Daskalakis E (2020) Information and communication technology solutions for the circular economy. Sustainability 12:7272. https://doi.org/10.3390/su12187272

Deng M, Menassa CC, Kamat VR (2021) From BIM to digital twins: a systematic review of the evolution of intelligent building representations in the AEC-FM industry. J Inf Technol Construct 26:58–83

Diyamandoglu V, Fortuna LM (2015) Deconstruction of wood-framed houses: material recovery and environmental impact. Resour Conserv Recycl 100:21–30. https://doi.org/10.1016/j.resconrec.2015.04.006

Eastman C, Teicholz P, Sacks R, Liston K (2011) BIM handbook: a guide to building information modeling for owners, managers, designers, engineers, and contractors. Wiley, Hoboken

Fathi H, Dai F, Lourakis M (2015) Automated as-built 3D reconstruction of civil infrastructure using computer vision: achievements, opportunities, and challenges. Adv Eng Inform 29:149–161. https://doi.org/10.1016/j.aei.2015.01.012

Gálvez-Martos JL, Styles D, Schoenberger H, Zeschmar-Lahl B (2018) Construction and demolition waste best management practice in Europe. Resour Conserv Recycl 136:166–178. https://doi.org/10.1016/j.resconrec.2018.04.016

Ge XJ, Livesey P, Wang J, Huang SH, He X, Zhang C (2017) Deconstruction waste management through 3d reconstruction and bim: a case study. Vis Eng 5:13. https://doi.org/10.1186/s40327-017-0050-5

Ghaffar SH, Burman M, Braimah N (2020) Pathways to circular construction: an integrated management of construction and demolition waste for resource recovery. J Clean Prod 244:118710. https://doi.org/10.1016/j.jclepro.2019.118710

Gordon M, Batallé A, De Wolf C, Sollazzo A, Dubor A, Wang T (2023) Automating building element detection for deconstruction planning and material reuse: a case study. Autom Constr 146:104697. https://doi.org/10.1016/j.autcon.2022.104697

Gorgolewski M (2008) Designing with reused building components: some challenges. Build Res Inf 36:175–188. https://doi.org/10.1080/09613210701559499

Gorgolewski M (2018) Resource salvation: the architecture of reuse. Wiley, Hoboken

Govindan K, Soleimani H (2017) A review of reverse logistics and closed-loop supply chains: a Journal of Cleaner Production focus. J Clean Prod 142(Part 1):371–384. https://doi.org/10.1016/j.jclepro.2016.03.126

Han D, Kalantari M, Rajabifard A (2021) Building Information Modeling (BIM) for construction and demolition waste management in Australia: a research agenda. Sustainability 13:12983. https://doi.org/10.3390/su132312983

Honic M, Kovacic I, Aschenbrenner P, Ragossnig A (2021) Material passports for the end-of-life stage of buildings: challenges and potentials. J Clean Prod 319:128702. https://doi.org/10.1016/j.jclepro.2021.128702

Hosseini MR, Chileshe N, Rameezdeen R, Lehmann S (2014) Reverse logistics for the construction industry: lessons from the manufacturing context. Int J Constr Eng Manag 3:75–90. https://doi.org/10.5923/j.ijcem.20140303.01

Hosseini MR, Chileshe N, Rameezdeen R, Lehmann S (2015a) Integration of design for reverse logistics and harvesting of information: a research agenda. Int J Logist Syst Manag 20:480–515. https://doi.org/10.1504/IJLSM.2015.068490

Hosseini MR, Rameezdeen R, Chileshe N, Lehmann S (2015b) Reverse logistics in the construction industry. Waste Manag Res 33:499–514. https://doi.org/10.1177/0734242X1558484

Iacovidou E, Purnell P (2016) Mining the physical infrastructure: opportunities, barriers and interventions in promoting structural components reuse. Sci Total Environ 557:791–807. https://doi.org/10.1016/j.scitotenv.2016.03.098

Jayasinghe LB, Waldmann D (2020) Development of a BIM-based web tool as a material and component bank for a sustainable construction industry. Sustainability 12:1766. https://doi.org/10.3390/su12051766

Jayasinghe RS, Rameezdeen R, Chileshe N (2019) Exploring sustainable post-end-of-life of building operations: a systematic literature review. Eng Constr Archit Manag 26:689–722. https://doi.org/10.1108/ECAM-08-2017-0148

Jongert J, Nelson N, Goosens F (2011) Recyclicity: a toolbox for resource-based design. Archit Design 81:54–61. https://doi.org/10.1002/ad.1320

Kang K, Besklubova S, Dai Y, Zhong RY (2022) Building demolition waste management through smart BIM: a case study in Hong Kong. Waste Manag 143:69–83. https://doi.org/10.1016/j.wasman.2022.02.027

Kibert CJ (2016) Sustainable construction: green building design and delivery. Wiley

Koutamanis A, van Reijn B, van Bueren E (2018) Urban mining and buildings: a review of possibilities and limitations. Resour Conserv Recycl 138:32–39. https://doi.org/10.1016/j.resconrec.2018.06.024

Kreider RG, Messner JI (2013) The uses of BIM: classifying and selecting BIM uses. The Pennsylvania State University, University Park

Lee S, Pan W, Linner T, Bock T (2015) A framework for robot assisted deconstruction: process, sub-systems and modelling. In: 32nd ISARC: proceedings of the international symposium on automation and robotics in construction, Oulu, Finland

Li CZ, Zhao Y, Xiao B, Yu B, Tam VWY, Chen Z, Ya Y (2020) Research trend of the application of information technologies in construction and demolition waste management. J Clean Prod 263:121458. https://doi.org/10.1016/j.jclepro.2020.121458

Luscuere LM (2017) Materials passports: optimising value recovery from materials. Proc Inst Civil Eng Waste Resour Manage 170:25–28. https://doi.org/10.1680/jwarm.16.00016

Mah CM, Fujiwara T, Ho CS (2016) Construction and demolition waste generation rates for high-rise buildings in Malaysia. Waste Manag Res 34:1224–1230. https://doi.org/10.1177/0734242x16666944

Marzouk M, Elmaraghy A (2021) Design for deconstruction using integrated lean principles and BIM approach. Sustainability 13:7856. https://doi.org/10.3390/su13147856

Ness D, Swift J, Ranasinghe DC, Xing K, Soebarto V (2015) Smart steel: new paradigms for the reuse of steel enabled by digital tracking and modelling. J Clean Prod 98:292–303

Ochmann S, Vock R, Wessel R, Klein R (2016) Automatic reconstruction of parametric building models from indoor point clouds. Comput Graph 54:94–103. https://doi.org/10.1016/j.cag.2015.07.008

Platform CB'23 (2022) Leidraad paspoorten voor de bouw deel 1+2: Werkafspraken en onderbouwing voor paspoorten in een circulaire bouwsector. https://platformcb23.nl/images/downloads/2022/final/Leidraad_Paspoorten-voor-de-bouw-3.pdf

Rašković M, Ragossnig AM, Kondracki K, Ragossnig-Angst M (2020) Clean construction and demolition waste material cycles through optimised pre-demolition waste audit documentation: a review on building material assessment tools. Waste Manag Res 38:923–941. https://doi.org/10.1177/0734242x20936763

Rogers DS, Tibben-Lembke RS (1999) Going backwards: reverse logistics trends and practices. Reverse Logistics Executive Council, Pittsburgh

Schultmann F, Sunke N (2007) Energy-oriented deconstruction and recovery planning. Build Res Inf 35:602–615. https://doi.org/10.1080/09613210701431210

Shojaei A, Ketabi R, Razkenari M, Hakim H, Wang J (2021) Enabling a circular economy in the built environment sector through blockchain technology. J Clean Prod 294:126352. https://doi.org/10.1016/j.jclepro.2021.126352

Su S, Li S, Ju J, Wang Q, Xu Z (2021) A building information modeling-based tool for estimating building demolition waste and evaluating its environmental impacts. Waste Manag 134:159–169. https://doi.org/10.1016/j.wasman.2021.07.025

Tak AN, Taghaddos H, Mousaei A, Bolourani A, Hermann U (2021) BIM-based 4D mobile crane simulation and onsite operation management. Autom Constr 128:103766. https://doi.org/10.1016/j.autcon.2021.103766

Tang P, Huber D, Akinci B, Lipman R, Lytle A (2010) Automatic reconstruction of as-built building information models from laser-scanned point clouds: a review of related techniques. Autom Constr 19:829–843. https://doi.org/10.1016/j.autcon.2010.06.007

Tennakoon GA, Rameezdeen R, Chileshe N (2022) Diverting demolition waste toward secondary markets through integrated reverse logistics supply chains: a systematic literature review. Waste Manag Res 40:274–293. https://doi.org/10.1177/0734242X211021478

Thomsen A, Schultmann F, Kohler N (2011) Deconstruction, demolition and destruction. Build Res Inf 39:327–332. https://doi.org/10.1080/09613218.2011.585785

Van den Berg M, Voordijk H, Adriaanse A (2020a) Information processing for end-of-life coordination: a multiple-case study. Constr Innov 20:647–671. https://doi.org/10.1108/CI-06-2019-0054

Van den Berg M, Voordijk H, Adriaanse A (2020b) Recovering building elements for reuse (or not) – ethnographic insights into selective demolition practices. J Clean Prod 256:120332. https://doi.org/10.1016/j.jclepro.2020.120332

Van den Berg M, Voordijk H, Adriaanse A (2021) BIM uses for deconstruction: an activity-theoretical perspective on reorganising end-of-life practices. Constr Manag Econ 39:323–339. https://doi.org/10.1080/01446193.2021.1876894

Volk R, Stengel J, Schultmann F (2014) Building Information Modeling (BIM) for existing buildings — literature review and future needs. Autom Constr 38:109–127. https://doi.org/10.1016/j.autcon.2013.10.023

Wahlström M, Castell-Rüdenhausen MZ, Hradil P, Hauge-Smith K, Oberender A, Ahlm M, Götbring J, Hansen JB (2019) Improving quality of construction & demolition waste: requirements for pre-demolition audit. Nordic Council of Ministers

Wassenberg F (2011) Demolition in the Bijlmermeer: lessons from transforming a large housing estate. Build Res Inf 39:363–379. https://doi.org/10.1080/09613218.2011.585104

Wijewickrama MKCS, Chileshe N, Rameezdeen R, Ochoa JJ (2021a) Information sharing in reverse logistics supply chain of demolition waste: a systematic literature review. J Clean Prod 280:124359. https://doi.org/10.1016/j.jclepro.2020.124359

Wijewickrama MKCS, Rameezdeen R, Chileshe N (2021b) Information brokerage for circular economy in the construction industry: a systematic literature review. J Clean Prod 313:127938. https://doi.org/10.1016/j.jclepro.2021.127938

Winch GM (2015) Project organizing as a problem in information. Constr Manag Econ 33:106–116

Wu Z, Yu ATW, Shen L, Liu G (2014) Quantifying construction and demolition waste: an analytical review. Waste Manag Res 34:1683–1692. https://doi.org/10.1016/j.wasman.2014.05.010

Wu Z, Yang K, Xue H, Zuo J, Li S (2022) Major barriers to information sharing in reverse logistics of construction and demolition waste. J Clean Prod 350:131331. https://doi.org/10.1016/j.jclepro.2022.131331

Xing K, Kim KP, Ness D (2020) Cloud-BIM enabled cyber-physical data and service platforms for building component reuse. Sustainability 12:10329. https://doi.org/10.3390/su122410329

Yu Y, Yazan DM, Junjan V, Iacob ME (2022) Circular economy in the construction industry: a review of decision support tools based on Information & Communication Technologies. J Clean Prod 349:131335. https://doi.org/10.1016/j.jclepro.2022.131335

Marc van den Berg is Assistant Professor in Circular Construction Management & Engineering. His award-winning PhD thesis was one of the first to apply circularity thinking to the construction industry. Through combining theoretical depth and empirical subtlety, he strives to develop urban mining futures in his research. With an active role in leading circularity community of practices (like Platform CB'23), he, finally, strives to co-create actionable transition pathways for the construction industry.

Chapter 12
Blockchain Technology for a Circular Built Environment

Alireza Shojaei and Hossein Naderi

Abstract The built environment fundamentally suffers from organisational fragmentation in various aspects, such as data flow, finance, and supply chains. Blockchain technology can be considered a transformative solution to the inherent fragmentation of this industry. This chapter first defines the basics of blockchain technology to show how a peer-to-peer network could enable a decentralised, traceable, and immutable information system across the life cycles of built assets. Then, an overview of blockchain literature within the context of a circular economy, with real-life examples and the current state of blockchain adoption in the circular built environment, is presented, and the role that this technology plays in addressing certain circular strategies is discussed. Afterward, implementation challenges and incentives are identified to set realistic expectations regarding the capabilities of blockchain technologies. Emerging concepts within blockchain technologies are then presented to give insights into prospects beyond current literature and use cases in the circular built environment. Finally, the future of blockchain technology in a circular built environment is discussed to present the applicability of blockchain and its possible integration with other emerging digitalisation tools, such as building information modelling (BIM) and material passports, in wider domains of circular, smart cities and communities.

Keywords Distributed ledger · Blockchain technology · Decentralised technologies · Circular economy · Built environment

12.1 What Is Blockchain Technology?

Blockchain technology is an advanced database that is dispersed across many computers (each called a node) and eliminates the need for a central authority or intermediaries. Since all nodes are equally privileged on the network, it creates a

A. Shojaei (✉) · H. Naderi
Myers-Lawson School of Construction, Virginia Tech, Blacksburg, VA, USA
e-mail: shojaei@vt.edu

C. De Wolf et al. (eds.), *A Circular Built Environment in the Digital Age*, Circular Economy and Sustainability, https://doi.org/10.1007/978-3-031-39675-5_12

peer-to-peer network that creates trust among all peers because each transaction is securely signed by cryptography algorithms that ensure consistency, immutability, and traceability (Ledger 2022). To add information to this database, a validation process, called consensus, is performed to cryptographically link each block of information to the previous one. As new blocks are added, older blocks become more difficult to modify. New blocks are replicated across copies of the ledger within the network, and any conflicts are resolved automatically using established rules of the network (Yaga et al. 2019).

In the following sections, two terms are frequently used: blockchain technology and blockchain network. The first term refers to a general concept of blockchain and its associated features, while the second is related to a decentralised network that is built using blockchain technology (see Fig. 12.1).

The structure behind blockchain technology provides it with some fundamental features that can revolutionise open issues across a variety of fields, including the built environment (Li et al. 2019). The first notable feature is immutability, which means that data cannot be changed once added to the blockchain network. Each block contains a cryptographic hash of the previous block, a timestamp, and transaction information. As a result, data stored in a block cannot be altered since all subsequent blocks would need to be changed as well (Atlam et al. 2018). Another feature is the consensus mechanism, which brings reliability to the blockchain network. A consensus algorithm (for example, Proof of Work in Bitcoin's case) ensures that all transaction data are identical between blocks. In simple words, the Proof of Work can be explained as the mechanism that requires nodes in the

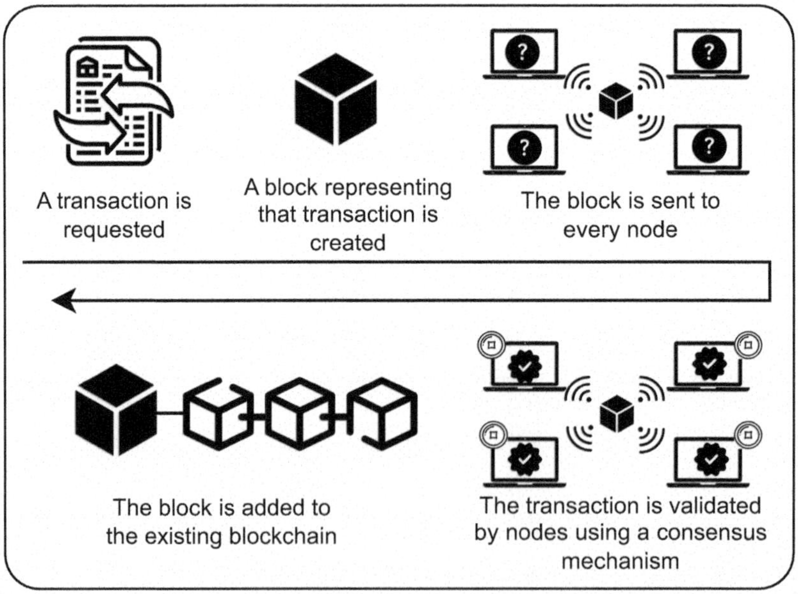

Fig. 12.1 Process of adding a block to the blockchain. (Adapted from Euromoney 2023)

blockchain network to solve a complex mathematical problem in exchange for cryptocurrency incentives (Euromoney 2023). Data traceability and integrity can also be considered other features of blockchain technology. By accessing any node in the blockchain's distributed network, users can easily trace previous transactions that have been validated and recorded on the blockchain (Perera et al. 2020). Moreover, all blocks link to the genesis block (the first block on the chain), ensuring the integrity of the blockchain (Nofer et al. 2017).

Nevertheless, while the general characteristics are the same, these features differ among different types of blockchain technologies, and it is important to note that each blockchain network has its own characteristics. Blockchain technology can generally be divided into two types, public and private, each with its own characteristics (Tasca and Tessone 2019). Bitcoin and Ethereum are two of the most well-known public blockchains. The public blockchain, also called permissionless blockchain, is accessible to everyone. Given this wide accessibility, more peers can participate and validate the network transactions, resulting in a more immutable, transparent, and traceable network with almost no downtime (Tapscott and Tapscott 2016). However, permissionless blockchain technologies have disadvantages: the fact that they are fully transparent and accessible makes them inappropriate in situations where nodes need to keep the transaction information protected from other users. As a constantly evolving technology, blockchains are gaining new capabilities. Ethereum introduced smart contracts, which allowed peers to execute codes and enforce terms of contracts on blockchain networks without reliance on a trusted third party (Han et al. 2020).

Unlike a public blockchain, a private blockchain has only a few participants (who are authorised by the owner of the network), and changes are made when the majority of nodes (or all of them unanimously, depending on the network structure) reach a consensus (Perera et al. 2020). Hyperledger Fabric platform, which is developed by Linux Foundation, is one real-life example of a private or permissioned blockchain platform. A private blockchain gives users control over the level of data transparency, which makes it an effective option in situations where users are reluctant to share data with other participants (Boucher 2017). Moreover, the limited number of participants not only makes this type of blockchain faster but also provides a more manageable ecosystem (Haritonova 2021). However, considering the limited number of nodes in this type of blockchain, they cannot provide a fully decentralised system compared to public blockchains. Additionally, few nodes can bring an additional risk of database downtime, which can result in disruption in the network operation.

12.2 Blockchain Technology in the Built Environment

The architecture, engineering, and construction (AEC) industry is frequently referred to as a fragmented industry that suffers from a lack of transparency (Bakis et al. 2007; Jiao et al. 2013). Projects within the built environment usually involve organisations from a number of disciplines, which can create conflicts due to a lack of control over data. Moreover, these projects have a long lifespan, which poses additional risks to the cybersecurity of data and data accessibility. In such projects, blockchain technology and its inherent features are frequently employed as possible solutions (Lee et al. 2021; Wu et al. 2022). One recent study reported a 192% annual increase in the average number of blockchain-based academic publications from their first appearance in 2017 to 2020 (Scott et al. 2021). However, most of these publications only propose a conceptual framework or review of blockchain technologies; practical implementation of blockchain technology within the built environment has remained less explored. The following sections offer an overview of blockchain technology applications in the AEC industry as they relate to the circular economy.

12.2.1 Supply Chain Management

Increasing transparency and reliability of the information in the supply chain is critical in achieving a circular economy in the built environment. Blockchain technology can potentially address this gap and thus facilitate a circular economy. For example, Wang et al. (2019) conducted an interview with 14 supply chain practitioners who validated the advantages of blockchain technology in the supply chain field.

It is virtually impossible to have smooth construction supply chain management in complex and hard-to-reach construction sites without real-time and reliable information among all parties. To address this issue, many researchers have focused on blockchain technology. For example, Wang et al. (2020) proposed a blockchain-based information management framework based on a model for real-time information sharing in a supply chain for precast components. Their results showed that the proposed model positively impacted the tracking of precast components and helped find the root causes of disputes about the precast supply chain.

Tracking materials or assets also plays an important role in the transition to a circular economy. In this situation, blockchain technology can be used to improve the traceability of materials in projects. For example, blockchain technology and radio-frequency identification (RFID) have been applied to track ready-mixed concrete in construction sites (Lanko et al. 2018).

12.2.2 BIM and Digital Twins

Since building information modelling (BIM) is the primary source of construction data and blockchain facilitates the handling of data, many researchers are focusing on the integration of these two technologies. Ye et al. (2018) identified the Internet of Things (IoT) and blockchain as two potential technologies for integration with BIM in terms of bringing a single source of truth within the context of digitalisation in the AEC industry. Shojaei et al. (2019) also proposed a blockchain solution based on the Hyperledger Fabric platform that can maintain a record of project progress and thus automatically govern construction contracts and avoid many potential disputes. For digital twin technology, Lee et al. (2021) proposed an integrated digital twin and blockchain model for communicating traceable data among project stakeholders. The integration of blockchain and BIM technologies will further ensure data availability and reliability for all stakeholders.

12.2.3 Cost Saving

Construction projects often struggle with a considerable number of disputes and transaction costs, mostly due to a lack of trust and transparency in the contract administration (Cheng et al. 2021). Blockchain technology is suggested to solve this issue by eliminating intermediaries in construction agreements. In this regard, Dakhli et al. (2019) examined the amount of cost savings after applying blockchain technology in a real estate company and found that deploying blockchain technology in residential construction could save 8.3% of total cost. Hamledari and Fischer (2021) also discussed the inappropriateness of current centralised workflows for automatic payments based on project progress. To tackle this problem, they proposed a decentralised smart contract framework enabling construction progress payments to be made automatically based on an unmanned vehicle-based progress monitoring process. These applications indicate how blockchain technology can enable industry actors to gain more value with fewer costs.

12.2.4 Information Management

Blockchain technology features are suitable to address the challenge of transparency, trust, and intellectual property protection in construction documents. A primary challenge of construction quality management is that, traditionally, quality information is recorded on paper by specialists, which can lead to data loss. To address this issue, Wu et al. (2021) proposed a conceptual framework based on Hyperledger Fabric and the consortium blockchain network that records all data on an immutable network making it more reliable than paper reports. The immutable feature of the blockchain network facilitated data integrity in the proposed document management system. In some cases, the large amount of data makes decision-making susceptible to error. To overcome this problem, Ciotta et al. (2021) proposed smart contracts with different levels of complexity, focusing on reducing human error and increasing the reliability and transparency of the decision-making processes in construction.

12.3 Circular Economy Through Blockchain Technology

Blockchain technology has been identified as a promising tool to support circular economy strategies in a variety of ways (Kouhizadeh et al. 2019). This section investigates how blockchain technology and its features can improve different circular economy strategies (particularly, *regenerate*, *slow*, *narrow*, and *close* strategies) in the built environment.

The *regenerate* principle contains efforts to transition from fossil fuels to renewable energy and materials, avoid the use of hazardous contents, and improve biodiversity (Çetin et al. 2021). To this end, traceability and immutability features in blockchain technology enable industry actors to manage and track energy and material flows from production lines to specific points of consumption. This brings about a level of transparency to the system that can promote renewable energy and material use. This strategy can also benefit from the decentralised structure of blockchain through leveraging material or energy trading on a peer-to-peer network without relying on third parties.

The *slow* strategy includes efforts to maximise value through sharing goods while minimising duplications and waste (Çetin et al. 2021). Utilising used goods or sharing assets can be classified in this category. An immutable and secure network of blockchain lays a solid foundation for sharing material and goods while tracking ownership and usage information. For example, blockchain technology has been utilised to provide a secure platform for car-sharing (Shrestha et al. 2020; Auer et al. 2022).

In addition, the slow strategy is defined as a circular path for remanufacturing goods or components instead of a linear path of make-use-dispose (Çetin et al. 2021). To this end, we need effective ways to track and trace materials, components, and products from the production point to the end of life. In this situation, blockchain technology can provide a reliable way to trace information for logistic activities, history of energy use, etc. In one example, a Hyperledger Fabric network (a type of private blockchain) is applied to provide a traceable network of material data. This framework allows preplanning for reusing materials in the built environment (Shojaei et al. 2021).

The *narrow* strategy aims to boost system performance by minimising non-value-added activities in processes of manufacturing, operating, and consuming (Çetin et al. 2021). The application of big data analytics in this strategy has drawn considerable attention as one of the most promising technologies (Marinakis 2020). Additionally, automation and enforceability of smart contracts in blockchain networks can facilitate the transition to more optimised business systems. Many existing manual workflows, which are highly reliant on centralised workflows, can be upgraded to automated workflows based on decentralised systems (Hamledari and Fischer 2021).

In addition, the narrow strategy includes dematerialising efforts by delivering utilities virtually (such as e-documents, online conferencing, etc.). Intrinsic features of blockchain technology can support this strategy in various ways. Trading goods, energy, and components with cryptocurrency can help reduce issues such as lack of trust, transparency, and the need for intermediary facilitators related to current trade

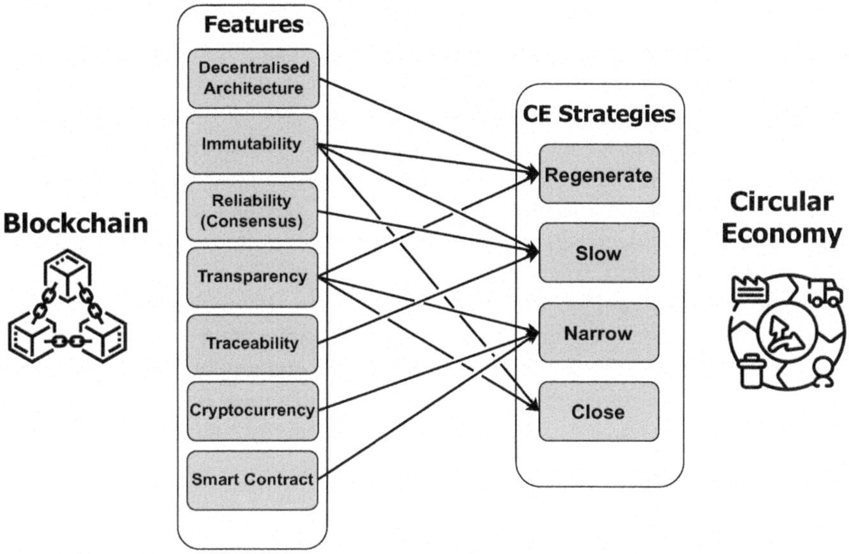

Fig. 12.2 Links between blockchain features and main circular economy strategies

practices. Blockchain technology's transparency feature can also help reduce the massive paperwork currently used in business practices without sacrificing trust and traceability. Furthermore, smart contracts can help automate manual and labour-intensive workflows.

Using the *close* strategy, buildings' end-of-life resources can be reintroduced to the economic cycle (Çetin et al. 2021). Traceability, transparency, and immutability are all features associated with blockchain technology that can improve this strategy. Blockchain-based systems can provide a more reliable platform for material flow during the whole life cycle of buildings through their applications in supply chain. In summary, Fig. 12.2 presents how blockchain features and advantages are aligned with the four discussed circular economy strategies.

12.4 Examples of Blockchain Applications for a Circular Built Environment

Although much attention has been given to conceptual frameworks based on intrinsic features of blockchain technology, practical services using blockchain features have remained mostly unexplored. In this section, we examine some real-life examples of blockchain technology adoption for a transition to a circular economy in the built environment.

Nowadays, a considerable number of distributed energy resources enable us to access renewable energy microgrids on a range of buildings. However, balancing loads from different inputs and outputs is a major barrier to this opportunity. In this situation, blockchain technology has been applied as a solution in some real-life services explained below. Immutability along with transparency in a blockchain network is, for example, being utilised along with artificial intelligence (AI) tools in the Port of Rotterdam to create an energy market for trading renewable microgrid energy on the buildings. This platform reduced user costs by 11% in 2021 and has attracted much attention for a promising future solution (Distro 2021). In a similar effort in Australia, Powerledger (a software and technology company) provides a decentralised market for renewable energy generation, storage, and purchasing in an optimal manner. A blockchain-based software has been developed based on this solution to trade rooftop solar power between international schools, apartment complexes, shopping centres, and dental hospitals in Bangkok. Therefore, these services support regenerating the resource loops. The traceability of the blockchain network is utilised in a pilot project by the Iberdrola company (a company working in the field of renewable energy) to certify sources of green energy to build trust among users and encourage them to adopt renewable energy sources. This is aligned with the *regenerate* strategy in the previous section.

Blockchain technology is not always utilised as a tool to add traceability and transparency values to old systems; it can be used to change the whole business model. The Brooklyn Microgrid in the United States is a good example of this change. Instead of using a central grid network, this platform redesigned the energy grid model by providing a community-based and decentralised network applying blockchain technology. According to this concept, the actual energy in the system is generated, stored, and traded locally by community users instead of using a third-party utility company. Users on this platform can contribute to electricity generation by solar panels in their own buildings or the energy stored in electric cars. Users who need more energy can buy energy directly from other users. Extra energy in the network is shared and stored among all users and allows them to be a step closer to implementing a circular economy in their community. This example utilises almost all strategies for reaching circularity at a community level.

A carbon credit market in China is based on a blockchain network, where individuals can track carbon emissions on a smartphone app in a reliable manner and trade with those who need carbon credits (ECO2 2015; Jackson 2022). (The same concept could be applied in the built environment to enable buildings with high energy efficiency or low carbon footprints to monetise their savings.) For another example, EZ Blockchain in the United States utilises waste natural gas that cannot be effectively sold as fuel to mine Bitcoin. Similar solutions could be applied to waste materials in the built environment to recreate value from waste. This alignment with the *narrow* strategy demonstrates additional real-life efforts to transition to a circular economy.

The potential to use blockchain for a more circular economy is not limited to examples within the energy context. A significant number of real-life examples utilises blockchain technology to create a more efficient supply chain. For example, Circularise partnered with the City of Amsterdam to improve sustainability within the construction procurement process by proposing a traceable and transparent platform for tracking material information. This platform answers the need for effective ways of tracing materials from the production point to the end of life.

The provided traceability enables companies to circulate material effectively and also mitigate risks across the supply chain. However, transparency, while offering benefits, can also create concerns about data privacy and confidentiality. Different types of blockchain technology can be utilised to address this issue. For example, Circularise developed a solution called 'Smart Questioning', based on a public blockchain, which not only provides transparency but also preserves companies' confidential data in a secure and reliable manner. Another example is BanQu, which created a supply chain platform to authenticate transactions during the whole extent of the supply chain. This service allows active players to manage records. These innovations are aligned with the *close* strategy.

Furthermore, material management during the construction phase of the built environment is critical. Despite academic efforts to apply blockchain technology as a solution to the construction supply chain (Tezel et al. 2020; Shemov et al. 2020), there are few real-life examples of these efforts. As one of the few examples, DigiBuild developed a blockchain-based solution to provide trusted material management among all parties involved in construction projects. Another example is the company Empower, which motivates transparent and traceable waste collection through the use of blockchain technology. ReCheck is another example of a company putting efforts into creating a material circularity passport for the built environment. Through a blockchain-enabled network, this platform records the material information in an immutable and transparent environment. Such information provides us with an easier process of recycling at the end of the building life cycle.

12.5 Challenges of Applying Blockchain Technology

Despite the advantages of applying it towards a circular economy, blockchain, like many other emerging technologies, faces considerable challenges for practical implementation. In this section, some of the most significant barriers are discussed to set realistic expectations for this technology. Furthermore, despite efforts noted in the previous section, real-life examples of blockchain technology for circular economy practices in the built environment are very limited. The reason for this scarcity can be explained by various challenges in implementing blockchain technology in practice.

First of all, although blockchain technology started in 2009 (the year of the first mined Bitcoin), it still can be considered an emerging technology, as it has experienced significant changes from its first appearance. Its constantly changing nature has led to many challenges for its implementation. The lack of sufficient experts is one of these challenges (Connolly 2021), which poses an additional risk when deciding on technologies for the relatively new concept of a circular built environment. Furthermore, the multi-stakeholder structure of a circular built environment makes it hard to get all parties on the same page, especially when they must decide whether to use a technology with only scant resources and few experts who understand it well. Moreover, AEC tends to be a risk-averse industry that is very slow to adopt new technologies (Oesterreich and Teuteberg 2016; Li et al. 2019). Creating a solution that brings all these stakeholders together in a unified platform is a challenging task.

Blockchain interoperability is another challenge in decentralised networks. This means that each blockchain network has a unique structure, making it incompatible with some or all other networks. For example, if the Ethereum blockchain, one of the most well-known public networks, is chosen for implementing a circular economy solution, any platform developed with it will be incompatible with many other blockchain networks.

The next challenge is the speed and performance (including costs) of blockchain networks. Blockchain technology is relatively slow when applied to massive amounts of everyday data, be it energy or any other type of data. Each transaction in Ethereum can take to several seconds, making applications function very slowly, especially in comparison to centralised common databases (Wang et al. 2020). However, it should be noted that many new blockchain networks have been developed to address this issue. For example, while Ethereum can perform 15 transactions per second, the Polygon blockchain can perform more than 65,000 transactions per second. Applying blockchain technology is also associated with implementation costs because each transaction has a transaction fee. It can make decision-makers reluctant to apply this technology early in their transition to a circular economy. However, some new blockchain networks, such as Polygon, offer more affordable transaction fees.

Another challenge is related to regulatory issues within and beyond organisations. For example, regulations in governmental agencies may not allow the implementation of a new technology like blockchain that can potentially change existing workflows. This is particularly evident in the public sector, where all workflows are governed by legislation. There is no solid and certain standard for the proper implementation of blockchain technology (Alaloul et al. 2020), which in part hinders the development of appropriate legislation for implementing blockchain technology in practice. In addition, a fully integrated adoption of blockchain technology in the various fragmented organisations of the AEC industry requires regulatory changes, which brings additional time and costs for organisations.

Kiu et al. (2019) have mentioned the challenge of developing a smart contract and its associated coding, especially when it comes to encoding the developing concepts in the circular economy field. Although the immutability and enforceability of smart contracts were mentioned as advantages above, they also can become a challenge as codes cannot be updated after final execution. This can be challenging in the field of circular economy, which is an emerging field that needs constant improvements.

The interaction between a blockchain network (on-chain) and technologies outside of the blockchain (off-chain) is also mentioned as one of the most serious challenges for developing an automated solution based on blockchain technologies. Oracles are middleware agents that bridge real-world, off-chain data to on-chain networks (Al-Breiki et al. 2020). However, the main problem with widely used oracles is that they are centralised services and are thus vulnerable to all the traditional problems associated with centralised systems, such as single points of failure and lack of transparency, cost, and dependency. This challenge can impact the functionality of digital twins and IoT when they are integrated with the blockchain technology. As a result, it can impact the performance of applications developed for building industry based on these tools.

It should also be noted that although there is a considerable number of open challenges for adopting blockchain technology, potential solutions are being provided on a daily basis, making the future of adoption more promising. For example, decentralised oracle networks were recently developed as a solution for avoiding the problems associated with centralised oracles that were mentioned above.

12.6 Future of Blockchain Technology in a Circular Built Environment

Blockchain technology has experienced numerous innovations since its introduction in 2008 (under the pseudonym Satoshi Nakamoto). In this section, we explore some of these concepts that can potentially influence the future adoption of blockchain in a circular built environment. Decentralised application (dApp) is one of these new concepts, which refers to a kind of application built over smart contracts. These kinds of applications are almost similar to web applications in how they look and are accessed (front-end), with the difference that they mostly use smart contracts as their functioning mechanism (back-ends) (Ethereum 2022). These applications are associated with intrinsic transparency and accessibility of the Ethereum network. Most real-life examples introduced in Sect. 12.4 were developed as dApps powered by smart contracts for the purpose of each platform. However, dApps hold much more potential for future developments. BIM and digital twin assets can be integrated with smart contracts for building dApps that improve the circular built environment based on strategies discussed in Sect. 12.3. For example, integrating BIM with blockchain technology allows us to create material passports in a secure and immutable manner integrating into the design workflow or maintenance operations to inform the users in decision-making processes. Furthermore, material information is stored on a blockchain network and then can be used by different parties to select the best strategy for a transition to a circular environment.

Smart contracts also can be applied to generate crypto tokens. Tokens are digital representations of assets or interests that have been tokenised on the blockchain of a cryptocurrency (Frankenfield et al. 2023). Basically, tokens can be divided into two categories: fungible tokens (FTs) and non-fungible tokens (NFTs). FTs are divisible and interchangeable tokens, each equivalent to another, such as cryptocurrencies. NFTs, however, are indivisible, verifiable tokens representing a piece of information on a given blockchain network, whether digital art or any other kind of information (Bal and Ner 2019). Although few examples use FTs to trade energy (see details in Sect. 12.4), to the best of our knowledge, NFT features have not yet been used in the transition to a circular built environment. For example, NFTs can be used to monitor, verify, and report building energy performance or dynamically present an asset status during its life cycle or circular economy strategies (regenerate, slow, narrow, close) implemented on it. To be more specific, the efforts to maximise sharing goods through its life cycle can be tracked using dynamic NFTs as a way for monitoring slow strategy in implementing circular economy in the built environment. This capability builds lost trust among stakeholders and enables them to clearly track, monitor, and manage the building energy performance or the status of an asset in a secure and immutable manner. This can increase the investments and buy-in on green buildings and circular strategies.

There are also other emerging concepts under the general term of blockchain that need more investigation. Decentralised autonomous organisation (DAO) is one of these concepts. Instead of relying on a central governmental component, DAOs are controlled by various users in a decentralised network (Reiff 2022). DAO is a community-based organisation that distributes decision-making without any intervention from a centralised power. This decentralised structure brings about a true bottom-up management approach, which can offer opportunities for transition to a circular built environment. Integrating DAO and digital twins in the built environment can bring a higher level of functionality. For example, a decentralised autonomous organisation (DAO) prototype to build a self-governing house (Hunhevicz et al. 2021). This integration can eliminate the reliance on a single authority for decision-making. This integration can be applied to create an autonomous built environment that runs its decisions automatically, optimising the use of resources and energy. This opportunity can save considerable time and cost in building operations and lead the built environment function with circular principles.

Another newly discussed concept, which is rarely explored, is decentralised finance (DeFi). This financial system is powered by smart contracts and a decentralised blockchain network without relying on intermediaries such as banks (Sharma 2022). DeFi can enable public users to fund circular building projects securely and quickly without being charged bank service fees. This opportunity can promote circular economy projects and lead to achieving sustainable goals quicker.

In conclusion, the potential for blockchain technology to revolutionise the circular built environment is immense, offering a myriad of opportunities for stakeholders to enhance transparency, efficiency, and collaboration. This technology can be particularly transformative in areas such as energy and resource management, waste reduction, and promoting sustainable practices throughout the entire life cycle of buildings. While there have been some promising real-life examples of blockchain technology being applied to the built environment, its full potential has yet to be realised due to the numerous challenges and barriers that must be addressed, including the emerging nature of the technology, interoperability, performance, regulatory issues, and integration with other technologies.

As we look to the future, emerging concepts such as dApp, NFTs, DAO, and DeFi present new avenues to explore for the integration of blockchain technology in the circular built environment. By fostering innovation and collaboration among stakeholders, addressing the challenges facing the technology, and promoting regulatory changes, the full potential of blockchain technology can be harnessed to create a more sustainable and circular built environment, ultimately contributing to the broader goal of a more sustainable and resilient society.

12.7 Key Takeaways

- Blockchain can create a transparent, traceable platform for tracking materials and managing waste in the built environment, supporting a circular economy.
- Blockchain-based solutions can protect data privacy while offering transparency and traceability.
- Blockchain technology faces challenges like lack of experts, interoperability, and regulatory issues that slow down its adoption in the circular built environment.
- Decentralised applications, tokens, and decentralised autonomous organisations can unlock new opportunities for blockchain in the circular built environment.
- Blockchain integration with digital twins, BIM, and decentralised finance can promote efficient resource use, optimise decision-making, and support a sustainable circular economy in the built environment.

References

Alaloul WS, Liew MS, Zawawi NAWA, Kennedy IB (2020) Industrial Revolution 4.0 in the construction industry: challenges and opportunities for stakeholders. Ain Shams Eng J 11: 225–230. https://doi.org/10.1016/J.ASEJ.2019.08.010

Al-Breiki H, Rehman MHU, Salah K, Svetinovic D (2020) Trustworthy blockchain oracles: review, comparison, and open research challenges. IEEE Access 8:85675–85685. https://doi.org/10.1109/ACCESS.2020.2992698

Atlam HF, Alenezi A, Alassafi MO, Wills GB (2018) Blockchain with Internet of Things: benefits, challenges, and future directions. Int J Intell Syst Appl 10:40–48. https://doi.org/10.5815/IJISA.2018.06.05

Auer S, Nagler S, Mazumdar S, Mukkamala RR (2022) Towards blockchain-IoT based shared mobility: car-sharing and leasing as a case study. J Netw Comput Appl 200:103316. https://doi.org/10.1016/J.JNCA.2021.103316

Bakis N, Aouad G, Kagioglou M (2007) Towards distributed product data sharing environments: Progress so far and future challenges. Autom Constr 16:586–595. https://doi.org/10.1016/J.AUTCON.2006.10.002

Bal M, Ner C (2019) NFTracer: a non-fungible token tracking proof-of-concept using hyperledger fabric. https://doi.org/10.48550/arxiv.1905.04795

BanQu. https://www.banqu.co/platform. Accessed 15 Feb 2023

Boucher P (2017) How blockchain technology could change our lives: in-depth analysis, European Parliament Research Service. Brussel

Brooklyn Microgrid. https://www.brooklyn.energy/about. Accessed 6 Dec 2022

Çetin S, De Wolf C, Bocken N (2021) Circular digital built environment: an emerging framework. Sustainability 13:6348. https://doi.org/10.3390/SU13116348/S1

Cheng M, Liu G, Xu Y, Chi M (2021) When blockchain meets the AEC industry: present status, benefits, challenges, and future research opportunities. Buildings 11. https://doi.org/10.3390/buildings11080340

Ciotta V, Mariniello G, Asprone D et al (2021) Integration of blockchains and smart contracts into construction information flows: proof-of-concept. Autom Constr 132:103925. https://doi.org/10.1016/J.AUTCON.2021.103925
Circularise. https://www.circularise.com/company. Accessed 6 Dec 2022
Connolly G (2021) Overcoming the obstacles to implementing blockchain technology. In: Forbes. https://www.forbes.com/sites/forbesbusinesscouncil/2021/09/29/overcoming-the-obstacles-to-implementing-blockchain-technology/?sh=618dbcb5d0ea
Dakhli Z, Lafhaj Z, Mossman A (2019) The potential of blockchain in building construction. Buildings 9:77. https://doi.org/10.3390/buildings9040077
DigiBuild. https://digibuild.com/. Accessed 6 Dec 2022
Distro (2021) Port of Rotterdam. In: Distro Energy. https://www.portofrotterdam.com/en/news-and-press-releases/successful-trial-port-rotterdam-electricity-platform. Accessed 6 Dec 2022
ECO2 (2015) ECO2 ledger: linking the global carbon market. https://www.eco2.cc/. Accessed 6 Dec 2022
Empower. https://www.empower.eco/tracking/. Accessed 15 Feb 2023
Ethereum (2022) Introduction to dapps. https://ethereum.org/en/developers/docs/dapps/. Edited 26 Sep 2022
Euromoney (2023) Blockchain explained: how does a transaction get into the blockchain? In: Euromoney Learning. https://www.euromoney.com/learning/blockchain-explained/how-transactions-get-into-the-blockchain. Accessed 13 Feb 2023
EZ Blockchain. https://ezblockchain.net/. Accessed 6 Dec 2022
Frankenfield J, Mansa J, Kvilhaug S (2023) What are crypto tokens, and how do they work? In: Investopedia. https://www.investopedia.com/terms/c/crypto-token.asp. Accessed 14 Feb 2023
Hamledari H, Fischer M (2021) Role of blockchain-enabled smart contracts in automating construction progress payments. J Leg Aff Disput Resolut Eng Constr 13:04520038. https://doi.org/10.1061/(asce)la.1943-4170.0000442
Han D, Zhang C, Ping J, Yan Z (2020) Smart contract architecture for decentralized energy trading and management based on blockchains. Energy 199:117417. https://doi.org/10.1016/J.ENERGY.2020.117417
Haritonova A (2021) Public vs. private blockchain: What's the difference? In: Pixelplex, 19 Mar 2021. https://pixelplex.io/blog/public-vs-private-blockchain/
Hunhevicz JJ, Wang H, Hess L, Hall DM (2021) no1s1: a blockchain-based DAO prototype for autonomous space. Proc 2021 Eur Conf Comput Constr 2:27–33. https://doi.org/10.35490/ec3.2021.185
Iberdrola. Blockchain in the energy market. https://www.iberdrola.com/innovation/blockchain-energy. Accessed 6 Dec 2022
Jackson J (2022) The blockchain projects making renewable energy a reality: In: Cointelegraph Magazine https://cointelegraph.com/magazine/blockchain-projects-making-renewable-energy-reality/
Jiao Y, Wang Y, Zhang S et al (2013) A cloud approach to unified lifecycle data management in architecture, engineering, construction and facilities management: integrating BIMs and SNS. Adv Eng Inform 27:173–188. https://doi.org/10.1016/j.aei.2012.11.006
Kouhizadeh M, Zhu Q, Sarkis J (2019) Blockchain and the circular economy: potential tensions and critical reflections from practice. Prod Plan Control 31:950–966. https://doi.org/10.1080/09537287.2019.1695925
Lanko A, Vatin N, Kaklauskas A (2018) Application of RFID combined with blockchain technology in logistics of construction materials. MATEC Web Conf 170:03032. https://doi.org/10.1051/MATECCONF/201817003032
Ledger (2022) What is blockchain? In: Ledger Academy. https://www.ledger.com/academy/blockchain/what-is-blockchain. Updated 23 May 2023
Lee D, Lee SH, Masoud N et al (2021) Integrated digital twin and blockchain framework to support accountable information sharing in construction projects. Autom Constr 127:103688. https://doi.org/10.1016/j.autcon.2021.103688

Li J, Greenwood D, Kassem M (2019) Blockchain in the built environment and construction industry: a systematic review, conceptual models and practical use cases. Autom Constr 102: 288–307. https://doi.org/10.1016/j.autcon.2019.02.005

Marinakis V (2020) Big data for energy management and energy-efficient buildings. Energies 2020 13:1555. https://doi.org/10.3390/EN13071555

Nofer M, Gomber P, Hinz O, Schiereck D (2017) Blockchain. Bus Inf Syst Eng 593(59):183–187. https://doi.org/10.1007/S12599-017-0467-3

Oesterreich TD, Teuteberg F (2016) Understanding the implications of digitisation and automation in the context of Industry 4.0: a triangulation approach and elements of a research agenda for the construction industry. Comput Ind 83:121–139. https://doi.org/10.1016/J.COMPIND.2016.09.006

Perera S, Nanayakkara S, Rodrigo MNN et al (2020) Blockchain technology: is it hype or real in the construction industry? J Ind Inf Integr 17:100125. https://doi.org/10.1016/j.jii.2020.100125

Powerledger. T77 & O77 – BCPG, Thailand. https://www.powerledger.io/clients/t77-bcpg-and-mea-thailand-thailand. Accessed 6 Dec 2022

Reiff N (2022) Decentralized autonomous organization (DAO): definition, purpose, and example. In: Investopedia. https://www.investopedia.com/tech/what-dao/. Updated 23 Sep 2022

Scott DJ, Broyd T, Ma L (2021) Exploratory literature review of blockchain in the construction industry. Autom Constr 132:103914. https://doi.org/10.1016/j.autcon.2021.103914

Sharma R (2022) What is decentralized finance (DeFi) and how does it work? In: Investopedia. https://www.investopedia.com/decentralized-finance-defi-5113835. Updated 21 Sep 2022

Shemov G, de Soto BG, Alkhzaimi H (2020) Blockchain applied to the construction supply chain: a case study with threat model. Front Eng Manag 7:564–577. https://doi.org/10.1007/S42524-020-0129-X

Shojaei A, Flood I, Moud HI et al (2019) An implementation of smart contracts by integrating BIM and blockchain. Adv Intell Syst Comput 1070:519–527. https://doi.org/10.1007/978-3-030-32523-7_36

Shojaei A, Ketabi R, Razkenari M et al (2021) Enabling a circular economy in the built environment sector through blockchain technology. J Clean Prod 294:126352. https://doi.org/10.1016/J.JCLEPRO.2021.126352

Shrestha AK, Vassileva J, Deters R (2020) A blockchain platform for user data sharing ensuring user control and incentives. Front Blockchain 3. https://doi.org/10.3389/FBLOC.2020.497985

Tapscott D, Tapscott A (2016) Blockchain revolution: how the technology behind bitcoin and other cryptocurrencies is changing the world. Penguin, New York

Tasca P, Tessone CJ (2019) A taxonomy of blockchain technologies: principles of identification and classification. Ledger 4:1–39. https://doi.org/10.5195/LEDGER.2019.140

Tezel A, Papadonikolaki E, Yitmen I, Hilletofth P (2020) Preparing construction supply chains for blockchain technology: an investigation of its potential and future directions. Front Eng Manag 7:547–563. https://doi.org/10.1007/S42524-020-0110-8

Wang Y, Singgih M, Wang J, Rit M (2019) Making sense of blockchain technology: how will it transform supply chains? Int J Prod Econ 211:221–236. https://doi.org/10.1016/J.IJPE.2019.02.002

Wang Z, Wang T, Hu H et al (2020) Blockchain-based framework for improving supply chain traceability and information sharing in precast construction. Autom Constr 111:103063. https://doi.org/10.1016/J.AUTCON.2019.103063

Wu H, Zhong B, Li H et al (2021) On-site construction quality inspection using blockchain and smart contracts. J Manag Eng 37:04021065. https://doi.org/10.1061/(ASCE)ME.1943-5479.0000967

Wu H, Zhong B, Li H et al (2022) On-site safety inspection of tower cranes: a blockchain-enabled conceptual framework. Saf Sci 153:105815. https://doi.org/10.1016/j.ssci.2022.105815

Yaga D, Mell P, Roby N, Scarfone K (2019) Blockchain technology overview. https://doi.org/10.6028/NIST.IR.8202

Ye Z, Yin M, Tang L, Jiang H (2018) Cup-of-water theory: a review on the interaction of BIM, IoT and blockchain during the whole building lifecycle. ISARC 2018 – 35th Int Symp Autom Robot Constr Int AEC/FM Hackathon Futur Build Things. https://doi.org/10.22260/ISARC2018/0066

Alireza Shojaei is an Assistant Professor of Building Construction in the Myers-Lawson School of Construction at Virginia Tech. Dr. Shojaei has a Bachelor of Science in civil engineering, a master's degree in construction management with distinction from the University of Reading, and a PhD from the University of Florida, where he also received a master of science in management. His research programme is focused on digital engineering, construction automation, digital twin and smart cities, construction informatics, computer-based modelling and simulation, artificial intelligence, and project management and economics.

Hossein Naderi received a Bachelor's degree in civil engineering from Semnan University in 2016 and a master's degree in construction engineering and management from Pars University, Tehran, in 2019. He is currently a doctoral candidate at the Myers-Lawson School of Construction at Virginia Tech. His main research interests include digital twins, machine learning, blockchain technology, construction automation, and construction informatics.

Chapter 13
The Role of Digital Building Logbooks for a Circular Built Environment

Joana Dos Santos Gonçalves, Wai Chung Lam, and Michiel Ritzen

Abstract Digital building logbooks (DBLs) are digital repositories of building-related data gathered throughout the full life cycle of a building. DBLs help increase transparency and access to information during the design, construction, operation, and end-of-life phase of a building. They thereby facilitate an efficient and cost-effective transition to a zero energy and circular built environment. DBLs could slow down resource loops by extending the service life of buildings through better coordination of maintenance and repair and close resource loops by promoting adaptability and reuse of the whole building and/or its components with multi-cycle approaches. This chapter analyses examples of DBLs developed in five countries to show that they are useful tools at different life stages of the building and for different stakeholders (homeowners, property managers, or building professionals). Challenges for establishing DBLs as a central tool for a circular built environment lie in improving the user experience and ease of implementation; enhancing interoperability; and effectively collecting, managing, and transforming data into actionable information for the management, maintenance, and reuse at building and district levels.

Keywords Digital building logbooks · Building passports · Whole life cycle data · Traceability · Data management

13.1 Introduction

Building logbooks are repositories of building-related information. They are also commonly referred to as building passports, electronic building files, and, in specific cases, building renovation passports. They provide a single source for inputting, accessing, and visualising all the information associated with a building that can be continuously monitored and updated (Hartenberger et al. 2021). As data is captured

J. D. S. Gonçalves (✉) · W. C. Lam · M. Ritzen
VITO (Flemish Institute for Technological Research – Vlaamse Instelling voor Technologisch Onderzoek)/EnergyVille, Genk, Belgium
e-mail: joana.goncalves@vito.be; waichung.lam@vito.be; michiel.ritzen@vito.be

C. De Wolf et al. (eds.), *A Circular Built Environment in the Digital Age*, Circular Economy and Sustainability, https://doi.org/10.1007/978-3-031-39675-5_13

and managed throughout a building's whole life cycle, DBLs facilitate transparency, trust, and informed decision-making in the construction sector and are considered enablers of a circular built environment (Dourlens-Quaranta et al. 2021). While material passports (MPs) (see Chap. 5 by Honic et al. on this topic) focus on the material-related data of a product and its underlying components, such as life cycle impacts or circular characteristics (van Capelleveen et al. 2023), digital building logbooks (DBLs) can include technical, spatial, and functional characteristics as well as environmental, social, and financial performance data of a building.

A DBL is intended to be a flexible repository of building-related information that can be accessed and managed in different ways by different stakeholders. These stakeholders should be able to manually enter, upload, and update information, import data from external sources, or link to external databases. DBLs have the potential to cover a wide range of building-related information: static data (such as administrative documents, building plans, bills of materials, etc.) and dynamic data (such as maintenance logs, operational energy consumption, etc.) (Hartenberger et al. 2021). DBLs allow centralised access to information and can cluster digital product passports (DPPs) and MPs at the component and material level, including information on energy performance certificates and renovation roadmaps towards minimum energy performance requirements.

13.2 Digital Building Logbooks (DBLs)

13.2.1 DBLs in the European Built Environment

Several European policy documents, ranging from European legislation to future recommendations, have been established to pave the way for a low carbon, digital, and circular Europe. Despite the emphasis on the 'energy efficiency first' principle, there is a clear trend towards the inclusion of embodied greenhouse gases and whole life carbon in order to meet climate targets and decouple growth from resource use (European Commission 2019). In this context, the review of the Ecodesign framework, foreseen by 2025, will establish mandatory DPPs to improve the traceability of products along the value chain (Directorate-General for Environment European Commission 2022), including construction-related products. At the building scale, DBLs are specifically referred to in the Energy Performance in Buildings Directive (EPBD) recast proposal (European Commission 2021a), as tools to promote circular economy principles throughout the life cycle of buildings. The EPBD recast proposal (European Commission 2021a) also outlines the concept of a building renovation passport as a customised action plan for a specific building to help it achieve a higher level of energy efficiency.

In 2020, the European Commission commissioned a study on the development of a European Union framework for the DBLs (Dourlens-Quaranta et al. 2021). In several European countries, DBLs are already in use or in the process of being introduced (Jansen et al. 2022; Gómez-Gil et al. 2022). Some of these can be

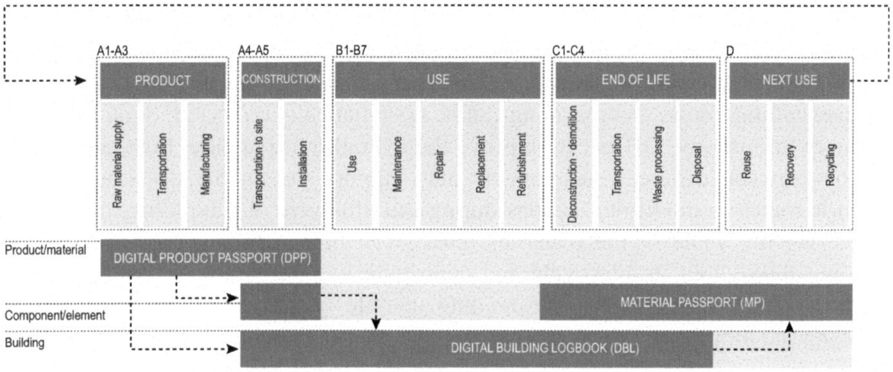

Fig. 13.1 Relation of DPPs, MPs, and DBLs across scales and life cycle stages, with the darker shade highlighting the focus life cycle stages

identified as a DPP or MP. The differences in scope between these digital tools are still a topic of discussion, as they can be related to the scale of implementation (from material and product to building), life cycle stages coverage, and scope of the contents. All three – DBLs, DPPs and MPs – are intended to be useful throughout the whole life cycle of a product. However, they have different focus: DPPs are created in the moment of production; MPs can be created during design and construction or at the end-of-life stage of a product or a building (Honic et al. 2021), but their most essential contribution lies at the beginning and at the end-of-life and next-use stages, enabling reuse and recovery of materials (see Chap. 5 by Honic et al.); and the main focus of DBLs is the use stage, as represented in Fig. 13.1. DBLs might 'nest' lower-level passports (such as DPPs or MPs), so that information can be inherited at a higher (building) level from underlying levels (components or materials) (Platform CB'23 2020). Furthermore, DBLs can be seen as 'living' logbooks that can be updated, automatically or not, during the life cycle stages, and they can include information related to energy performance, health and comfort, and operational management, while DPPs and MPs tend to be more static.

13.2.2 DBLs for a Circular Economy

The EU Parliament's Strategy for a Sustainable Built Environment, aiming to set the legislative priorities for the built environment regarding the implementation of the European Green Deal, refers to the importance of DBLs to increase material efficiency and to reduce the climate impact of the built environment, in particular by promoting circularity principles throughout the life cycle of buildings (European Parliament 2023). DBLs are an 'important means to achieve a more circular construction sector, as they promote reuse at the material, product, element, and building scale' (Platform CB'23 2020). In addition, DBLs can promote the principles of durability, adaptability, and circularity principles throughout the life cycle of a building (Hartenberger et al. 2021). As with DBLs, information on the installed construction elements, components and materials, their lifespan, and the possibilities for dismantling, reuse, and recycling can be systematically collected, organised, and updated. In this way, DBLs can improve the overall transparency, trust, and cooperation between different stakeholders and support sustainable decision-making when it comes to modifying actions during the life cycle of a building, ultimately preserving the value of the materials. DBLs can help maintain the value of the building throughout its life cycle and contribute to smarter use of materials and products (narrowing the loop), extending the life of buildings and components (slowing the loop) and ensuring beneficial end of life (closing the loop).

13.2.3 Examples of DBLs in Europe

This section presents examples of DBLs developed in five geographical contexts with different local drivers and normative frameworks: France (CLÉA), the UK (Residential Logbook Association/Chimni), Germany (CAPSA), Belgium (De Woningpas), and the Netherlands (CIRDAX). These DBLs were chosen to showcase the wide variety of legal and market backgrounds, level of maturity, functionalities, and target audiences. The DBLs are analysed in terms of functionalities, data management, data fields, and contribution to circularity strategies.

CLÉA, France

Since January 2023, French regulations have made the 'Carnet d'Information du Logement' (dwelling information file) mandatory for all new buildings; however, digitalisation is not mandatory. In this context, Qualitel, a French certification body, has developed CLÉA – a DBL that was launched to the market in October 2020. Currently, CLÉA is used in 50,000 dwellings (45,000 privately owned multi-family homes and 5000 single-family homes). This business-to-consumer (B2C) DBL is intended for building owners and tenants, either directly or through real estate managers. The CLÉA DBL is divided into different information categories, namely: general dwelling information (cadastre information); documents (repository of pdf files with invoices, rules, or minutes of residents' association); equipment (user guides and maintenance alerts for heating, ventilation, and air conditioning (HVAC) equipment); news (blog); and energy monitoring (connected to smart meters).

Residential Logbook Association, the United Kingdom

In the UK, there is no specific building logbook legislation. In 2021, the Coalition for the Energy Efficiency of Buildings (CEEB) developed a standardised framework for

Building Renovation Passports in the UK to help finance a net-zero carbon built environment. In this context, the Residential Logbook Association brings together several DBL companies to contribute to the regulatory process. Approximately 250,000 homes have a DBL verified by the RLBA. Chimni is one of these business-to-business (B2B) and B2C DBLs. It is tailored for homeowners, estate agents, and house builders for existing and new buildings. Information categories currently included in this DBL are pictures and floor plans; geolocation; document storage (deeds, certificates, etc.); utility dashboard (connecting to gas, electricity, and water companies); and property history timeline.

De Woningpas, Belgium

The Woningpas (De Woningpas, 2023) is a DBL owned and developed by the Flemish government as part of the implementation trajectory for the renovation wave and the regional decree on building passports (Vlaams Overheid, 2018). It makes building passports available for all building units in Flanders. The Woningpas was launched in December 2018 as a B2C DBL for residential building units, and an extension of the DBL to all non-residential buildings is planned for the end of 2023. The DBL data is linked to external platforms via Application Programming Interfaces (APIs), connecting all available information from public authorities or other institutions (e.g. inspection organisations of energy network operators). In this way, the Woningpas is automatically fed with data and made freely available to the building owners of all 4 million individual building units in Flanders. A building owner can add information about work carried out and certificates (pdf files) in a digital environment and also can share the information in this DBL with the public. The information categories currently offered by this DBL include building information (cadastre information); energy (energy performance certificate, renovation advice, renovations work); insulation, glazing, and installation characterisation; soil characterisation; building permits; dwelling quality; mobility; water and sewage; flood sensitivity; biodiversity level; and asbestos.

CAPSA, Germany

Chillservices is a commercial company that has been providing building logbooks for large food retailers since 2016. In 2021, the company launched a new variant for office and residential buildings – CAPSA, which is currently applied to 50,000 apartments in Germany, but also in smaller test cases in Scotland, the Netherlands, and Italy. CAPSA is a B2B DBL to support housing owners and facility managers. It consists of a smartphone app to collect primary data, supported by geo-positioning and image recognition. The collected data is stored in a cloud-based platform and interpreted with the support of external data sources. Functionalities currently offered by this DBL include the following information categories: calculation of energy performance; surface area; material catalogue and embodied carbon; asset management (condition assessment, monitoring, and maintenance advice); and semi-automated calculation of decarbonisation roadmaps.

CIRDAX, the Netherlands

CIRDAX (2023) is a commercial materials management system launched in 2016 in the Netherlands by the company Re-Use Materials. As the focus of this DBL is mostly on materials, it can be considered an MP (see Chap. 5 by Honic et al. on material passports). However, CIRDAX is also an example of an integrative approach of digitalisation, as it combines the inventory of materials and components in a building scale with a digital twin (see Chap. 1 by Koutamanis on BIM and digital twinning) and includes building management functionalities, and should therefore be considered as a DBL. The data collected by 3D-scanning or manual inputs are aggregated in a DBL, linked to a blockchain to provide verifiable information about the ownership of materials for future transactions. CIRDAX is a B2B DBL, currently used by governmental organisations and real estate organisations for in-depth digitalisation of existing real estate portfolios. This DBL currently includes material passport; 3D Digital Twin; CO_2 balance calculator; management and maintenance (condition assessment and maintenance alerts); performance dashboards (circular potential, financial value, and CO_2 emissions); and material marketplace.

13.3 Data Fields Supporting Circular Strategies

Table 13.1 presents a summary of the most relevant data fields enabling circularity in the built environment present in the analysed DBLs. All analysed tools include geolocation of the building, a data field that can be linked to GIS (see Chap. 2 by Tsui et al) to optimise distances in the construction and end-of-use stages, encourage smart use of available space, track, and trace available resources (from materials to energy, including space), and encourage excess resource exchange.

The focus of most DBLs is on energy in the use stage of the buildings: information on maintenance and use of HVAC equipment (CLÉA, Chimni, Woningpas, CAPSA), links to energy certificates (Woningpas), invoices and consumption data from utilities (Woningpas, Chimni), or live monitoring through smart meters (CLÉA, CAPSA). Thus, they support the narrowing of resource loops in the use stage, by improving and tracing energy efficiency in buildings, with energy renovation roadmaps (such Woningpas and CAPSA) and encouraging the reduction of primary energy inputs, by integrating renewable energy sources and analysis of solar potential (Woningpas).

Slowing resource loops is also an important aspect tackled by the analysed DBLs. By integrating data about the heritage values of the building, tools like Chimni and Woningpas reinforce the emotional connection with the users so that the users feel attached to their buildings (Çetin et al. 2021). Together with information about user

Table 13.1 Data fields related to circularity strategies

			Digital building logbook				
Category	Data field	Strategy	CLÉA	CHIMNI	WONINGPAS	CAPSA	CIRDAX
Plot	Geolocation	ALL	×	×	×	×	
	Soil characterisation	Regenerate			×		
	Flood sensitivity	Regenerate			×		
	Water & sewage	Narrow/regenerate			×		
	Blue-green level	Regenerate			×		
	Solar potential	Narrow			×		
	Mobility	Narrow/regenerate			×		
Building	Construction date	Slow	×	×	x		
	Heritage listing	Slow		×	×		
	Building timeline	Slow		×			
	Home quality	Regenerate			×		
	Surface area	Slow		×		×	x
	Architectural characteristics	Slow/close		×	×	×	×
Services	HVAC systems	Narrow	×	×	×	×	
	HVAC user guides	Slow	×	×			
	HVAC maintenance alerts	Slow	×	×		×	
	Energy performance	Narrow	×	×	×	×	
	Energy consumption	Narrow	×	x	×		
	Energy monitoring	Narrow	×			×	
Components and materials	Characterisation	Slow/close				×	×
	Embodied carbon	Narrow/close				×	×
	Circular potential	Close					×
	Marketplace	Close					×
	Residual financial value	Close					×

(continued)

Table 13.1 (continued)

			Digital building logbook				
Category	Data field	Strategy	CLÉA	CHIMNI	WONINGPAS	CAPSA	CIRDAX
Maintenance	Condition assessment	Slow				×	×
	Monitoring hotspots	Slow				×	
	Maintenance strategies	Slow		×		×	×
Roadmaps	Energy renovation	Narrow			x	×	
	Project templates	Slow		×			

guidance, condition assessment, and maintenance (for instance, in CAPSA and CIRDAX), these strategies contribute to redesign strategies that extend the service life of the building.

By including modules related to materials, such as an inventory of materials and components and analysis of embodied carbon, CIRDAX and CAPSA show the potential of DBLs to contribute to closing resource loops, avoiding waste, and bringing resources back into the economic cycle. CIRDAX's MP links to a circular potential analysis (Potting et al. 2017) and the residual value of the building and connects supply and demand for material reuse with blockchain technology (see Chap. 12 by Shojaei and Naderi on the topic). Woningpas is the only DBL analysed to include information on the plot and city level, such as soil characterisation, mobility, and blue-green levels. It also includes information related to the quality of the indoor environment (home quality assessment), making this the only analysed tool already targeting the regeneration of natural and human systems, promoting biodiversity, healthy environment, and exchange of resources at the community level.

13.4 Business Models for DBLs

As identified in the European Commission study on building logbooks, several European approaches to DBLs do not yet have a clear business model that can be easily be replicated (Carbonari 2020). For the stakeholders involved, the lack of definition of business models is a significant barrier to the development of a DBL or its replication (Carbonari 2020). The analysis of the five DBLs identified some common benefits highlighted by all the DBLs analysed: the centralisation of information, which becomes easier to find and to share, resulting in streamlined workflows, the reduction of sectoral fragmentation, and the reduction of administrative burden. At the same time, the availability of reliable information contributes to greater transparency in all the processes, reducing risks, speeding up transactions, and, ultimately, increasing the property value. Despite the very different market groups, the benefits presented by the different DBLs tend to be overarching and thus may miss the unique value proposition for each specific stakeholder.

Three business models were identified in the five DBLs analysed: a B2B sale (product-oriented), where DBLs are sold to real estate promotors for a limited period of time; a B2C sale, where DBLs are sold directly to individual end-users; and B2B (use-oriented) commercial licence, where DBLs are offered as a service. The B2C approaches (CLÉA and Chimni) are currently free for individual users as an experimental approach to attract new users but are likely to gradually become 'freemium' services, combining some free features with more advanced features available only for a fee. The commercial licence fees are associated with the use of a software tool and are targeted at real estate owners and housing corporations with larger real estate portfolios. For example, access to the full list of functionalities identified for

CIRDAX requires the payment of a premium licence per month, per user, and per building.

Depending on the objective of the DBL, business models should be based on the clearly defined added value of using a DBL. This could result, on the one hand, in a single unique selling proposition such as a B2B/B2C opportunity that by using a DBL in which a maintenance company has access to building-related data, actual maintenance and operational costs decrease. For example, a technical installer with a maintenance contract with a private homeowner could timely plan maintenance because the operational efficiency of the installation decreases with collaterally higher energy consumption (and thus costs). On the other hand, business models based on the clear added value of DBLs could result in a multitude of B2C opportunities, for instance, for covering flood-related insurance costs, assessing photovoltaic potential, or estimating the costs of asbestos removal. A pre-condition for these business models would be a certain level of data sharing between the different parties involved.

13.5 Discussion

13.5.1 Future Developments for DBLs

Ambitions for developing DBLs and increasing the contribution to a circular built environment vary widely depending on the current level of complexity and stakeholders targeted. Most of the current DBLs still have a one-dimensional focus on the use phase and operational energy consumption, with little coverage of the whole cycle (Hartenberger et al. 2021). DBLs such as Woningpas are aiming to integrate external data from smart meters to monitor real performance, and CAPSA has already done so. Some of the DBLs presented already provide users with automated renovation advice (Woningpas) or detailed decarbonisation roadmaps (CAPSA), which can support the renovation of the building stock, investment decision-making, and access to EU funding, green financing, and insurance products.

According to the European Commission, the automatic input of data from a BIM model (see Chap. 1 by Koutamanis) is considered important for the majority of stakeholders (Dourlens-Quaranta et al. 2021), as it would contribute to speeding up the processes and reducing costs – two major barriers to the implementation of DBLs (Dourlens-Quaranta et al. 2021). This is not yet common practice, as the analysis of cases demonstrated, with only CIRDAX offering that possibility, and Chimni actively working on its integration with the DBL.

Collaboration is an essential strategy in the transition towards a circular built environment (Çetin et al. 2021), which will require integrating needs and expectations of multiple stakeholders at multiple scales. Understanding the building as a part of a larger complex system shaped by social, economic, and environmental forces is important for identifying flows of material products and waste across different scales. DBLs contribute to a better overview of the existing building stock and can enhance collective approaches that significantly reduce impacts at the neighbourhood and urban levels. DBLs, together with GIS technologies (see Chap. 2 by Tsui et al.), can support community-driven decarbonisation and the decentralisation of water, energy, and waste flows and simultaneously establish urban mining networks with information on the location and availability of materials.

To improve the contribution towards a circular built environment, the next generation of DBLs needs to go beyond energy and support sustainable flows throughout the entire life cycle of the building and beyond. In the study of the European Commission, participating stakeholders identify the building material inventory as one of the most important features (Dourlens-Quaranta et al. 2021). However, the analysis of the practical cases in this chapter shows that DBLs integrating this feature are still the exception and not the common practice. Requiring a bill of materials could increase the completeness and accuracy of the DBLs (Platform CB'23 2020) in the early stages and, later on, facilitate the traceability of embodied carbon and life cycle costing (Hartenberger et al. 2021). It also would offer an opportunity to integrate DBLs with current policy frameworks, such as LEVEL(s), by providing the necessary information to assess resource efficiency and material life cycles (European Commission 2021b), as soon required by the EPBD (European Commission 2021a).

13.5.2 Market Uptake

To ensure that DBLs are effectively useful tools, a more systematic and aligned approach to data collection, storage, and exchange is needed. Passports should allow comparison and interchange of information, and 'it is important that everyone uses the same technical terms and uses the same definitions' (Platform CB'23 2020). The five practical cases analysed show that the same functionalities may mean different things in the different DBLs. This was clear in the data fields related to general cadastre information and building characterisations, for instance, and in the integration of maintenance advice or environmental product declarations (EPDs). Future developments need to establish protocols and tools to ensure interoperability and compatibility of information so that DBLs are effective tools for information sharing and not obstacles to access. A harmonised framework of minimum requirements and protocols for DBLs is essential to ensure that accurate and correct data is available while still allowing for a diverse range of DBLs to meet different market needs and local drivers. Standardisation of minimum requirements goes hand in hand with the financing of the development of DBLs (Dourlens-Quaranta et al. 2021): certain mandatory aspects can be developed by the public sector (such as Woningpas), ensuring transparency and harmonisation, while more advanced features can be developed with commercial purposes, targeting stakeholders' specific needs (such as CIRDAX). The highest value for the end-users will be achieved when both approaches can be combined.

User-friendliness is a key factor determining the success of DBLs. Greater market uptake depends on the extent to which governments impose obligations (Platform CB'23 2020), but also on a better understanding of users' needs, attitudes, and personal motivations (Gonçalves et al. 2021), as there is no 'one-size-fits-all' solution for DBLs. Despite the overarching benefits of implementing DBLs identified in the findings, not all levels of information are relevant to all stakeholders. Therefore, DBLs, despite their role as an information hub, need to allow for different levels of granularity and user roles to avoid overburdening stakeholders with additional work and costs for data storage and management (Hartenberger et al. 2021). A key issue for the successful development and large-scale application of DBLs will depend on the business model. While the overarching objectives of DBLs are in line with the EU and national ambitions, it does not seem to be the case here as well; there is no one-size-fits-all solution for DBLs. Some will be based on a B2B model, B2C model, or fully supported by governments. For the B2B and B2C models, it will be key to define clear unique selling propositions that generate value for the customer.

13.5.3 DBLs as Enablers of Circular Economy

DBLs have the potential to contribute to three main circularity goals: (1) measuring achieved circularity; (2) management and maintenance in the use phase; and (3) facilitating future reuse and value retention (Platform CB'23 2020). Despite the different levels of complexity and detail, all the five DBLs presented in this chapter contribute to the second goal, facilitating the maintenance of the existing building stock; CAPSA and CIRDAX include some functionalities that contribute to the first goal, namely the material inventory and calculation of embodied carbon, but only CIRDAX actively aims at future circularity, value retention, and circular potential. Future developments should integrate renovation advice with MPs (see Chap. 5) and reuse marketplaces with blockchain technology (see Chap. 13). This would allow to balance achievements on operational and embodied carbon and make the most of the resources already existing in the building or its surroundings to avoid disposal and loss of value and enable multiple life cycles.

The development of DBLs presents challenges ahead, but the practical cases of DBLs already implemented demonstrate the potential of DBLs to enable a circular economy in the four strategies proposed by Çetin et al. (2021). They facilitate the upgrade and improvement of energy efficiency in buildings in the use phase (narrowing resource loops); contribute to extending buildings' lifetime through maintenance and repair, and enabling smart reuse of space (slowing resource loops); enable tracking, tracing, and bringing material resources back into the economic cycle in the next-use phase (closing resource loops); and contribute to a net positive impact when including indicators on biodiversity, surplus resources, and environmental quality (regenerating resource loops).

13.6 Key Takeaways

- Digital building logbooks (DBLs) provide transparency and access to building-related data throughout the full life cycle of a building.
- Reliable data can help improve the design, construction, and management of buildings, increase market transparency, create innovative services and business models, and lead to more effective policymaking.
- DBLs have the potential to promote a circular economy by facilitating energy efficiency of buildings, lifespan extension, intelligent reuse, and tracking and tracing material resources for future use.
- DBLs can contribute to creating a net positive impact if they include indicators related to biodiversity, surplus resources, and environmental quality at the building and neighbourhood level.
- A life cycle thinking approach to DBLs can support decision-making based on resource optimisation and circularity principles.
- Future developments should integrate renovation advice with material passports and marketplaces to balance achievements on operational and embodied carbon performance.
- Despite the benefits of DBLs to support a circular built environment, a successful business model has not yet been proved on the market with fully defined unique selling propositions in a challenging context with high expectations, policy requirements, and a competitive environment with more and more DBL developers.

Acknowledgements This work was supported by Demo-BLog and ICEBERG projects. Demo-BLog is co-funded by the European Union's Horizon Europe research and innovation programme under Grant Agreement No.101091749. ICEBERG is funded by the European Union's Horizon 2020 Research and Innovation Programme under Grant Agreement No: 869336. Views and opinions expressed are however those of the author(s) only and do not necessarily reflect those of the European Union. Neither the European Union nor the granting authority can be held responsible for them.

References

CAPSA (n.d.). https://capsa-building.com/. Accessed 7 Feb 2023

Carbonari G (2020) Building logbook state of play: report 2 of the study on the development of a European Union framework for buildings' digital logbook. European Commission Executive Agency for Small and Medium-sized Enterprises, Brussels

Çetin S, De Wolf C, Bocken N (2021) Circular digital built environment: an emerging framework. Sustainability 13:6348. https://doi.org/10.3390/su13116348

Chimni (n.d.). https://chimni.com. Accessed 7 Feb 2023

CIRDAX (2023). https://www.cirdax.com. Accessed 7 Feb 2023

CLÉA (n.d.). https://clea.qualitel.org. Accessed 7 Feb 2023

Directorate-General for Environment European Commission (2022) Proposal for a Regulation establishing a framework for setting ecodesign requirements for sustainable products, amending Regulation (EU) 2019/1020 and repealing Directive 2009/125/EC

Dourlens-Quaranta S, Carbonari G, De Groote M et al (2021) Study on the development of a European Union framework for digital building logbooks: final report. European Commission Executive Agency for Small and Medium-sized Enterprises, Brussels

European Commission (2019) Communication from the Commission to the European Parliament, The European Council, The Council, The European Economic and Social Committee and the Committee of the Regions "The European Green Deal", COM(2019) 640 final, Brussels, Belgium. https://eur-lex.europa.eu/legal-content/EN/TXT/?uri=COM:2019:640:FIN

European Commission (2021a) Proposal for a directive of the european parliament and of the council on the energy performance of buildings (recast), COM(2021) 802 final, Brussels, Belgium. https://eur-lex.europa.eu/legal-content/EN/TXT/?uri=CELEX:52021PC0802

European Commission (2021b) Level(s), Putting whole life carbon into practice. Directorate-General for the Environment (European Commission), Luxembourg. https://data.europa.eu/doi/10.2779/79139

European Parliament (2023) Legislative Train 01.2023: A European green deal. Strategy for a sustainable built environment. https://www.europarl.europa.eu/legislative-train/theme-a-european-green-deal/file-strategy-for-a-sustainable-built-environment

Gómez-Gil M, Espinosa-Fernández A, López-Mesa B (2022) Review and analysis of models for a European digital building logbook. Energies 15:1994. https://doi.org/10.3390/en15061994

Gonçalves JDS, Mateus R, Silvestre JD, Roders ARP (2021) Beyond good intentions: the role of the building passport for the sustainable conservation of built heritage to behavioural change. Sustain Switz 13. https://doi.org/10.3390/su13158280

Hartenberger U, Ostermeyer Y, Lützkendorf T (2021) The building passport: a tool for capturing and managing whole life data and information in construction and real estate – practical guideline. Global Alliance for Buildings and Construction, United Nations Environment Programme

Honic M, Kovacic I, Aschenbrenner P, Ragossnig A (2021) Material passports for the end-of-life stage of buildings: challenges and potentials. J Clean Prod 319:128702. https://doi.org/10.1016/j.jclepro.2021.128702

Jansen M, Gerstenberger B, Bitter-Krahe J et al (2022) Current approaches to the digital product passport for a circular economy: an overview of projects and initiatives. Wuppertal Institut für Klima, Umwelt, Energie

Platform CB'23 (2020) Guide: passports for the construction sector – Working agreements for circular construction. Version 2.0, 2 July 2020. Platform CB'23. https://platformcb23.nl/images/downloads/Platform_CB23_Guide_Passports_for_the_construction_sector_2.0.pdf

Potting J, Hekkert M, Worrell E, Hanemaaijer A (2017) Circular economy: measuring innovation in the product chain. PBL Netherlands Environmental Assessment Agency, The Hague

RLBA (Residential Logbook Association) (n.d.). https://www.rlba.org.uk/. Accessed 7 Feb 2023

van Capelleveen G, Vegter D, Olthaar M, van Hillegersberg J (2023) The anatomy of a passport for the circular economy: a conceptual definition, vision and structured literature review. Resour Conserv Recycl Adv 17:200131. https://doi.org/10.1016/j.rcradv.2023.200131

Joana Dos Santos Gonçalves is a researcher on circularity in the built environment at VITO in Belgium. She graduated as an architect from the School of Architecture of University of Minho, Portugal in 2014, with a master's thesis on the contributions of vernacular architecture for sustainability. In 2022 she concluded her PhD on 'Beyond good intentions: building passport for sustainable conservation' at the University of Minho. Between 2019 and 2021 Joana was a guest scholar in the UNESCO chair on Heritage and the Reshaping of Urban Conservation for Sustainability, and between 2020 and 2022, she was a lecturer on heritage and sustainability in the Faculty of Architecture and the Built Environment at TU Delft.

Wai Chung Lam obtained her master's degree in 2008 at the Department of Built Environment of the Eindhoven University of Technology. She worked as a consultant sustainable development at a commercial engineering consultancy company. In 2014, Wai Chung joined the group Sustainable Built Environment at VITO as a researcher. At VITO she focuses on construction-related projects, such as life cycle assessments of buildings and the further development of the building evaluation framework MMG (Environmental profile of building(s) (elements)), i.e. the underlying method of the Belgian building assessment webtool TOTEM (Tool to Optimise the Total Environmental impact of Materials), and projects on ecodesign and design for change/circular building.

Michiel Ritzen received a master's degree in architecture with a specialisation in technology in sustainable design from Delft University of Technology in 2004. He received a PhD in the environmental impact assessment of building integrated photovoltaics from Eindhoven University of Technology in 2017. Until 2010, Michiel Ritzen worked for several architectural firms as an architect, internal auditor (ISO 9001), sustainable energy expert, and environmental impact assessor. Between 2010 and 2021 he worked as a senior researcher at Zuyd University of Applied Sciences in Heerlen, the Netherlands and was the programme leader for circular building technology, covering the interaction between the energy transition and circular economy in the built environment. Currently, he is a project manager and senior expert in circularity in the built environment in the unit Smart Energy and Built Environment of VITO. Michiel Ritzen is the technical coordinator of the European H2020 research project 'DRIVE 0' and a member of the steering group of the European H2020 research project 'HOUSEFUL', both of which focus on a circular built environment and the energy transition.

Chapter 14
Circular Business Models for Digital Technologies in the Built Environment

Julia Nussholz, Ingvild Reine Assmann, Philip Kelly, and Nancy Bocken

Abstract Business model innovation enabled by novel digital technologies can accelerate the impact and upscaling of the circular economy in the built environment. Digital technologies not only enable highly impactful new business models but also enable innovation of existing business models. Considering the disruptive power of digital technologies, rethinking business models in the construction sector for the circular economy is vital to manage risks and capture opportunities. This chapter presents 12 real-life cases of emerging business models enabled by digital technologies that successfully narrow, slow, close, or regenerate resource loops in the construction sector. Cases are analysed regarding how they create, deliver, and capture value and how they enable circularity. Findings present different types of business models for digital technologies prevalent for narrowing, closing, slowing, and regenerating resource loops and that enabling capabilities for circularity, such as tracking, monitoring, control, optimisation, design evolution, and information exchange, are at the core of their value propositions. Industry practitioners can use findings to familiarise themselves with emerging business models and innovation opportunities.

Keywords Circular business models · Business model innovation · Digital technologies · Built environment · Building industry

J. Nussholz (✉)
Ramboll Sustainability Management Consulting, Copenhagen, Denmark
e-mail: junu@ramboll.com

I. R. Assmann
Ramboll Corporate – Sustainability & CR, Copenhagen, Denmark

Centre for Technology Entrepreneurship, Danish Technical University (DTU), Lyngby, Denmark

P. Kelly
Ridge and Partners LLP, Birmingham, UK

N. Bocken
Maastricht Sustainability Institute, School of Business and Economics, Maastricht University, Maastricht, The Netherlands

C. De Wolf et al. (eds.), *A Circular Built Environment in the Digital Age*, Circular Economy and Sustainability, https://doi.org/10.1007/978-3-031-39675-5_14

14.1 Introduction

A business model is a useful management tool to analyse and design a firm's business logic (Casadesus-Masanell and Ricart 2010) and how a company delivers, creates, and captures value (Osterwalder and Pigneur 2010). During this process, it helps managers to focus on the most relevant building blocks for the creation of commercial value (Osterwalder et al. 2005). The business model concept in management literature originates from the time when the Internet proliferated, and companies' blueprints to creating value diversified and became more complex compared with traditional business models (Osterwalder et al. 2005, Amit and Zott 2010). Business models are considered a strong indicator of competitive advantage (Magretta 2002) because changes are harder to replicate than product innovations (Amit and Zott 2010). Thus, in order to stay successful, companies must adapt their business model over time to changing business environments (Demil and Lecocq 2010).

Technological advances since the diffusion of the Internet in the 1990s have enabled new digital business models, which are now transforming industrial-age industries, such as the media, retail, financial services, and logistics sector (Veit et al. 2014). Digital business models can be defined as those that rely on digital technology and leverage the effects of digitalisation (Guggenberger et al. 2020; Bärenfänger and Otto 2015). Veit et al. (2014, p. 48) define a business model as digital "if changes in digital technologies trigger fundamental changes in the way business is carried out, and revenues are generated" of which Uber in the transport sector and AirBnB in the hospitality sector are prominent examples that have caused major disruption of previous business practices.

Even though the adoption of digital technologies in the building sector is slow compared with other sectors (ESCO 2021), an increasing number of digital technologies and business models are proliferating. Business models are paramount for the market introduction and uptake of these technologies. Only if technologies are embedded in business models that create superior customer and business value, the technology-enabled offers can be commercialised and scaled. This is the case, for instance, in platform models, such as those operated by the Norwegian company Loopfront that enables material or second-hand product exchanges (Loopfront 2022). Given the enormous challenges, such as stagnating productivity, high construction costs, resource intensity, and scarcity paired with pending ambitious environmental regulation in national legislation (ESCO 2021; JRC 2019), new digital business models could provide unforeseen solutions to challenges and serve customers in radically superior ways. Digital business models are understood as innovations in business models that transform analogue, physical objects, processes, or content into primarily digital formats (Trischler and Li-Ying 2022).

Digital technologies, such as platforms or building information modelling (BIM), enable a plethora of benefits, such as improved collaboration, easier transactions, and greater control of the value chain. The Internet of Things (IoT) increases data availability and enables data-driven decision-making for more efficient operations.

These developments are fundamentally changing traditional ways companies approach operations, procurement, design, and construction and engage with value chain partners (McKinsey 2020). For example, Boston Consulting Group estimates that 10–17% of total annual spending can be saved in the operation of buildings and 13–21% in the construction phase from full digitalisation (BCG 2016). Considering the disruptive power of digital technologies, rethinking business models and technological capabilities is vital to manage risks and capturing opportunities.

To provide an overview of the developments of digital business models in the circular built environment, this chapter presents 12 real-life cases of emerging business models enabled by digital technologies that successfully narrow, slow, close, or regenerate resource loops in the built environment. Cases are identified through desk research focusing on Europe, particularly the Netherlands, due to the authors' familiarity with this geographical context and proliferation of commercialised circular solutions in the built environment. Cases are analysed regarding how they create, deliver, and capture value, the digital technologies used, and their level of maturity. Also, their enabling capabilities to help narrow, slow, close, and regenerate resource loops in the built environment are presented. Based on the product and service offers of the case studies, several types of digital business models for the circular built environment are identified.

This chapter proceeds with outlining the theoretical background of the circular business model concept (Sect. 14.2), the presentation of the case studies for narrowing, slowing, closing, and regenerating resource loops (Sect. 14.3), and the discussion and conclusion (Sect. 14.4).

14.2 Circular Business Model Innovation

A business model can be described as a conceptual tool that can assist in understanding how a company conducts business to create and capture economic value (Schaltegger et al. 2012). This chapter defines business models by three main elements: value proposition, value creation and delivery, and value capture (Bocken et al. 2014; Richardson 2005).

The value proposition concerns product/service offerings, customer segments, and customer relationships of a company's business model (Boons and Lüdeke-Freund 2013). Value creation and delivery mechanisms are concerned with the activities, resources, partners, and distribution channels of a company's business model. Value capture is about the cost and revenue model, and in the case of a circular or sustainable model, it also concerns the positive value for society and the natural environment.

Business model innovation is considered to be a holistic approach that can function as an enabler to fulfil radical changes in a company's offers and value chains (Wells and Seitz 2005; Bocken et al. 2016; Tunn et al. 2019). Innovating the business model involves either reconfiguring the main elements of the company's existing business model or developing new business models (Zott and Amit 2010).

In the context of circular economy, business models have received substantial attention in literature and industry as an avenue to achieve increased sustainability in organisations across industries. Circular business models aim to create, deliver, and capture value while implementing circular strategies that can close material loops and extend the useful life of products and parts (Nussholz 2018). Adopting circular strategies usually requires radical and holistic alterations to a company's offers and value chains (Bocken and Geradts 2022; Wells and Seitz 2005; Nussholz 2018).

14.3 Digital Business Models to Enable Circularity

Twelve business model cases enabled by digital technologies were selected to exemplify business models that are narrowing, slowing, closing, and regenerating resource loops in the built environment. The following sections describe the companies' offers, how they enable circularity, and the main elements of their business models. It should be noted that all cases are examples of new business models, sometimes operated through daughter companies or spin-offs, and that not necessarily the whole company associated with the example is fully circular.

14.3.1 Digital Business Models for Narrowing Resource Loops

This section discusses the companies Parametric Solutions, Philips Lighting, and EDGE Next as examples of digital business models for narrowing resource loops.

The Swedish company Parametric Solutions offers an analytics app based on a parametric design method for architectural teams to create and compare design options (Parametric Solutions 2022a). Designs are developed based on the client's criteria and downloadable into design tools such as Revit. Optimisation criteria are, for example, space efficiency, energy efficiency, and reduced embodied carbon. As such, the main enabling capabilities of Parametric Solution's business models for narrowing resource flows are optimisation and design evolution. Parametric Solutions, for instance, partnered with the engineering consultant COWI and architect Arkitema to generate options for building volumes for a respective site (Parametric Solutions 2022b). Parametric Solutions creates value through the development of the parametric method and customised app based on the client's design criteria. Value is captured through users' payments for the app licence (Table 14.1).

Philips Lighting, with its headquarters in the Netherlands, offers an interactive IoT and Big Data System for lighting solutions. Sensors in the lighting panels are connected to interactive app-based systems that measure the occupancy, movement, and lighting levels to adjust and distribute energy usage where needed (Philips

Table 14.1 Examples of business models for narrowing resource loops enabled by digital technologies

Company	Parametric solution	Philips lighting	EDGE next
Sector	Building design	Lighting	Smart buildings
Country	Sweden	The Netherlands	The Netherlands
Business model type	Analytics app developer	Light as a service	Service provision platform
Digital technologies	Artificial intelligence	IoT, big data, and analytics	Digital twin, digital platform, IoT, big data analytics
Enabling capabilities	Optimisation, design evolution	Tracking, monitoring, control, optimisation	Tracking, monitoring, control, optimisation
Value proposition	Instant creation and comparison of design options. Design optimisation based on architectural teams' criteria. Optimisation of sustainability criteria, e.g. efficient space use, embodied carbon, energy consumption and efficiency, biodiversity.	Improved lighting quality. Adjustments based on user preferences. Reduction of energy use. Real-time data on operations and activities for facility managers to streamline operations.	Based on sensors, delivering data and insights for corporate real estate, portfolio managers, and human resources to optimise building performance. Optimisation of space utilisation, operational efficiency, employee Well-being, sustainable performance.
Value creation	Developing algorithms, front and back end by a team of architects and coders. Customisation of backend to customers' needs.	Developing lighting panels, sensor systems, big data system and analytics, and user apps. Maintenance of lighting system.	Developing sensor systems, software, platform, and dashboards apps for different optimisation targets.
Value capture	Payments for licence for app	Payment for products of lighting system and services	Payment per package
Company type	Start-up	Multi-national	Scale-up

Lighting 2022). As a result, increased user comfort is achieved and combined with a significant energy reduction for lighting. For example, energy usage decreased by 70% in the office building The Edge Amsterdam (Philips Lighting 2022). The control application provides building managers with real-time data on operations and activities to optimise operational efficiency and provides users with the possibility of adjusting the lighting. The main enabling capabilities of Philips Lighting's business model for narrowing resource loops are tracking, monitoring, control, and optimisation. Value is created by developing lighting panels, sensors, and a software system to monitor and control the lighting. Value is captured through the sale of the lighting system and services, while apps are offered free to users.

EDGE Next is a Netherlands-based real estate developer that also operates a service provider platform based on digital twin, sensor-based solutions, and big data analytics. EDGE Next offers different service and technology packages for various optimisation purposes, such as improved space utilisation, operational efficiency, and indoor comfort (EDGE Next 2022a). EDGE Next's business model's main enabling capabilities for narrowing resource loops are tracking, monitoring, control, and optimisation. For the Swedish power company Vattenfall, EDGE Next developed a 22,000 m^2 office building in Berlin, using their technologies to achieve a significant reduction in energy use (EDGE Next 2022b). Value is created through the development of the sensor systems, platform applications, and user dashboards, with targeted customers being corporate real estate, portfolio managers, and human resources. Value is captured through continuous payments for different service packages.

14.3.2 Digital Business Models for Slowing Resource Loops

This section discusses the companies Madaster, Rehub, and Excess Material Exchange as examples of digital business models for slowing resource loops.

The Netherlands-based Madaster operates as a digital platform offering a registry of all materials and products used in real estate and infrastructure. Madaster bases its registry on material passports developed for the objects. Amsterdam Metropolitan Area has, for instance, been involved in using Madasters's material passport to stimulate the regional circular economy (Madaster 2022). The enabling capability of Madaster's business model is information exchange. Value is created by linking the registry to material databases of partner companies to facilitate data entry and quality. Value is captured through offering a licence for use.

Rehub is a Norwegian start-up offering a material bank platform that connects the supply and demand side for the reuse of construction materials (Rehub 2022). Rehub's business model offers the enabling capabilities of optimisation and information exchange. The value proposition is about the database for reusable materials and warranties, environmental impact analyses, and assistance. Value is created through the development of the platform and data registry of the materials, and value is captured via subscription-based payments for access to the platform.

Excess Material Exchange (EME) is a Dutch start-up operating as a digital marketplace platform focused on allowing clients to find new high-value reuse options for their end-of-use materials and products (Excess Materials Exchange 2022). EME's tools are, for instance, applied in the European carpet industry to ensure that recyclable carpet tiles are matched with the demand side. The carpet tiles are given a product identification to gather all product information and allow for recyclability (Excess Materials Exchange 2019). The business model's enabling capabilities involve optimisation and information exchange. The company's value proposition is about the offering of an online material matching platform focused on selling B2B. Value is created through developing the platform, and value is captured by selling subscriptions to access the platform. Examples are given in Table 14.2.

Table 14.2 Examples of business models for slowing resource loops enabled by digital technologies

Company	Madaster	Rehub	Excess material exchange
Sector	Buildings and infrastructure	Construction materials	Cross-industries
Country	The Netherlands	Norway	The Netherlands
Business model type	Material passport platform provider	Material bank platform provider	Marketplace platform provider
Digital technologies	Digital platform, material passports	Digital platform	Digital platform, blockchain, artificial intelligence
Enabling capabilities	Information exchange	Optimisation, information exchange	Optimisation, information exchange
Value proposition	Registry of information on all materials and products in a building project. Circularity, embodied carbon, or toxicity assessment. Material passport for optimised end-of-use and end-of-life value management for construction materials and products.	Database for reusable material. Warranties on the material. Documenting CO_2 savings.	Online marketplace for all excess material. B2B sale by matching the supply and demand across industries.
Value creation	Acquiring partner companies to facilitate data entry and data quality	Development of digital platform	Development of platform
Value capture	Licence for use	Subscription-based payment for platform access	Subscription-based payment for platform access
Company type	SME	Start-up	Start-up

14.3.3 Digital Business Models for Closing Resource Loops

This section discusses the companies MetroPolder, Circularise, and Loopfront as examples of digital business models for closing resource loops.

The Dutch company MetroPolder (2022a) offers a green roof with a rainwater storage system. Storage and discharge are controlled through a sensor-based software system allowing for controlled discharge of rainwater to prevent flooding and enable reuse, thereby preventing the use of drinking water. This system for control and optimisation helps close resource loops for rainwater. Through its biodiversity and cooling benefits, the green roof also fits the regenerate principle. In Amsterdam, MetroPolder's water storage system is used on the roof park/garden Babylon providing a 1500 m^2 park with a water storing capacity of 50,000 l. Water is used for plant irrigation, for example, for the vegetable and fruit garden, enabling suitable irrigation levels (Metropolder 2022b). MetroPolder's business model creates value by developing a sensor and software system, green roof technology, an operating system, and a dashboard for users, e.g. facility managers. Value is captured through the sale of the water capture system technology and services such as construction and maintenance.

The company Circularise, based in the Netherlands, offers a blockchain-enabled software platform to help companies track products and materials and allow information exchange to enable closing loops of materials (Circularise 2022a). Circularise partnered with the City of Amsterdam to increase traceability and transparency in their construction procurement process and gather data on environmental impact, enabling information sharing without risking sensitive data. Circularise also partnered with a concrete product company to help trace materials end-to-end throughout the supply chain, and that information can be shared without risking sensitive data (Circularise 2022b). Circularise's business model creates value through the development of blockchain technology and the creation of data, product passports, and other certificates. Value is captured through selling services and payment for licences for software solutions.

Loopfront is a Norwegian company that offers clients working across the built environment access to a reuse platform. The digital platform offers material passports, a material bank, and a survey tool and assists in closing resource loops through its enabling capabilities of optimisation, tracking, monitoring, and control (Loopfront 2022). The value is created through the development of the digital platform and is captured through selling membership packages on four different levels (Starter, Basic, Standard, or Enterprise). Examples are given in Table 14.3.

14.3.4 Digital Business Models for Regenerating Resource Loops

This section discusses the companies WASP, Lo3Energy, and AUAR as examples of digital business models for regenerating resource loops.

WASP is an Italian firm specialised in designing, developing, and selling 3D printers (WASP 2022a). The company has succeeded in 3D printing structures that are entirely developed using reusable and recyclable bio-based materials from local soil. Specialist software allows for two printing arms to be synchronised for the construction, which allows for avoiding collisions and ensuring simultaneous operation. WASP recently created an installation for Dior in which they 3D printed two pop-up stores on Jumeirah beach in Dubai from all-natural materials (WASP 2022b). WASP's business model creates value through the development of advanced 3D printers, whereas value is captured through the sale of 3D printers and 3D printing services.

Lo3Energy is an American company that has developed a front-end blockchain-powered platform called Pando that enables suppliers and clean energy operators to support 24/7 load matching and offers intelligent incentives to drive renewable energy use (LO3Energy 2022). The Pando software solution has, for instance,

Table 14.3 Examples of business models for closing resource loops enabled by digital technologies

Company	MetroPolder	Circularise	Loopfront
Sector	Roofs and water management	Manufacturing and recycling	Platform developer
Country	The Netherlands	The Netherlands	Norway
Business model type	Software system provider	Software platform provider	Digital platform and surveying tool
Digital technologies	IoT	Blockchain technology, digital platform, material passports and databank	Digital platform
Enabling capabilities	Control and optimisation	Tracking and information exchange	Tracking, monitoring, control, optimisation
Value proposition	Sensor-equipped roof system with rainwater storage, e.g. developers or facility managers. Controlled discharge of water to prevent flooding and enable rainwater use. Biodiversity benefits and cooling effects.	Blockchain technology to trace products and materials and verify their origins. Creation of product passport and certificates.	Survey tool. Material cards. Marketplace. Material passports.
Value creation	Developing sensor and software system, green roof technology, operating system and dashboard	Developing software and platform solutions, including back end and dashboards	Developing and piloting material bank and material passport system
Value capture	Sale of roof systems and services, e.g. planning, construction, maintenance	Sale of services and licences for software solutions	Sale of membership to access platform
Company type	SME	Start-up	Pilot project

been installed in a shopping centre in New South Wales, Australia, where it will be used to optimise renewable energy production. The company's business model's main enabling capabilities are monitoring, optimisation, and information exchange, helping to regenerate resource loops. The business model is capturing value through developing a grid-edge accounting service platform that can match the production and consumption of clean energy at defined time intervals. The value is captured through payment by grid operators and energy utilities to promote their offers on the app.

AUAR is a British start-up, which develops dwelling units through robotic manufacturing using bio-based materials with a zero-carbon life cycle (AUAR 2022). It has been used in an installation at The Building Centre in London to show how it can act as a home, office, and co-working station solution (Design Boom 2020). AUAR's business model's enabling capabilities consist of optimisation and design evolution. Value is created through the development of robotically assembled dwelling units, and value is captured through the payment for customised

Table 14.4 Examples of business models for regenerating resource loops enabled by digital technologies

Company	WASP	Lo3Energy (Pando)	AUAR
Sector	3D printed construction	Renewable energy	Automated architecture
Country	Italy	USA	UK
Business model type	3D printer manufacturer	Web platform provider for energy retail	Automation developer
Digital technologies	3D printer manufacturing	Blockchain technology	Additive/robotic manufacturing
Enabling capabilities	Optimisation, design evolution	Optimisation, monitoring, information exchange	Optimisation, design evolution
Value proposition	Optimising construction to be more time and resource-efficient. Use of 100% bio-based materials.	Software platform allowing clients to forecast the availability of cheap and clean energy.	Modular dwelling units with installation that can be developed according to clients' specific needs. Zero-carbon life cycle.
Value creation	Development of 3D printers or building constructions with 100% bio-based materials for reuse and recycling	Development of grid-edge accounting service to match production and consumption of clean energy at specific time intervals	Development of robotically assembled and customised dwelling units
Value capture	Sale of 3D printers and 3D printing services	Payment by grid operators and energy utilities to promote their offers on the app	Payment for dwelling units on demand
Company type	SME	Start-up	Start-up spinout

dwelling units on demand. The prices are dependent on the dwelling unit size and amounts of units needed. Examples are given in Table 14.4.

14.4 Discussion and Conclusions

This chapter has presented 12 cases of business models enabled by digital technologies that help narrow, slow, close, and regenerate resource flows in the built environment. The analysed companies were active in various sectors within the built environment, such as smart buildings, interiors, building design, and construction. Business models were found to use a variety of technologies, often pairing multiple technologies such as digital twins, digital platforms, IoT, and big data analytics. No emerging business models, however, were found based primarily on BIM and Geoinformation Systems (GIS) technologies. A reason could be that these types of software are available through licences of established companies, widely used, but in the case of GIS, also accessible open source. Both technologies however have the potential to track stocks and locations of components and materials suitable for reuse and recycling (see Chap. 2 for industry use cases of GIS).

Through developing and using digital technologies and thinking of resource efficiency and circularity in their business models, the analysed case companies make significant contributions to enabling circularity in the built environment through their offers. They capitalised on several enabling capabilities of digital technologies to realise circular resource flows. In particular, tracking, monitoring, control, optimisation, and information exchange and optimisation were prominent examples of how digital technologies help enable different strategies for circularity. It should be noted that some of the presented cases explicitly define themselves or their services as circular (e.g. Circularise) while most of them do not (e.g. EDGE Next).

Based on the overview of several case studies, various business model types were identified, summarising commonalities of companies' offers. Types identified were 3D printer manufacturer, platform provider (e.g. material registry, marketplace, service provision, retail), automation developer, product manufacturer, light as a service model, and analytics app developer. Specifically, service offers facilitated through platforms were common even though they had a lot of variation in terms of their use and offerings. For narrowing resource loops, business model types based on software for optimisation were the most common. For slowing and closing resource loops, business model types based on platforms were dominant. For regeneration, manufacturers or providers of automation and 3D printing machinery or services dominated.

Many of the identified cases were in the Netherlands. The Netherlands has a progressive circular economy policy (Ministerie van Infrastructuur en Waterstaat 2021) and ranks high in the Global Innovation Index (GII 2021), which might be an explanation for the proliferation of circular start-ups in this country. However, the fact that the authors of this chapter have better insights into the developments in the Dutch built environment and might have missed cases in other countries, for example, if company websites were not available in English or less emphasis was put on communication outside of the national market, might have contributed to the dominance of Dutch case studies.

Most of the studied cases were start-ups. Some companies are already small to medium-sized enterprises, such as the digital twin and optimisation platform provider EDGE Next or the 3D printing company WASP. Many of the identified start-ups are daughter companies or spin-offs of incumbent multinationals (e.g. PolderRoof by Wavin, Rehub by Ramboll). Certainly, many digital technologies, such as parametric design, BIM, and GIS are also already used by incumbents. This study presented companies with circular business models enabled by digital technologies, offering their benefits to other actors in the sector. Future research is needed to investigate potential pitfalls and uncertainties associated with digital business models for enabling circularity in the built environment that might stem from a higher dependence on critical materials, data and technology, or environmental rebound effects. Despite these pitfalls, these developments in the uptake of digital technologies are critical as wide adoption is a prerequisite to capitalise on the improvement potential of digital technologies for circularity and other sustainability benefits (JRC 2019).

14.5 Key Takeaways

- Considering the disruptive power of digital technologies, rethinking business models in the construction sector for the circular economy is vital for companies to manage risks and capture opportunities.
- Companies considering resource efficiency and circularity in their business models and developing offers based on digital technologies can make significant contributions to enable circularity in the built environment.
- Emerging business model examples for the circular economy include 3D printer manufacturers, platform providers (e.g. material registry, marketplace, service provision, retail), automation developers, product manufacturers, light as a service models, and analytics app developers.
- Different business model types (e.g. digital marketplaces, platforms, etc.) are suitable for enabling different circular principles (i.e. narrowing, slowing, closing, and regenerating resource loops).

References

Amit R, Zott C (2010) Business model innovation: creating value in times of change. Long Range Plan 43(2–3):216–226

AUAR (2022) Automated Architecture. https://automatedarchitecture.io/. Accessed 01 Dec 2022

Bärenfänger R, Otto B (2015) Proposing a capability perspective on digital business models. IEEE 17th Conference on Business Informatics 1:17–25

Bocken N, Geradts T (2022) Designing your circular business model. Stanf Soc Innov Rev 20(2): 34–39

Bocken NM, Short SW, Rana P et al (2014) A literature and practice review to develop sustainable business model archetypes. J Clean Prod 6:42–56

Bocken NM, De Pauw I, Bakker C et al (2016) Product design and business model strategies for a circular economy. J Ind Prod Eng 33(5):308–320

Boons F, Lüdeke-Freund F (2013) Business models for sustainable innovation: state-of-the-art and steps towards a research agenda. J Clean Prod 45:9–19

Boston Consulting Group (2016) How technology is revolutionizing construction. https://www.bcg.com/publications/2016/how-technology-is-revolutionizing-construction. Accessed 01 Dec 2022

Casadesus-Masanell R, Ricart JE (2010) From strategy to business models and onto tactics. Long Range Plan 43(2–3):195–215

Circularise (2022a) We create software to make products traceable. Circularise. Accessed 22 Dec 2022

Circularise (2022b) Case study: city of Amsterdam tracks environmental impact of construction procurement. Circularise. Accessed 22 Dec 2022

Demil B, Lecocq X (2010) Business model evolution: in search of dynamic consistency. Long Range Plan 43(2–3):227–246

Design Boom (2020) Architecture. AUAR uses robotically fabricated timber blocks to build temporary home-office. https://www.designboom.com/architecture/auar-robotically-fabricated-timber-blocks-build-temporary-home-office-london-03-16-2020/. Accessed 22 Dec 2022

EDGE Next (2022a) Home. Optimizing your building performance. Home – EDGE Next (edge-next.com). Accessed 22 Dec 2022

EDGE Next (2022b) EDGE delivers EDGE Suedkreuz Berlin to Vattenfall. EDGE delivers EDGE Suedkreuz Berlin to Vattenfall – EDGE Next (edge-next.com). Accessed 22 Dec 2022
ESCO (2021) Digitalisation in the construction sector. Analytical Report 1–159
Excess Materials Exchange (2019) EME pilot report. EME+Pilot+Report+compressed.pdf (excessmaterialsexchange-pilot-report.s3.eu-central-1.amazonaws.com). Accessed 22 Dec 2022
Excess Materials Exchange (2022) About us. Excess Materials Exchange. Accessed 22 Dec 2022
GII (2021) Global innovation index 2021. https://www.wipo.int/edocs/pubdocs/en/wipo_pub_gii_2021/nl.pdf. Accessed 01 Dec 2022
Guggenberger T, Möller F, Boualouch K et al (2020) Towards a unifying understanding of digital business models. PACIS, p 70
JRC (2019) Digital transformation in transport, construction, energy, government and public administration. https://ec.europa.eu/jrc/en/publication/eur-scientific-and-technical-research-reports/digital-transformation-transport-construction-energygovernment-and-public-administration. Accessed 01 Dec 2022
Lo3Energy (2022) Empowering companies and communities through new DER incentives and renewable energy accounting. The Future of Energy | Pando | LO3 Energy. Accessed 01 Dec 2022
Loopfront (2022) Loopfront makes reuse easy. Loopfront. Accessed 01 Dec 2022
Madaster (2022) Amsterdam metropolitan area uses material passports to boost the circular economy in the region. Madaster. Accessed 01 Dec 2022
Magretta J (2002) Why business models matter. Harv Bus Rev:4–8
McKinsey & Company (2020) The next normal in construction: How disruption is reshaping the world's largest ecosystem. How disruption is reshaping construction | McKinsey. Accessed 01 Dec 2022
Metropolder (2022a) About. Metropolder. Accessed 01 Dec 2022
Metropolder (2022b) Roof park/roof garden B. Bylon. Metropolder. Accessed 01 Dec 2022
Ministerie van Infrastructuur en Waterstaat (2021) Uitvoeringsprogramma Circulaire Economie 2021–2023. https://open.overheid.nl/repository/ronl-669a180a-7f09-4336-890c-633cf2c3b852/1/pdf/uitvoeringsprogramma-circulaire-economie.pdf. Accessed 01 Dec 2022
Nussholz JL (2018) A circular business model mapping tool for creating value from prolonged product lifetime and closed material loops. J Clean Prod 197:185–194
Osterwalder A, Pigneur Y (2010) Business model generation: a handbook for visionaries, game changers, and challengers, vol 1
Osterwalder A, Pigneur Y, Tucci CL (2005) Clarifying business models: origins, present, and future of the concept. Commun Assoc Inf Syst 16(1):1
Parametric Solutions (2022a) Home. Parametric Solutions. Accessed 01 Dec 2022
Parametric Solutions (2022b) Tools to explore architecture. Parametric solutions – Tools to explore architecture. Accessed 01 Dec 2022
Philips Lighting (2022) Cases. Office life with IoT enabled connected lighting. The Edge – Philips lighting. Accessed 22 Dec 2022
Rehub (2022) Connecting supply & demand of reusable building materials in a hub of supporting features. Home – Rehub. Accessed 01 Dec 2022
Richardson JE (2005) The business model: an integrative framework for strategy execution. University of Hawaii at Manoa
Schaltegger S, Lüdeke-Freund F, Hansen EG (2012) Business cases for sustainability: the role of business model innovation for corporate sustainability. International journal of innovation and sustainable development 6(2):95–119
Trischler MFG, Li-Ying J (2022) Digital business model innovation: toward construct clarity and future research directions. Rev Manag Sci:1–30
Tunn VS, Bocken NM, van den Hende EA et al (2019) Business models for sustainable consumption in the circular economy: an expert study. J Clean Prod 212:324–333
Veit D, Clemons E, Benlian A et al (2014) Business models. Bus Inf Syst Eng 6(1):45–53
WASP (2022a) Discover our 3D printers Solutions. 3D printers for sale online | WASP (3dwasp.com). Accessed 01 Dec 2022

WASP (2022b) WASP 3D prints a unique concept store in collaboration with Dior. A 3D printed pop-up store by WASP for Dior (3dwasp.com). Accessed 01 Dec 2022

Wells P, Seitz M (2005) Business models and closed-loop supply chains: a typology. Supply Chain Manag 10(4):249–251

Zott C, Amit R (2010) Business model design: an activity system perspective. Long Range Plan 43(2–3):216–226

Julia Nussholz is a Senior Consultant at Ramboll's Strategic Sustainability Consulting team in Denmark and holds a PhD in Circular Business Model Innovation from Lund University's International Institute of Industrial Environmental Economics. She has more than 8 years of experience working with circular innovation in positions in research, policy, and industry, with a focus on the building sectors.

Ingvild Reine Assmann is a sustainability consultant in Ramboll Group's Sustainability & CR function. She is currently undertaking an Industrial PhD degree at the Technical University of Denmark (DTU) focused on circular business model innovation in the built environment. She has a background as an Industrial Engineer from DTU and University College London.

Philip Kelly is partner at Ridge and Partners LLP leading the area of Net Zero and Circularity. Previously Philip was director and the head of Sustainability & Building Physics in Ramboll's Building Services Group in the UK. He has extensive experience in technical analysis and environmental assessment methodologies. With over 12 years of experience in delivering low-energy buildings, Philip works with CIBSE, the UK Green Building Council (UKGBC), and the Association of Consulting Engineers' (ACE) Sustainability working groups to influence the industry towards a more sustainable future. Working predominantly in the higher education sector over the last 7 years, Philip has led the energy and sustainability strategy for some of the most complex buildings in the sector.

Nancy Bocken is a Professor in Sustainable Business and Circular Economy at Maastricht University, Maastricht Sustainability Institute (MSI) and is a leading researcher on topics such as sustainable business models, business experiments for sustainability, circular economy, and sufficiency. She is also a Fellow at Cambridge Institute for Sustainability Leadership, advisor to TNO (Dutch association for applied scientific research) and a Board member of the Philips Foundation. Before going into academia, she held positions in the logistics, banking, and consulting sectors. She holds a PhD from the Department of Engineering, University of Cambridge and is a co-founder of her own circular and sustainable business, HOMIE.

Chapter 15
Digital Transformation of the Built Environment Towards a Regenerative Future

Catherine De Wolf and Nancy Bocken

Abstract The concept of regeneration and its application in the built environment is crucial when considering how digital technologies contribute to the transition towards a circular economy. Regeneration in the built environment fosters economic, social, and environmental prosperity for all stakeholders involved, through coevolution, adaptation, knowledge and skill exchange, diversity of ecosystems, harmonisation, and reconciliation. These advantages extend to building users and owners, businesses, local governments, the environment, and the community as a whole. The regenerative design, construction, and maintenance of buildings and infrastructure enhances the economic, social, and environmental aspects of a region. This chapter discusses examples and business models that showcase the implementation of regenerative practices in the built environment and examines how the digital technologies discussed in the book can contribute to regeneration.

Keywords Regeneration · Resilience · Regenerative business models · Regenerative design

15.1 The Relevance of Regeneration

The concept of regeneration has gained significant attention in recent years as a powerful approach to creating thriving socio-ecological systems (Konietzko et al. 2023). By embracing regenerative principles, we can effectively tackle global environmental challenges through minimising harm, restoring and revitalising ecosystems, and achieving a net positive impact (Morseletto 2022). Regeneration as a

C. De Wolf (✉)
Circular Engineering for Architecture, Institute of Construction & Infrastructure Management, Department of Civil, Environmental and Geomatic Engineering, ETH Zurich, Zurich, Switzerland
e-mail: dewolf@ibi.baug.ethz.ch

N. Bocken
Maastricht Sustainability Institute, School of Business and Economics, Maastricht University, Maastricht, The Netherlands

C. De Wolf et al. (eds.), *A Circular Built Environment in the Digital Age*, Circular Economy and Sustainability, https://doi.org/10.1007/978-3-031-39675-5_15

concept goes beyond narrow interpretations of sustainability and resilience. While sustainability usually focuses on meeting present needs without compromising the future (Brundtland 1987), and resilience aims to withstand and recover from disturbances (Sayer et al. 2013; Standish et al. 2014; Capdevila et al. 2021; Wyss et al. 2022), regeneration as an approach seeks to have a continuous net positive impact on the environment, health, society, and the economy (Polman and Winston 2021; Hahn and Tampe 2021). While a narrow interpretation of sustainability may concern the mitigation of negative impacts ('less bad'), regeneration endeavours to go beyond that (Reed 2007) by actively reversing past damage through renewal, nurturing the ecosystem, and enhancing well-being ('more good'). Regenerative approaches recognise humans as active participants in the broader ecosystem, transforming the concept of sustainability into a more comprehensive and impactful paradigm that emphasises holistic engagement and collaboration (Mang et al. 2016).

Inclusive definitions of the circular economy go beyond the efficient use of resources to also include the active improvement of the natural environment (Ellen McArthur Foundation 2013; Konietzko et al. 2020; Bocken and Geradts 2022). Regeneration has been identified as a key strategy of the circular economy by the editors of this book (Çetin et al. 2021). While other circular economy strategies typically involve minimising waste and closing material loops, regeneration adds an extra layer of value by actively restoring and revitalising ecosystems, enhancing biodiversity, and promoting the well-being of both the environment and communities. This is necessary to prevent further environmental damage, already evident in the significant decline in biodiversity (Almond et al. 2022; Naeem et al. 2022) and the effects of climate change, such as extreme weather patterns and the melting of ice caps (IPCC 2022).

Regeneration goes beyond environmental considerations as it offers comprehensive and interconnected design and construction practices that empower us to generate societal and economic benefits. Through recognising the interdependencies among domains such as finance, agriculture, design, ecology, economy, sustainability, and broader societal issues (Wahl 2016), we have the opportunity to be inspired by nature and harness its self-healing and self-organising abilities to foster symbiotic relationships with natural ecosystems (Mang and Reed 2012).

In the built environment, examples of regenerative approaches include buildings as carbon sinks, self-repairing or pollution-cleaning envelopes, green facades and roofs, the use of regenerative materials, and building approaches that support biodiversity and renewable energy generation (Konietzko et al. 2023; Churkina et al. 2020). At the 18th International Architecture Exhibition (Venice Biennale), for example, the Belgian 'In Vivo' pavilion showcased an innovative application of mycelium as a regenerative building material, defining regeneration as the 'process of reversing programmed obsolescence' (Fakharany 2023) (Fig. 15.1).

Regeneration in the built environment, however, is not limited to biological and ecological approaches. It must also actively improve both the natural environment and human activities, recognising the collaborative role of humans in the ecosystem and applying it at various scales (Attia 2018). To integrate environmental restoration, social development, economic revitalisation, and urban transformation into the built environment, we need to address social inequalities, stimulate green economic

Fig. 15.1 The In Vivo Pavilion at the 18th International Architecture Exhibition in Venice. Mycelium, used as a regenerative building material, is kept alive so that the walls can self-repair

growth (Terzi 2022), and build inclusive cities in which people can co-evolve (Mang et al. 2016). Regenerative design fosters symbiosis between human activities and the natural environment, promoting ecological balance and resilience in order to create a harmonious future in which humans and nature thrive together (Watson 2019).

While models for nature regeneration have long been discussed, the consideration of regeneration as a tenable model for adoption by business and policymakers is much more recent. Konietzko et al. (2023: 1) have proposed a comprehensive definition and framework for regenerative business models to enable organisations to focus on planetary health and societal well-being. They suggest that businesses 'create and deliver value at multiple stakeholder levels—including nature, societies, customers, suppliers and partners, shareholders and investors, and employees—through activities promoting regenerative leadership, co-creative partnerships with nature, and justice and fairness'. By doing so, businesses can aim for a net positive impact.

Although regeneration may currently seem distant from being a mainstream business practice, investing in regenerative innovations holds the potential to enhance resource security, lower costs, and gain a competitive advantage in the long run. However, achieving equitable and resilient systems requires collaboration across diverse fields (Bocken and Geradts 2022; Polman and Winston 2021). By forming partnerships and leveraging collective knowledge, we can enhance positive feedback loops to put into action principles of repair, renewal, flexibility, adaptability, harmonisation, reconciliation, and resilience within self-organising local systems. To effectively tackle uncertain challenges, it is crucial to embrace complexity, employ new tools, and consider context-specific interventions. Adopting a regenerative approach can help us address these complex problems. Such an approach has the potential to both facilitate adaptation to climate change (e.g. through the cooling effects of green roofs) and serve as a means of mitigating and repairing environmental damage (e.g. through emissions-capturing facade materials or natural habitat-improving design).

Designing buildings in a regenerative manner is crucial not only for the welfare of the natural environment but also for human health (Coady 2020). Climate change-induced heat waves, storms, air pollution, and contamination directly affect human well-being. Incorporating nature-inspired elements and strategies into building design, such as natural landscapes, natural ventilation systems, and regenerative materials, can improve air quality, reduce pollution, and mitigate the urban heat island effect. These design approaches create healthier indoor and outdoor environments, ultimately enhancing the overall well-being of occupants. By prioritising symbiotic (i.e. mutually beneficial) relationships with nature, buildings can protect and enhance human health while fostering a regenerative future where both the environment and humanity can thrive.

15.2 Examples of Regeneration in the Built Environment

Several strategies can be implemented to embody the principle of regeneration in the circular built environment. These include stimulating human–nature co-habitation and local biodiversity through the creation of shared spaces. By adopting these strategies, the circular built environment can embrace regeneration as a core principle. Many examples of regenerative architecture already exist at the material, product, building, neighbourhood, and community scale.

Regenerative approaches at the *material scale* involve not only using sustainable and renewable materials in construction and infrastructure projects, but also developing self-repairing or environment-improving materials. The concrete of the Pantheon in Rome provides inspiration for developing materials that can heal themselves: its 'lime clasts' create mineral deposits, which give the concrete self-healing properties (Seymour et al. 2023). Regenerative materials should use healthy (non-hazardous) and renewable (e.g. bio-based materials) resources, such as the mycelium used for the Venice Biennale in 2023 (Heisel 2017; Bitting et al. 2022).

Regenerative strategies at the *product scale* focus on designing and creating products that use or generate renewable materials, turn waste into resources, and enhance the natural habitat for plants and animals. Green roofs, also referred to as living roofs or vegetated roofs, are examples of regenerative building products. Their vegetation improves air and water quality, reduces the urban heat island effect, minimises stormwater runoff, improves energy efficiency by providing insulation, and enhances aesthetics (Wang et al. 2022; Calheiros et al. 2022). Vertical gardens, green facades, and urban farming have similar regenerative benefits (Rodrigues do Amaral 2020). One example is building materials that capture greenhouse gas emissions from the air (Dring and Schwaag 2021). Through the conversion of wood waste into biochar, a negative emissions technology, CO_2 stored by trees remains permanently locked in a stable form. Such materials replace environmentally harmful substances in a true end-of-life solution, as materials can safely return to the earth or be transformed into biochar after decades of use. Finally, facades and roofs can also be products that generate and store renewable energy locally in communities (e.g. through solar power). In fostering sustainable and equitable access to clean energy, these products empower communities and contribute to the restoration, self-sufficiency, and well-being of both the natural environment and its inhabitants.

Regenerative practices at the *building scale* aim to create structures that actively contribute to environmental enhancement and improve the quality of life of the occupants, for example, through on-site renewable energy generation, water conservation and treatment, natural lighting, and ventilation. Vernacular architecture often incorporates passive systems that achieve regeneration goals, while integrating digital technologies has the potential to further enhance outcomes, if done well. While achieving a completely regenerative building is challenging, there are notable instances in which buildings have incorporated regenerative systems, such as integrating on-site renewable energy generation, using smart sensors to adjust lighting and climate based on occupancy, harvesting rainwater, and generally setting new standards for environmental responsibility in terms of design and operation. One example of such a building is The Edge in Amsterdam (Wakefield 2016; Jalia et al. 2022).

Regenerative approaches at the *neighbourhood scale* enhance social equity, resilience, and environmental well-being through natural landscape design, walkability, public transportation, and regenerative infrastructure. Regeneration involves holistic urban planning through the integration of interconnected systems and regenerative principles into governance. Regenerative cities prioritise inclusivity, environmental restoration, and economic prosperity for all residents. For example, Singapore's Garden City vision emphasises biodiversity, air quality, and water quality; Copenhagen's infrastructure vision incorporates wind energy generation, walkability, cyclability, and inclusive public spaces; and Medellin's 'Corredores Verdes' (Green Corridors) and electric transportation vision fosters biodiversity, inclusivity, emissions sequestration, and air pollution reduction (Newman 2014; Reflow 2022; Zingoni dec Baro 2022; Copenhagen City 2014; Future of Cities 2023).

Other regenerative strategies consider the natural environment at the *community scale*. Improving outdoor spaces can transform misused or unused areas into public spaces that benefit local communities. Cleaning wastewater through regenerative design strategies, for example, promotes the restoration and enhancement of ecosystems and the preservation of water resources. The East Kolkata Wetlands exemplify the regenerative design strategy of sewage management by integrating indigenous practices of aquaculture (Watson 2019). By channelling sewage through a network of interconnected ponds, these wetlands utilise the natural purification capacity of aquatic plants and microorganisms to clean the water, while simultaneously providing a fertile habitat for fish farming (Saha 2019). Other wastewater

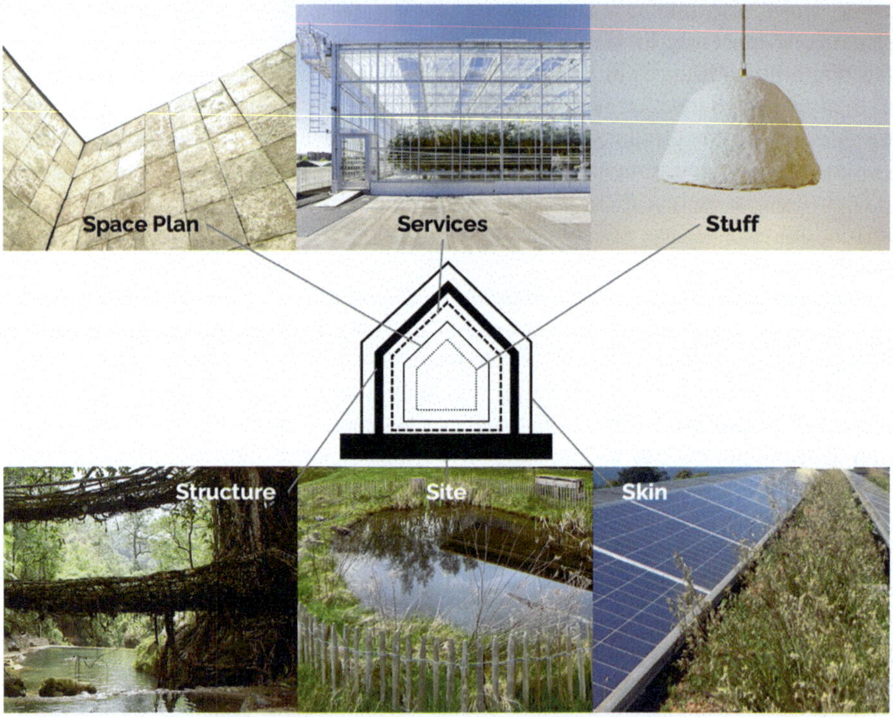

Fig. 15.2 Examples of regenerative strategies for the built environment, connected to the shearing layers of Brand, clockwise starting from the upper left: (**a**) mycelium as interior finishes as shown in the 2023 Venice Biennale's Belgian pavilion, (**b**) urban farming systems (© BIGH), (**c**) mycelium equipment such as lamps (© PermaFungi), (**d**) solar and green roofs (© Biosolar Roofs), (**e**) aquaculture (© Anku), and (**f**) living root bridges (© Arshiya Urveeja Bose)

treatment technologies employ a series of tanks that support vegetation and diverse organisms. These innovative systems mimic natural wetland processes to effectively treat wastewater, fostering ecological regeneration and promoting sustainable water management practices (Watson 2019).

In the context of a regenerative built environment, Fig. 15.2. establishes a connection between the above-mentioned examples and the layers of change. Inspired by the concept of shearing layers, introduced by architect Frank Duffy and further elaborated by Stewart Brand (Brand 1995), a regenerative built environment is understood to be composed of multiple layers that can be dynamically transformed and adapted over time to enhance their environmental, social, and economic performance. The concept of shearing layers can be applied to the various components and systems of regenerative buildings, emphasising the importance of considering different rates of change and adaptability in their design and operation.

Reinstating the symbiotic coexistence between nature and the man-made world involves dissolving the boundaries that separate the two realms (Sayer et al. 2013; Watson 2019; Wyss et al. 2022). Throughout history, ancient buildings have harmoniously responded to climate, ecology, culture, and location (Wahl 2016). Such structures use natural methods of heating, cooling, ventilation, and construction that have stood the test of time. These buildings became an expression of their communities, reflecting their unique surroundings. However, by contrast, modern architecture often neglects this vital relationship with the environment, resulting in a cookie-cutter approach that disregards local context. Rather than reverting to the lifestyle of our ancestors, we can embrace systems thinking to enhance traditional practices by integrating digital technologies in the construction industry, thereby improving efficiency, streamlining processes, and optimising outcomes for better project management and delivery (Binder 2007; Wyss et al. 2022).

Smart and sustainable city technologies strengthen a symbiotic relationship with the local environment and community while scaling up circular economy principles (Allam and Takun 2022; Hota et al. 2023). Net positive buildings equipped with advanced technologies can share surplus resources like energy, water, and food with their surroundings. Smart grid technologies enable 'prosumers' to trade surplus energy within their neighbourhoods, promoting a localised and sustainable energy ecosystem. Smart contracts, along with digital platforms like Pando, facilitate the purchase and receipt of local renewable energy within communities (Kirli et al. 2022). Smart cities can enable the efficient scaling up of regenerative architectural practices by leveraging digital technologies like smart grids, the Internet of Things (IoT), artificial intelligence (AI), and blockchain technology (Gligoric et al. 2019; Bugaj et al. 2022).

15.3 Digital Technologies Towards a Regenerative Built Environment

The wide array of digital technologies explored in this book has the potential to significantly contribute to a regenerative future for the built environment. Technologies that offer opportunities for stakeholders to collaborate in bringing about a more environmentally conscious and regenerative future include building information modelling (BIM), digital twinning, geographic information systems (GIS), smart cities, the Internet of Things (IoT), reality capture and scanning technologies, artificial intelligence (AI), data templates and material passports, computational design tools, digital fabrication, extended reality (XR), and blockchain, among others.

BIM and digital twins (see Chap. 1) contribute to regenerative architecture by providing a collaborative platform for stakeholders to optimise sustainable designs. By integrating data on materials, energy consumption, and life cycle analysis, BIM enables real-time monitoring, predictive maintenance, and resource optimisation. Digital twinning enhances this by accommodating past and present states of the building, providing temporal insights for informed decision-making. Together, BIM and digital twins support a data-driven and holistic approach to creating buildings that contribute to a regenerative future for the built environment.

GIS (see Chap. 2) supports the development of regenerative cities and regions by analysing spatial data to optimise resource flows and to conduct infrastructure planning for regenerating natural areas. By mapping and visualising natural resources, waste streams, material flows, and transportation networks, GIS enables policymakers and urban planners to locate recycling centres, optimise waste collection routes, design decentralised renewable energy systems, and predict patterns for building materials' reuse and regeneration (Raghu et al. 2022). GIS may play a crucial role in understanding global riverbed drying, aiding in the development of improved water management strategies (Yao et al. 2023). GIS can support the implementation of regenerative practices for safeguarding and restoring watersheds in an increasingly urbanised built environment (Cotler et al. 2022). Smart cities can play a crucial role in shaping a regenerative future by leveraging intelligent infrastructure, such as smart grids, efficient transportation systems, and smart buildings. Additionally, the implementation of IoT sensors and data analytics enables real-time monitoring and analysis of various parameters, facilitating informed decision-making for resource management and urban planning. By fostering citizen engagement and participation through digital platforms, smart cities can empower individuals to actively contribute to regenerative practices and promote a sense of community ownership.

Reality capture and scan-to-BIM technologies (see Chap. 3) allow for the cataloguing of existing buildings and construction sites, including their materials and components. By creating accurate digital representations of physical assets, these technologies enable effective inventory management, as well as the monitoring and tracking of regenerative materials. Reality capture technologies also enable the precise documentation and preservation of historic structures. LiDAR scanning, for instance, can create highly detailed 3D models of heritage buildings, capturing intricate architectural details. This data can aid in the restoration process, ensuring the accurate replication of original features and materials while supporting sustainable preservation practices. Moreover, facility managers can use reality capture and scan-to-BIM to track maintenance needs and simulate scenarios to optimise building performance. This helps identify opportunities for energy savings, predictive maintenance, and resource allocation.

AI algorithms (see Chap. 4) can analyse vast amounts of data related to material properties, market demand, and life cycle analysis to predict the potential for reuse, recycling, and regeneration in the built environment (Raghu et al. 2022). By identifying patterns and trends, AI enables stakeholders to make informed decisions regarding material selection, design for disassembly, and end-of-life strategies. AI-powered systems can also optimise supply chains and enhance circularity by matching supply with demand, thus facilitating the exchange of reusable and regenerative materials and products. AI can play a significant role in creating matchmaking algorithms that facilitate regenerative architecture and infrastructure. By analysing enormous quantities of information regarding project requirements, resources, and stakeholders, AI can identify potential synergies and connections that align with regenerative principles. For example, AI can analyse geographical data – such as climate conditions, available resources, and local regulations – to tailor specific regenerative design strategies – such as integrating renewable energy systems, optimising water management techniques, or promoting biodiversity – to the unique characteristics of a location. AI can leverage its data-processing capabilities to facilitate collaboration and knowledge sharing among stakeholders. By creating platforms or algorithms that connect architects, engineers, developers, material suppliers, governments, and communities, AI can foster an exchange of ideas, best practices, and innovative solutions that drive regenerative design.

Data templates and material passports (see Chap. 5) provide standardised formats for capturing and sharing information about building materials, their composition, and their performance characteristics. By enabling the transparent exchange of information, they promote the reuse, recycling, and responsible sourcing of materials. Through the implementation of material passports, the regenerative potential of a building can be maximised, as materials can be tracked, maintained, and repurposed. Material passports can facilitate the exchange of information and thus the promotion of regenerative materials and practices. Material passports play a crucial role in maintenance and renovation by providing information about the specific materials and components used in a building. This simplifies the process of finding suitable replacements or performing repairs when needed. Material passports ensure that the building's regenerative qualities are preserved over time, promoting long-term sustainability and minimising unnecessary waste. Material passports for regenerative materials should capture not only the initial state and properties of the material but also its regenerative capabilities. This includes information on how the material grows and responds to damage as well as its ability to self-repair or regenerate over time. The material passport should include details about monitoring and tracking the growth of regenerative materials over their lifespan. Since regenerative materials may require specific maintenance or nurturing to support their growth and regeneration, material passports should also provide guidelines for proper care and maintenance. This can include information on optimal environmental conditions, moisture levels, and any necessary interventions to support the regenerative capabilities of the material.

Computational tools (see Chap. 6), such as parametric design software and generative algorithms, empower architects and designers to optimise building designs for circularity and, in particular, regeneration. These tools enable the exploration of various design options, considering factors such as material efficiency, adaptability, and end-of-life scenarios. By integrating circular design and regenerative design principles from the early stages of a project, computational tools facilitate the creation of buildings that are easily disassembled, have low embodied carbon, and can accommodate future adaptations and regeneration. For more information regarding digital tools towards regenerative design, readers are referred to the extensive discussions by the Rethink Sustainability Towards a Regenerative Economy or RESTORE group (Naboni et al. 2019). In the book *Regenerative Design in Digital Practice*, the authors delve into the application of regenerative design principles to buildings and cities, showcasing digital computational design approaches and emphasising the importance of integrating science, big data, and multidisciplinary digital tools into the design process in order to reverse the effects of climate change and enhance the built environment.

Additive manufacturing, or 3D printing (see Chap. 7), offers opportunities to produce customised building components on demand while eliminating material waste. The 3D-printed materials can be recycled or bio-based. For example, tailored fibre placement (TFP) and coreless filament winding (CFW) techniques enable the creation of lightweight architectural solutions using natural fibre-reinforced polymers (NFRP), integrating moulds and frames as active structural elements and leveraging regenerative materials in innovative digital approaches Cutajar et al. 2020).

Robotics (see Chaps. 8 and 9) serve as transformative digital technologies for creating a regenerative built environment as they enable precision, efficiency, waste minimisation, resource use optimisation, customisation, flexibility, and adaptability to unique designs. Robotic systems facilitate scalability and replicability, promoting the widespread adoption of unique regenerative strategies. Robotic fabrication excels in handling complex geometries, enabling the realisation of intricate regenerative designs with regenerative materials. In urban farming, robotics enables a symbiotic coexistence of humans, plants, and robots in cities, integrating robotic agents with edible plants to enhance the urban environment by generating fresh food, improving the microclimate, and promoting local biodiversity (IaaC Robotic Urban Farmers 2022).

Extended reality (XR) (see Chap. 10) serves as a valuable digital technology for creating a regenerative built environment through the integration of regenerative materials and design strategies. XR technologies, including virtual reality (VR) and augmented reality (AR), offer immersive and interactive experiences that enable stakeholders to visualise and experience regenerative designs before they are constructed. This enhances design collaboration, facilitates informed decision-making, harmonisation, and reconciliation within a community, and reduces the risk of costly errors. XR can simulate the performance and behaviour of regenerative materials, allowing designers to assess their impact on energy efficiency, environmental sustainability, and occupant well-being. By providing a virtual testing ground, XR enables iterative design processes and the exploration of innovative regenerative solutions. Combining this with strategy gaming can enable an inclusive community decision-making process for a regenerative future. Indeed, XR can enhance user engagement and education, fostering a deeper understanding and appreciation of regenerative principles and practices, ultimately supporting the transition towards a regenerative built environment.

Blockchain technology (see Chap. 12) can help create a regenerative built environment by facilitating scalable socio-economic-ecological interactions along three lines of inquiry: rethinking data governance, reassessing stakeholder roles and responsibilities, and developing a new approach to the governance of value and ownership (Wang et al. 2023). Blockchain technology can enable secure and transparent data collection, distribution, maintenance, and evaluation in alignment with regenerative principles. To reassess stakeholder roles and responsibilities, blockchain technology can provide a decentralised platform for collaboration and decision-making among diverse stakeholders. Additionally, this technology can facilitate the governance of value and ownership of non-human entities such as buildings and nature, enabling a harmonious coexistence. Overall, the blockchain has the potential to transform governance structures in the built environment by uniting social, economic, and technological aspects to achieve effective regenerative development.

The adoption and effective use of the various digital technologies discussed above can contribute to a regenerative built environment. While incorporating all the digital technologies discussed can contribute to advancing regeneration in the built environment, it is not necessarily mandatory to use all of them simultaneously. The adoption and effective use of any combination of these technologies, depending on the specific needs and goals, can still bring significant benefits and help advance regenerative practices. The key is to identify the most relevant and impactful technologies that align with the objectives of the project or initiative at hand on a case-by-case basis – as is often leading to better design and construction in the built environment. These technologies enable collaboration, optimisation, real-time monitoring, and new methods of design and fabrication, supporting the integration of regenerative principles throughout the life cycle of buildings and infrastructure. It is important to note that the extent to which these digital technologies contribute to a regenerative future depends on how they are implemented and used. While they have the potential to assist in regeneration, their effectiveness relies on thoughtful application and strategic use. Strategic use of digital technologies for regeneration involves thoughtful application based on a needs assessment, careful technology selection, trade-offs assessment, integration planning, stakeholder engagement, and monitoring. A regenerative approach maximises effectiveness, aligns with project objectives, optimises resource allocation, and mitigates risks, ultimately driving positive and measurable change in the built environment. By purposefully leveraging the power of digital technologies in a calculated manner, we have the potential to create a built environment that actively restores ecosystems, fosters well-being for all, promotes biodiversity, enhances social equity and inclusive collaboration, improves community resilience, and cultivates circular building practices more generally.

15.4 Business Models for a Regenerative Built Environment

Organisations that embrace regenerative business models place a dominant emphasis on the health of the planet and the welfare of society, aiming to create value for multiple stakeholders, including the natural environment and society (Stubbs and Cocklin 2008; Muñoz and Branzei 2021). Regenerative business models share design principles with sustainable and circular models but diverge in their objectives by emphasising planetary health and societal well-being above profit-making and by recognising the dependency of business and society on nature (Konietzko et al. 2023). Successfully implementing regenerative business models necessitates robust policy frameworks that recognise the rights of animals and nature and incorporate true pricing (Fullerton 2015).

Regenerative business models have the potential to revolutionise the built environment by creating and delivering value across multiple stakeholder levels. These models emphasise regenerative leadership and co-creative partnerships with nature, as well as justice and fairness (Konietzko et al. 2023). In this context, digital technologies play a crucial role in enabling and supporting regenerative business models. One key aspect of digital transformation is facilitating effective communication and collaboration among stakeholders in the built environment. Digital platforms and tools can enhance transparency and information sharing, enabling stakeholders to work together towards regenerative goals. Digital technologies help create value that goes beyond individual organisations and benefits the entire ecosystem. Moreover, digital transformation enables the implementation of multi-capital accounting (Fullerton 2015), which allows organisations to capture and account for the various forms of capital involved in the built environment, including natural, social, and economic capital. By adopting a holistic approach to accounting, organisations can measure and optimise their impact across multiple dimensions, aiming for a net positive outcome (Hahn and Tampe 2021).

However, to bring about a regenerative built environment, new governance approaches are needed. This entails rethinking data governance, ensuring the responsible collection, management, and utilisation of data to inform decision-making and drive positive change. Additionally, reassessing stakeholder roles and responsibilities is crucial to foster collaboration and ensure that all relevant parties are actively engaged in the regenerative process. Furthermore, the governance of value and ownership needs to be reimagined in alignment with regenerative principles. This involves exploring new mechanisms that enable the equitable distribution of benefits and promote sustainable ownership models. By leveraging digital technologies, such as blockchain, organisations can achieve transformative governance that integrates social, economic, and technological aspects to drive effective regenerative development. Digital technology could be used as a governance tool to facilitate scalable socio-economic-ecologic interactions, addressing the three lines of inquiry mentioned earlier (Wang et al. 2023). By leveraging blockchain's decentralised and transparent nature, stakeholders can collaborate, track, and verify sustainability initiatives and ensure accountability throughout the built environment ecosystem.

Finally, we need ways of assessing the effects of new business models and building project designs. The Living Building Challenge label is an example of an international sustainable building certification programme setting higher standards in the realm of regenerative architecture. It certifies buildings that actively restore and regenerate the environment, promoting self-sufficiency, renewable materials, and healthy indoor spaces. Note that it is essential to recognise that true regenerative outcomes extend beyond a checklist approach due to the holistic nature of regeneration.

In summary, business models for a regenerative built environment embrace digital transformation to create and deliver value at multiple stakeholder levels. Digital technologies enable effective communication, support multi-capital accounting, contribute to the development of new governance approaches, and enable us to tackle complex challenges. More work is needed to truly incorporate intrinsic notions of value beyond financial capital and avoid misleading claims of regeneration and net positive impact. Such claims can be misleading and create a perception of regeneration that may not align with the actual impact or practices employed. To avoid this, it is important for stakeholders, regulators, and consumers to exercise caution, conduct thorough assessments, and demand transparency and accountability in verifying the legitimacy of regeneration claims. Additionally, ongoing efforts are being made by organisations, industry associations, and certifications to establish standards and frameworks that ensure the credibility and integrity of regeneration initiatives.

15.5 Conclusion

Regeneration goes beyond narrow interpretations of sustainability and resilience, offering a holistic approach to addressing socio-ecological systems. In recent years, there has been a growing recognition and adoption of regenerative practices in the built environment. Many projects are incorporating strategies that go beyond sustainability and aim to restore, rejuvenate, and enhance ecosystems and communities. These practices often involve holistic design approaches, renewable energy integration, resource-efficient systems, circular economy principles, and community engagement. A regenerative built environment requires collaboration among experts from various fields and aims to create positive impacts on the environment, health, society, and the economy. Embracing complexity, uncertainties, and context-dependent interventions is essential for implementing effective regenerative practices in the built environment. It is important to recognise the diverse perspectives and interpretations of regeneration in different fields and contexts, which allow for the development of comprehensive and inclusive solutions.

The adoption of various digital technologies (including BIM, digital twins, GIS, reality capture, AI, data templates, material passports, computational tools, additive manufacturing, robotic fabrication and construction, XR, and blockchain) holds immense potential for enabling regeneration in the built environment. Regenerative business models take a holistic approach to circularity by prioritising net positive value creation and delivery across different stakeholder levels, including society and the natural environment. Digital technologies thus play a crucial role in supporting regenerative business models by facilitating communication, multi-capital accounting, and new governance approaches. These technologies provide opportunities for collaboration, optimisation, visualisation, customisation, resource efficiency, circularity, and interdisciplinary integration.

If used well and for these purposes, digital technologies can contribute to the creation of a built environment that embraces regenerative practices and design strategies. While regenerative practices are gaining momentum, there is still a need for widespread implementation and continued innovation to fully realise its potential in the built environment. The technologies discussed in this book should be put to use to upscale the regenerative potential of the built environment.

15.6 Key Takeaways

- Regeneration offers a holistic approach to creating positive impacts in the built environment by combining collaboration, technology, and diverse expertise to foster a sustainable and thriving future.
- Regenerative architecture and infrastructure actively create positive change by nurturing ecosystems and replenishing the environment, the economy, and society towards an inclusive built environment.
- Digital technologies have the potential to revolutionise the built environment by enabling collaboration, optimisation, and customisation for a regenerative future.
- Regenerative business models create value, foster collective leadership, and drive positive impact across all stakeholders.

Acknowledgements The authors would like to acknowledge enriching discussions with Prof. Jaboury Ghazoul (Ecology) and Prof. Johan Six (Agroecosystems) of the Department of Environmental Systems Science, Prof. Ralph Müller (Biomechanics) of the Department of Health Sciences and Technology, and Dr. Chris Luebkeman and Dr. Johannes Heck of the Strategic Foresight Hub at ETH Zurich, as well as Daniel Wahl (Author of *Designing Regenerative Cultures*), and Bill Reed (Regenesis, Inc).

References

Allam Z, Takun YR (2022) Smart cities as an urban regeneration avenue: redefining the efficiency and performance of cities. In: Rethinking smart cities. Edward Elgar Publishing, pp 66–87

Almond REA, Grooten M, Juffe Bignoli D, Petersen T (2022) Living planet report 2022: building a nature-positive society. WWF, Gland. https://wwf.ca/wp-content/uploads/2022/10/lpr_2022_full_report_en.pdf

Attia S (2018) Regenerative and positive impact architecture: learning from case studies. Springer

Binder CR (2007) From material flow analysis to material flow management part I: social sciences modeling approaches coupled to MFA. J Clean Prod 15(17):1596–1604. https://doi.org/10.1016/j.jclepro.2006.08.006

Bitting S, Derme T, Lee J, Van Mele T, Dillenburger B, Block P (2022) Challenges and opportunities in scaling up architectural applications of mycelium-based materials with digital fabrication. Biomimetics 7:44

Bocken N, Geradts T (2022) Designing your circular business model. Stanf Soc Innov Rev 20(2): 34–39

Brand S (1995) How buildings learn: what happens after they're built? Penguine Ed

Brundtland GH (1987) Our common future. Report of the world commission on environment and development. United Nations. https://sustainabledevelopment.un.org/content/documents/5987our-common-future.pdf

Bugaj M, Machova V, Bratu S (2022) The geopolitics of smart sustainable city governance and simulated 3D environments. Geopolit Hist Int Relat 14(2):152–167

Calheiros CSC, Castiglione B, Palha P (2022) Nature-based solutions for socially and environmentally responsible new cities: the contribution of green roofs. In: Stefanakis A, Nikolaou I (eds) Circular economy and sustainability. Elsevier, pp 235–255

Capdevila P, Stott I, Oliveras Menor I, Stouffer DB, Raimundo RLG, White H, Barbour M, Salguero-Gómez R (2021) Reconciling resilience across ecological systems, species and subdisciplines. J Ecol 109(9):3102–3113. https://doi.org/10.1111/1365-2745.13775

Çetin S, De Wolf C, Bocken N (2021) Circular digital built environment: an emerging framework. Sustainability 13(11):6348. https://doi.org/10.3390/su13116348
Churkina G, Organschi A, Reyer CPO, Ruff A, Vinke K, Liu Z, Reck BK, Graedel TE, Schellnhuber HJ (2020) Buildings as a global carbon sink. Nature Sustain 3(4):269–276. https://doi.org/10.1038/s41893-019-0462-4
Coady T (2020) Rebuilding earth: designing ecoconscious habitats for humans. North Atlantic Books
Copenhagen City (2014) "Copenhagen: solutions for sustainable cities", report, January 2014. Available at: http://www.cleancluster.dk/wp-content/uploads/2017/06/5942682600c24.pdf
Cotler H, Cuevas ML, Landa R, Frausto JM (2022) Environmental governance in urban watersheds: the role of civil society organizations in Mexico. Sustainability 14(2):988. https://doi.org/10.3390/su14020988
Cutajar S, Martins VC, Hoven Cvd BP, Dahy H (2020) Towards modular natural fiber-reinforced polymer architecture. In: Proceedings of the 40th annual conference of the association of computer aided design in architecture. pp 564–73
Dring A, Schwaag D (2021) Carbon negative building materials: reversing climate change through urbanization. In: Mendelez F, Diniz N (eds) Data, matter, design. Routledge
Ellen McArthur Foundation (2013) Towards the circular economy. 3 vols. https://ellenmacarthurfoundation.org/towards-the-circular-economy-vol-1-an-economic-and-business-rationale-for-an
Fakharany N (2023) "In vivo": the Belgian pavilion investigates relationship of architects with resources at the 2023 Venice biennale. ArchDaily 25 April. www.archdaily.com/999937/in-vivo-the-belgian-pavilion-investigates-relationship-of-architects-with-resources-at-the-2023-venice-biennale
Fullerton J (2015) Regenerative capitalism. Greenwich, CT, USA, Capital Institute
Future of Cities (2023) Regenerative placemaking in action. Available at: https://focities.com/regenerative-placemaking-in-action/
Gligoric N, Krco S, Hakola L, Vehmas K, De S, Moessner K, Jansson K, Polenz I, van Kranenburg R (2019) SmartTags: IoT product passport for circular economy based on printed sensors and unique item-level identifiers. Sensors 19(3):586. https://doi.org/10.3390/s19030586
Hahn T, Tampe M (2021) Strategies for regenerative business. Strateg Organ 19(3):456–477. https://doi.org/10.1177/1476127020979228
Heisel F (2017) Design of a load-bearing mycelium structure through informed structural engineering: the MycoTree at the 2017 Seoul biennale of architecture and urbanism. In: Proceeding of the world congress on sustainable technologies (WCST)
Hota A, Gopal Krishna Patro S, Obaid AJ, Khatak S, Kumar R (2023) Constructed wetland challenges for the treatment of industrial wastewater in smart cities: a sensitive solution. Sustain Energy Technol Assess 55:102967. https://doi.org/10.1016/j.seta.2022.102967
IaaC Robotic Urban Farmers (2022) Tallinn Architecture Biennale 2022. Available at: https://iaac.net/the-project-robotic-urban-farmers-takes-part-to-tallinn-architecture-biennale-2022/
IPCC (2022) Climate change 2022: impacts, adaptation and vulnerability. Contribution of working group II to the sixth assessment report of the intergovernmental panel on climate change (Pörtner H-O, Roberts CD, Tignor M, Poloczanska ES, Mintenbeck K, Alegría A, Craig M, Langsdorf S, Löschke S, Möller V, Okem A, Rama B (eds)). Cambridge University Press, Cambridge/New York, 3056 pp. https://doi.org/10.1017/9781009325844
Jalia A, Bakker R, Ramage M (2022) The edge. Showcasing an exemplary IoT building, Amsterdam. https://www.cdbb.cam.ac.uk/system/files/documents/TheEdge_Paper_LOW1.pdf
Kirli D, Couraud B, Robu V, Salgado-Bravo M, Norbu S, Andoni M, Antonopoulos I, Negrete-Pincetic M, Flynn D, Kiprakis A (2022) Smart contracts in energy systems: a systematic review of fundamental approaches and implementations. Renew Sustain Energ Rev 158:112013. https://doi.org/10.1016/j.rser.2021.112013
Konietzko J, Bocken N, Hultink EJ (2020) A tool to analyze, ideate and develop circular innovation ecosystems. Sustainability 12(1):417
Konietzko J, Das A, Bocken N (2023) Towards regenerative business models: a necessary shift? Sustain Prod Consum 38:372–388. https://doi.org/10.1016/j.spc.2023.04.014

Mang P, Reed B (2012) Regenerative development and design. In: Meyers RA (ed) Encyclopedia of sustainability science and technology. Springer, New York. https://doi.org/10.1007/978-1-4419-0851-3_303

Mang P, Haggard B, Regenesis Group, I (2016) Regenerative development and design: a framework for evolving sustainability. Wiley. https://books.google.ch/books?id=ZjyXAQAACAAJ

Morseletto P (2022) Environmental principles for modern sustainable economic frameworks including the circular economy. Sustain Sci 17(5):2165–2171. https://doi.org/10.1007/s11625-022-01208-w

Muñoz P, Branzei O (2021) Regenerative organizations: introduction to the special issue. Organ Environ 34(4):507–516

Naboni E, Havinga L, Brown M, Loder A, Altomonte S, Peters T, Finocchiaro L, Chokhochian A, De Souza CB, De Wolf C, Marvuglia A (2019) Regenerative design in the digital practice. In: A handbook for the built environment. COST – European Cooperation in Science and Technology "rethink sustainability towards a regenerative economy" (RESTORE)

Naeem S, Lu Y, Jackson J (2022) Curtailing the collapse of the living world. Sci Adv 8(49): eadf9317. https://doi.org/10.1126/sciadv.adf9317

Newman P (2014) Biophilic urbanism: a case study on Singapore. Aust Plan 51(1):47. https://doi.org/10.1080/07293682.2013.790832

Polman P, Winston A (2021) Net positive: how courageous companies thrive by giving more than they take. Harvard Business Press

Raghu D, Marengo M, Markopoulou A, Neri I, Chronis A, De Wolf C (2022) Enabling component reuse from existing buildings: using Google Street View and machine learning to enhance building databases. In: Proceedings of the 27th international conference of the Association for Computer-Aided Architectural Design Research in Asia (CAADRIA) 2022. CAADRIA

Reed B (2007) Shifting from 'sustainability' to regeneration. Build Res Inf 35(6):674–680. https://doi.org/10.1080/09613210701475753

Reflow (2022) https://reflowproject.eu. Accessed 23 May 2023

Rodrigues do Amaral A (2020) BIGH: a case of an urban aquaponics farm in the metropolis Centre. Dissertation. University of Porto

Saha S (2019) Sewage management to fish culture: an age old eco-practice at East Kolkata wetlands. Environ Iss Approach Pract 1:10–19

Sayer J, Sunderland T, Ghazoul J, Pfund J-L, Sheil D, Meijaard E, Venter M, Boedhihartono AK, Day M, Garcia C, Van Oosten C, Buck LE (2013) Ten principles for a landscape approach to reconciling agriculture, conservation, and other competing land uses. Proc Natl Acad Sci U S A 110(21):8349–8356. https://doi.org/10.1073/pnas.1210595110

Seymour LM, Maragh J, Sabatini P, Di Tommaso M, Weaver JC, Masic A (2023) Hot mixing: mechanistic insights into the durability of ancient Roman concrete. Sci Adv 9(1). https://doi.org/10.1126/sciadv.add1602

Standish RJ, Hobbs RJ, Mayfield MM, Bestelmeyer BT, Suding KN, Battaglia LL, Eviner V, Hawkes CV, Temperton VM, Cramer VA, Harris JA, Funk JL, Thomas PA (2014) Resilience in ecology: abstraction, distraction, or where the action is? Biol Conserv 177:43–51. https://doi.org/10.1016/j.biocon.2014.06.008

Stubbs W, Cocklin C (2008) Conceptualizing a "sustainability business model". Organ Environ 21(2):103–127. https://doi.org/10.1177/1086026608318042

Terzi A (2022) Growth for good. Harvard University Press, Reshaping capitalism to save humanity from climate catastrophe

Wahl DC (2016) Designing regenerative cultures. Triarchy Press

Wakefield J (2016) Tomorrow's buildings: is world's greenest office smart? BBC. 6 Apr 2016. https://www.bbc.com/news/technology-35746647

Wang X, Li H, Sodoudi S (2022) The effectiveness of cool and green roofs in mitigating urban heat Island and improving human thermal comfort. Build Environ 217:109082. https://doi.org/10.1016/j.buildenv.2022.109082

Wang H, Hunhevicz J, Hall D, Meier G, De Wolf C (2023) Blockchain for regenerative built environment governance. CISBAT (International scientific conference cycle for a Sustainable Energy Transition in the Built Environment): The Built Environment in Transition, 13–15 September 2023, Lausanne, Switzerland

Watson J (2019) Lo—TEK. Design by Radical Indigenism. Taschen

Wyss R, Luthe T, Pedoth L, Schneiderbauer S, Adler C, Apple M, Erazo Acosta E, Fitzpatrick H, Haider J, Ikizer G, Imperiale J, Karanci N, Posch E, Saidmamatov O, Thaler T (2022) Mountain resilience: a systematic literature review and paths to the future. Mt Res Dev 42(2):A23–A36. https://doi.org/10.1659/MRD-JOURNAL-D-21-00044.1

Yao F, Livneh B, Rajagopalan B, Wang J, Crétaux J-F, Wada Y, Berge-Nguyen M (2023) Satellites reveal widespread decline in global lake water storage. Science 380(6646):743–749. https://doi.org/10.1126/science.abo2812

Zingoni dec Baro ME (2022) Regenerating cities: reviving places and plane. Springer International Publishing, Cham

Catherine De Wolf is an Assistant Professor of Circular Engineering for Architecture at the Swiss Federal Institute of Technology of Zurich (ETH Zurich), where she conducts research on digital innovation towards a circular built environment. She has a dual background in civil engineering and architecture and completed her PhD at the Massachusetts Institute of Technology (MIT). She is a faculty member in the ETH Centre for Augmented Computational Design in Architecture, Engineering and Construction (Design++), the ETH AI Center, the Swiss Federal Laboratories for Materials Science and Technology (Empa), the Circular Future Cities group of the Future Cities Lab, and the National Centre of Competence in Research on Digital Fabrication (DFAB). She has worked in academia (e.g., University of Cambridge, University of Technology Delft, Nanjing University), in industry (e.g., Arup, Elioth, Ney & Partners), and in governmental institutions (e.g., Joint Research Centre of the European Commission). She also co-founded the circular construction company Anku, the Structural Engineers 2050 Commitment, and the Digital Circular Economy (DiCE) Lab.

Nancy Bocken is a Professor in Sustainable Business and Circular Economy at Maastricht University, Maastricht Sustainability Institute (MSI) and is a leading researcher on topics such as sustainable business models, business experiments for sustainability, circular economy, and sufficiency. She is also a Fellow at Cambridge Institute for Sustainability Leadership, advisor to TNO (Dutch association for applied scientific research) and a Board member of the Philips Foundation. Before going into academia, she held positions in the logistics, banking, and consulting sectors. She holds a PhD from the Department of Engineering, University of Cambridge and is a co-founder of her own circular and sustainable business, HOMIE.

Concluding Perspective

Alessio Terzi

Avoiding a climate catastrophe and environmental collapse will require decarbonising our economies by mid-century and an overall reduction in material resource use (Terzi 2022). The challenge is immense for a variety of sectors, including energy production and manufacturing, but also the built environment, given buildings and construction contribute nearly 40% of overall global greenhouse gas emissions (Gates 2021; Byng 2022). How should we go about it? Posed in another way, is modern technology the solution or ultimately part of the problem (Wainwright 2020)?

To some, the herculean task of making humanity compatible with planetary boundaries takes the shape of a past-looking focus that is at high risk of slipping into a rejection of modernity (Pawlyn and Ichioka 2022). Innocuous acts at the individual level, such as favouring handicrafts or local town markets, do-it-yourself products or home gardening, are all part of a general trend of rejection of the standardisation that came with mechanisation and the Industrial Revolution (Westacott 2016). The label 'traditional', whether for agricultural techniques, food, medical practices or handicrafts, takes on a positive connotation, to be opposed to 'modern'. All of this builds on the implicit perception that life was better in the past, which is effectively the very negation of the concept of progress (Mokyr 2016).

This attachment to the idea of an idyllic past, and the rejection of the belief in progress, perhaps should not surprise, as it is not without precedent. Throughout the Middle Ages and Renaissance, for instance, much of Europe looked back to the classical era of the Roman Empire and Ancient Greece as a golden age that could never be matched (Mokyr 2016). The environmentalist movement itself has a long tradition of escapism from modernity, which draws its roots in classics like Henry David Thoreau's *Walden* (1854), the 'back-to-the-land' movement of the 1960/70s, and in today's manifestations such as eco-villages, which radical degrowth thinkers like Mattias Schmelzer, Tim Jackson or Jason Hickel have used as prime examples of how society should be reshaped to face the climate crisis (Hickel 2020; Jackson 2021; Schmelzer et al. 2022). In their view, society should

A. Terzi
Sciences Po, European Commission, HEC Paris, Paris, France
e-mail: alessio.terzi@sciencespo.fr

C. De Wolf et al. (eds.), *A Circular Built Environment in the Digital Age*, Circular Economy and Sustainability, https://doi.org/10.1007/978-3-031-39675-5

embrace a mental framework of scarcity, downscaling back to where humanity used to be before the Industrial Revolution, and do so in a voluntary manner.

The very separation between an idyllic past and the unsustainable present, ancient practices, and modern technology has a vein of artificiality, which generally creates a temporal division between before and after the Industrial Revolution. When focusing on the built environment, modernity is often associated with the building blocks of our civilisation – brick, glass, concrete and steel (Dartnell 2015) – even though the reality is that these very materials were not new at all. The Industrial Revolution only combined them with improvements in energy production and mechanisation to make them cheaper, more widely available and therefore useful to serve purposes that were seen as desirable at the time. Sun-dried bricks were foundational to the very first cities of antiquity, such as the Sumerian capital Uruk, while kiln-burned bricks were used in ancient Mesopotamia, including in the construction of Ziggurats, the Roman Empire and Han China (Smil 2017). Lime mortars have been used for millennia, including in Ancient Egypt. The Romans mixed slaked lime with volcanic ash (*pozzolana*) in what they called 'cementum'. When that was further mixed with crushed stones it created a form of concrete, which made it possible to build things like the Colosseum or the Pantheon: the largest single-piece dome in the world to this day. Glass, one of the first synthetic materials known to humanity, was invented in Mesopotamia in the third millennium BCE. Crucible steel itself was known to many preindustrial societies, although only in small amounts, making it very expensive and therefore available only for special purposes. Traditional East African steel-makers, for instance, used cone-shaped charcoal-fuelled slag and mud furnaces since the early centuries of the Common Era. In China during the Han dynasty, steel was obtained by removing carbon from cast iron through oxygenation and used for specific applications such as chains for suspension bridges. More broadly, steel was also used for ploughshares, body armour or sword and sabre blades at least since the third century CE (Feuerbach 2006).

While we artificially create a boundary between before and after, ancient and modern, human history can be better described through a pattern of continuity (Galor 2022). More broadly, the very development of know-how throughout human history has been based on experimentation, imitation of more successful practices, combination with pre-existing local knowledge, and adaptation to local context (Henrich 2016; Poskett 2022). While developing and spreading new technologies and innovations, ancient societies were generating value, expanding the realm of possibilities, and effectively growing the size of their economies. In other words, economic growth is not a recent phenomenon and did not start with the Industrial Revolution (Brooke 2014; Fouquet and Broadberry 2015).

Which brings us back to our modern challenge of decarbonising and reducing the overall environmental impact of the built environment in the twenty-first century. As architects, engineers and designers work towards this societal goal, it is only natural to expect them do what humanity has done all along when developing innovation: address the challenge of the moment. As they do so, they will be laying the foundations of a new economy – one that attaches value to solutions that are compatible with nature. In a nutshell, they will be contributing to a new growth model or green growth.

Ancient techniques and know-how will represent a great starting point for thinking of low-energy, fossil-free solutions, but they should then be combined with the state-of-the-art scientific and technological know-how currently available. The interplay between tradition and innovation is one that needs to be carefully managed (Nunn 2022). This implies that the ancient Achaemenid know-how of wind-catching towers for passively cooling buildings in ancient Persia will not be replicated identically in the twenty-first century, but rather perhaps serve as inspiration to be then combined with sensors and artificial intelligence technology to optimise airflow inside buildings, as they are in Masdar City (UAE). If the tree bridges of the Khasis hill tribe of north-eastern India ever prove to be useful in the current context, they will perhaps be combined with the most advanced biotechnology techniques to accelerate plant growth and resistance. If raw-earth buildings ever experience a comeback, given their strong thermal inertia and limited impact in construction and demolition, their structure could be optimised using digital modelling, and they might be 3D printed.

Designer, activist, and academic Julia Watson remarks that in the face of climate change and environmental degradation a lot of the answers are already there and have been for generations (Watson 2020). Her movement – Lo-TEK (traditional ecological knowledge) – imagines that part of the knowledge of how to survive the future is already embedded in low-energy, often ancient practices, to be then combined with modern technology (Savak 2020). As Italian architect Mario Cucinella correctly concludes, reducing the use of non-renewable energy requires us to look to the past, not out of nostalgia but to rediscover and reinterpret the way things were done in light of contemporary challenges and technology (Cucinella 2022).

It is through these lenses that the important role of digitalisation should be interpreted, meaning as one of the general-purpose technologies at the frontier of human know-how that will need to be leveraged when engaging in the unprecedented planetary challenge ahead. Chapters in this book have illustrated practical examples of how this can play out in practice by using blockchain, building information modelling (BIM) or LiDAR scanning technology to improve the circularity of building materials. In the process, experimentation and exchange of knowledge will be crucial. The New European Bauhaus was launched by the European Commission in 2021 precisely with the objective of catalysing actors that are re-imagining a new lifestyle that matches sustainability with good design, that needs less carbon and that is inclusive and affordable (European Commission 2021).

More broadly, governments around the world are developing an eclectic mix of policies to fast-track the green transition and promote a circular economy. These include but are not limited to the use of carbon pricing, which will need to be extended progressively to the whole economy, in line with the efforts of the recent 'Fit for 55' package approved by the European co-legislators for 'delivering the European Green Deal' (Council of the EU 2023). These could even include progressive bans on selected technologies that are harmful to the climate or environment, such as internal combustion engine cars or possibly conventional aviation fuels in the near future (Terzi 2023). Moreover, governments will need to lay the foundations for a green economy by leveraging public finances to roll out the infrastructure to accompany the transition, such as by providing charging points for electric vehicles. Their role as a large purchaser of goods and services in the economy enables them also to mandate that public sector buildings, for instance, respect certain standards of energy efficiency or circularity of materials. Finally, policies will need to be put in place to progressively embed so-called 'extended producer responsibility' in a variety of sectors, effectively making producers responsible for the disposal of their products. Concerning the built environment, this could very well extend to cement and steel, as already discussed in the European Commission's Circular Economy Action Plan (European Commission 2020). More broadly, as a leader in regulation for a circular economy, the EU is adopting an ambitious set of policy measures that include the right to repair, ecolabelling for textiles, and aggressive curtailing of packaging. Furthermore, it envisions a revision of the Construction Product Regulation to create a harmonised framework to assess and communicate the environmental and climate performance of construction products, including digital solutions such as a construction products database and a Digital Products Passport (European Commission 2022).

Parts of modernity will need to give way because of their unsustainable nature, but innovation and technology are, and always have been, humanity's greatest assets when facing its challenges. Achieving sustainability in the twenty-first century marks no exception.

Disclaimer The views expressed here are those of the author and do not necessarily reflect those of the institution to which he is affiliated.

Alessio Terzi is an economist at the European Commission and an Adjunct Professor at Sciences Po and HEC Paris. He is also the author of *Growth for Good: Reshaping Capitalism to Save Humanity from Climate Catastrophe* (Harvard University Press, 2022). His area of expertise centres on sustainable growth, and he was involved in the development of the European Green Deal. His work has been featured in several high-profile media outlets, including the Financial Times, The Wall Street Journal and Bloomberg. Alessio holds a PhD from the Hertie School and was a Fulbright Fellow at the Harvard Kennedy School of Government.

References

Brooke JL (2014) Climate change and the course of global history. Cambridge University Press, Cambridge

Byng M (2022) Can indigenous knowledge help us design a more sustainable future? Financial Times

Council of the EU (2023) "Fit for 55": Council adopts key pieces of legislation delivering on 2030 climate targets. In: Press Release. https://www.consilium.europa.eu/en/press/press-releases/2023/04/25/fit-for-55-council-adopts-key-pieces-of-legislation-delivering-on-2030-climate-targets/

Cucinella M (2022) The future is a journey to the past: ten stories about architecture. Quodlibet, Macerata
Dartnell L (2015) The knowledge: how to rebuild our world from scratch. Vintage, London
European Commission (2020) Circular economy action plan. European Commission, Brussels
European Commission (2021) New European Bauhaus beautiful, sustainable, together. European Commission, Brussels
European Commission (2022) Proposal for a regulation laying down harmonised conditions for the marketing of construction products. European Commission, Brussels
Feuerbach A (2006) Crucible Damascus steel: a fascination for almost 2,000 years. JOM 58:48–50. https://doi.org/10.1007/s11837-006-0023-y
Fouquet R, Broadberry S (2015) Seven centuries of European economic growth and decline. J Econ Perspect 29:227–244. https://doi.org/10.1257/jep.29.4.227
Galor O (2022) The journey of humanity: the origins of wealth and inequality. Dutton, Boston
Gates B (2021) How to avoid a climate disaster: the solutions we have and the breakthroughs we need. Random House, New York
Henrich J (2016) The secret of our success: how culture is driving human evolution, domesticating our species, and making us smarter. Princeton University Press, Princeton
Hickel J (2020) Less is more: how degrowth will save the world. Random House, London
Jackson T (2021) Post growth: life after capitalism. Polity, Cambridge
Mokyr J (2016) A culture of growth: the origins of the modern economy. Princeton University Press, Princeton
Nunn N (2022) On the dynamics of human behavior: the past, present, and future of culture, conflict, and cooperation. AEA Pap Proc 112:15–37. https://doi.org/10.2139/ssrn.4045382
Pawlyn M, Ichioka S (2022) Flourish: design paradigms for our planetary emergency. Triarchy Press, Bridport
Poskett J (2022) Horizons: a global history of science. Viking Press, New York
Savak N (2020) Design for the future when the future is bleak. New York Times Style Mag
Schmelzer M, Vetter A, Vansintjan A (2022) The future is degrowth: a guide to a world beyond capitalism. Verso, London
Smil V (2017) Energy and civilization: a history. MIT Press, Cambridge
Terzi A (2022) Growth for good: reshaping capitalism to save humanity from climate catastrophe. Harvard University Press, Cambridge
Terzi A (2023) Enrolling capitalism in the fight against climate change. ProMarket
Thoreau HD (1854) Walden. Ticknor and Fields, Boston
Wainwright O (2020) Living bridges and supper from sewage: can ancient fixes save our crisis-torn world? Guardian
Watson J (2020) How to build a resilient future using ancient wisdom. https://www.ted.com/talks/julia_watson_how_to_build_a_resilient_future_using_ancient_wisdom?language=enIf
Westacott E (2016) The wisdom of frugality: why less is more – more or less. Princeton University Press, Princeton

Conclusion

This book arises from the urgent need to shift from a linear to a circular model in order to lower the construction sector's greenhouse gas emissions, resource use, and waste generation. A circular economy approach in the construction sector can bring numerous benefits, ranging from environmental to economic and social advantages. By prioritising slowing, narrowing, closing, and regenerating the loops in buildings and infrastructure, we can reduce waste and environmental impacts while creating economic opportunities. Ultimately, the circular economy in the built environment can lead to the development of more sustainable and resilient communities while fostering innovation, creating jobs, and enhancing social value. We need to shift the construction sector from a linear take-make-waste model to a circular economy narrow-slow-close-regenerate to be able to meet the building needs now and in the future. While pioneering circular building projects are promising, large-scale global implementation of circular economy principles in the built environment has not yet been achieved due to the fragmentation of stakeholders and risk aversion in the construction industry.

Digital innovation can accelerate the transition to a circular economy by improving efficiency, promoting collaboration, and creating new business models. Bridging the latest developments in the fields of digitalisation and circularity is crucial to address our global challenges of climate change and a growing population that needs more buildings. Digitalisation has the potential to address barriers to circular innovation by enabling coordination and risk assessment. Circular construction management balances the benefits of automation with the value of human skills and creativity towards Industry 5.0 creating a more sustainable, efficient, and people-centred construction industry. This book has shown how different disciplines – such as civil engineering, computer science, architecture, mechanical engineering, finance, and management – can come together to harness digital transformation in the service of a better, more circular built environment. By leveraging the twin transition of the digital and circular economy, we can transform our approach to building and managing resources and turn buildings into material depots and banks, thus reducing waste and creating a more sustainable future.

C. De Wolf et al. (eds.), *A Circular Built Environment in the Digital Age*, Circular Economy and Sustainability, https://doi.org/10.1007/978-3-031-39675-5

As we strive to create a circular built environment, we need to consider the potential rebound effects of digitalisation in terms of resource use. While digital technologies can enable more sustainable and efficient use of resources, they are also often energy intensive and require rare earth elements, which could offset the benefits of circular strategies. To address this challenge, we need to take a holistic approach that considers the entire life cycle of digital technologies, from raw material extraction to disposal. This involves assessing the environmental impacts of each stage and identifying opportunities to reduce resource use and waste. As in the built environment, we should prioritise the use of renewable, reusable, and recyclable materials and resources in the production of digital technologies and machines. We should also design digital technologies with circular principles in mind, such as modular design, which can facilitate repair and upgrade, extend the life of products, and reduce the need for new materials. Ultimately, taking the potential rebound effects of digitalisation into account is crucial for creating a truly circular built environment. By carefully considering the environmental impacts of digital technologies and adopting circular design principles, we can ensure that digitalisation contributes to a sustainable and regenerative future.

This book provided a comprehensive overview of digital innovations that emerged in the past years and have the potential to enable a circular built environment. Part I of the book explored the role of data by investigating how capabilities of data-driven digital technologies can help transition to a circular built environment. Building information modelling (BIM), geographical information systems (GIS), reality capture technologies, artificial intelligence (AI), data templates, and material passports are essential for data collection, integration, and analysis towards circular construction. The first three chapters discussed how BIM, GIS, scanning, and scan-to-BIM can be used to promote circularity in the built environment. BIM and digital twinning can provide comprehensive information necessary for circular construction, while GIS can optimise the location of facilities for circular activities. Scanning and scan-to-BIM technologies can help digitise and identify circular opportunities, while AI can help predict the potential for building components to be reused in future projects. To upscale and promote circular building strategies, we need to create material passports and digital product passports, which require a standardised data infrastructure and stakeholder collaboration for effective development and implementation.

Part II of the book introduced digital technologies for design and fabrication. Computational design algorithms, additive manufacturing, cooperative robotic fabrication, on-site robotic construction, and extended reality were explored. These digital innovations can help optimise material use, facilitate material retrieval for reuse, and help predict how buildings might be used in practice. The chapters discussed the importance of circularity indicators and material passports as design parameters for reducing waste and promoting resource reuse. The use of additive manufacturing and cooperative robotic fabrication, which can generate less waste and require no formwork, was also explored. Robots can also be used for dis- and reassembly, as well as material sorting and separation, while extended reality technology can be used to simulate and visualise building designs and support building maintenance, repair, and reuse. These tools and techniques can help reduce waste, minimise extraction, and promote the reuse, recycling, renewal, and regeneration of resources in the construction industry, moving it towards a more circular model.

Part III of the book provides insights into business models, supply chain management strategies, and policies that can enable the transition to a more sustainable circular built environment, as well as novel strategies such as regeneration that are yet underexplored. Digital technology use cases for deconstruction and reverse logistics, blockchain technology, digital logbooks, and especially business model innovation were explored for accelerating the uptake of circular economy strategies in the built environment. By providing transparency and traceability, blockchain-based solutions can help manage waste, promote resource efficiency, and enable circular business models. While adoption is slowed down by challenges such as lack of expertise, interoperability, and regulatory issues, decentralised applications, tokens, and decentralised autonomous organisations offer new opportunities for blockchain in the circular built environment. Digital building logbooks provide reliable data for improving building design, construction, and management and have the potential to promote a circular economy by facilitating energy efficiency, lifespan extension, and tracking and tracing of material resources. Different types of business models are suitable for enabling different circular principles, and companies can make significant contributions to the transition by considering resource efficiency in their business models. Finally, the digital age offers innovative opportunities for regenerative strategies, which are crucial in the process of creating revitalised, resilient, and sustainable urban areas, buildings, and infrastructure.

While we have provided a comprehensive starting point to explore a twin transition of digital and circular innovation for the built environment, this should only be seen as the start. New, yet unimagined possibilities will emerge from existing digital technologies, and digital technologies not yet known to us will provide new opportunities and challenges. It is clear, however, that if digital technologies are used strategically and with positive intent, they will provide an essential lever for change to fundamentally transform the way we build.

Digital transformation offers a promising pathway to achieving a circular built environment by enabling transparency, traceability, and efficient resource use. Today, we are already witnessing the emergence of various digital tools and platforms that support circularity in the built environment, such as digital building logbooks, AI-based circular solutions, and digital twins. These technologies can help optimise resource use, reduce waste, and enable closed-loop systems. In the near future, we can expect to see the widespread adoption of digital tools and platforms that enable circularity in the built environment. For example, digital building logbooks could become a standard practice for tracking building-related data throughout the entire life cycle of a building. Blockchain-based solutions could be used to track and manage material resources and waste, facilitating a circular economy in the construction industry. While some technologies may just be a hype and others may still only exist at a conceptual level, we are already seeing practical applications and tangible benefits of the digital transformation in the built environment. For example, digital tools have already been used to optimise building design and construction, reduce material waste, and improve energy efficiency. Despite challenges and limitations, we can expect to see continued progress and innovation in this field. As companies and policymakers recognise the benefits of circularity and digitalisation, we can anticipate a shift towards more sustainable and regenerative built environments.

We hope this book will inspire new connections between different players in the supply chain. To this end, we included interdisciplinary examples where data-driven innovation is being used in the advancement of circular construction. We hope that these examples will help move circular projects out of niche spaces and into a world where the principles of a circular economy can be digitally upscaled in the construction sector. Augmenting the skills of architecture, engineering, and construction (AEC) players with digital technologies has the potential to completely disrupt the way we design, build, and look at our buildings. Digital technologies, if used well, have the power to enable the AEC industry to shift towards a circular future. Exploring digital transformation can advance not only the building industry but also all other sectors that struggle with implementing digital innovation for circularity.

Catherine De Wolf
Sultan Çetin
Nancy Bocken

SPRINGER NATURE

GPSR Compliance

The European Union's (EU) General Product Safety Regulation (GPSR) is a set of rules that requires consumer products to be safe and our obligations to ensure this.

If you have any concerns about our products, you can contact us on ProductSafety@springernature.com

In case Publisher is established outside the EU, the EU authorized representative is:

Springer Nature Customer Service Center GmbH
Europaplatz 3
69115 Heidelberg, Germany

The manufacturer's authorised representative in the EU is Springer Nature Customer Service Centre GmbH, Europaplatz 3, 69115 Heidelberg, Germany. If you have any concerns regarding our products, please contact ProductSafety@springernature.com

Printed and bound by CPI Group (UK) Ltd, Croydon, CR0 4YY
01/10/2026
02225855-0007